EPIGENETICS, ENVIRONMENT, AND GENES

EPIGENETICS, ENVIRONMENT, AND GENES

Edited by
Sun Woo Kang, MD, PhD

Apple Academic Press
TORONTO NEW JERSEY

Apple Academic Press Inc.
3333 Mistwell Crescent
Oakville, ON L6L 0A2
Canada

Apple Academic Press Inc.
9 Spinnaker Way
Waretown, NJ 08758
USA

First issued in paperback 2021

Exclusive worldwide distribution by CRC Press, a member of Taylor & Francis Group
No claim to original U.S. Government works

ISBN 13: 978-1-77463-274-1 (pbk)
ISBN 13: 978-1-926895-25-3 (hbk)

Library of Congress Control Number: 2012951950

Library and Archives Canada Cataloguing in Publication

Epigenetics, environment, and genes/edited by Sun Woo Kang.

Includes bibliographical references and index.
ISBN 978-1-926895-25-3
1. Epigenetics. 2. Gene expression. 3. Genes. 4. Chromatin. 5. Nature and nurture.
I. Kang, Sun Woo

QH450.E65 2013 572.8'65 C2012-906390-8

Apple Academic Press also publishes its books in a variety of electronic formats. Some content that appears in print may not be available in electronic format. For information about Apple Academic Press products, visit our website at **www.appleacademicpress.com** and the CRC Press website at **www.crcpress.com**

ABOUT THE EDITOR

SUN WOO KANG, MD, PhD

Sun Woo Kang, MD, PhD, is Associate Professor in the Division of Nephrology of the Department of Internal Medicine at the College of Medicine at Inje University, Korea. Dr. Kang earned his PhD and MD degrees at the Kyung Hee University Medical School in Seoul, Korea. He received a National Kidney Foundation Fellowship at the University of California at San Diego (Division of Nephrology, Department of Medicine) in 2009–2010. He has received several research grants and has several years of clinical trial experience. He is also a published author and presenter at professional meetings.

Normally a reserved and modest man, National Kidney Foundation researcher Sun Woo Kang's paternal pride overrides all shyness when the topic turns to his two boys. "They look just like me, " he says of 9-year-old Seung-Oh Kang and 4-year-old Minseong Kang. "Though they are really much more handsome than me!"

What keeps Kang working long hours in the laboratory at the Center for Human Genetics and Genomics, at the University of California at San Diego, is the fear that with his handsome genes, he also may have passed along a more lethal legacy to his sons: a precursor to kidney disease.

"My grandfather has had hypertension for 45 years," says the South Korean-born nephrologist and PhD researcher. My father has had diabetes and hypertension for 25 years, and had bypass surgery in 2002. When it was recently discovered that I had certain precursors for cardiovascular or kidney disease similar to my father and my father's father, I became even more interested in human genetics as it relates to cardiovascular disease and kidney diseases."

The irony is that Kang, whose research is funded with a National Kidney Foundation Fellowship, was passionate about his work long before it became entwined with his personal history. Now, however, there is an added urgency to his quest. He believes that testing blood samples and genomic DNA of patients with kidney failure or End Stage Renal Disease, recruited from dialysis units at three southern California medical centers, will help him discover if any genetic factors exist when it comes to cardiovascular and kidney disease.

"Of people suffering from hypertension there is a small group that is genetically at a high risk of developing hypertensive kidney disease (HKD)", he says. "I'm here to find out why and to help them." That group, Kang adds, would get more intensive treatment, shown to reduce the onset of HKD. Those without the genetic markers won't have to undergo that level of treatment, which also carries serious side effects.

When not in the lab, Kang loves playing basketball with his boys at Carmel Valley Park. "I would very much like them to grow up strong." He says. "And healthy."

CONTENTS

LIST OF CONTRIBUTORS

Paul A. Berry
YCR Cancer Research Unit, Department of Biology, University of York, Wentworth Way, York, UK

Misha Bilenky
Department of Medical Genetics, Life Sciences Institute, University of British Columbia, 2350 Health Sciences Mall, Vancouver, BC, Canada

Karol Bomsztyk
Institute for Stem Cell and Regenerative Medicine, University of Washington School of Medicine, 815 Mercer St., Seattle, Washington, USA

Pilib Ó. Broin
Institute for Stem Cell and Regenerative Medicine, University of Washington School of Medicine, 815 Mercer St., Seattle, Washington, USA

Jeremy P. Brown
Department of Medical Genetics, Life Sciences Institute, University of British Columbia, 2350 Health Sciences Mall, Vancouver, BC, Canada

Jörn Bullwinkel
Department of Medical Genetics, Life Sciences Institute, University of British Columbia, 2350 Health Sciences Mall, Vancouver, BC, Canada

Stephanie D. Byrum
University of Arkansas for Medical Sciences, 4301 West Markham Street, Little Rock, Arkansas, USA

Nick Cercone
Department of Computer Science and Engineering, York University Toronto, Canada

Anne T. Collins
YCR Cancer Research Unit, Department of Biology, University of York, Wentworth Way, York, UK

Tariq Enver
MRC Molecular Haematology Unit, Weatherall Institute of Molecular Medicine, University of Oxford, Oxford, UK

Fiona M. Frame
YCR Cancer Research Unit, Department of Biology, University of York, Wentworth Way, York, UK

David Garrick
MRC Molecular Haematology Unit, Weatherall Institute of Molecular Medicine, University of Oxford, Oxford, UK

Richard J. Gibbons
MRC Molecular Haematology Unit, Weatherall Institute of Molecular Medicine, University of Oxford, Oxford, UK

Nicholas Goardon
MRC Molecular Haematology Unit, Weatherall Institute of Molecular Medicine, University of Oxford, Oxford, UK

Marco De Gobbi
MRC Molecular Haematology Unit, Weatherall Institute of Molecular Medicine, University of Oxford, Oxford, UK

Aaron A. Golden
Institute for Stem Cell and Regenerative Medicine, University of Washington School of Medicine, 815 Mercer St., Seattle, Washington, USA

Irving I. Gottesman
Department of Psychiatry and Psychology, University of Minnesota, Minneapolis, Minnesota, USA

Preeti Goyal
Department of Medical Genetics, Life Sciences Institute, University of British Columbia, 2350 Health Sciences Mall, Vancouver, BC, Canada

John M. Greally
Institute for Stem Cell and Regenerative Medicine, University of Washington School of Medicine, 815 Mercer St., Seattle, Washington, USA

Anna Gréen
Division of Cell Biology, Department of Clinical and Experimental Medicine, Linköping University, SE-58185 Linköping, Sweden

Henrik Gréen
Division of Cell Biology, Department of Clinical and Experimental Medicine, Linköping University, SE-58185 Linköping, Sweden

Douglas R. Higgs
MRC Molecular Haematology Unit, Weatherall Institute of Molecular Medicine, University of Oxford, Oxford, UK

Jim R. Hughes
MRC Molecular Haematology Unit, Weatherall Institute of Molecular Medicine, University of Oxford, Oxford, UK

Sten Eirik W. Jacobsen
Haemopoietic Stem Cell Laboratory, Weatherall Institute of Molecular Medicine, University of Oxford, Oxford, UK

Ruby Jiang
Department of Medical Genetics, University of British Columbia, 2329 West Mall, Vancouver, BC, Canada

Nora Khaldi
UCD Conway Institute of Biomolecular and Biomedical Research, School of Medicine and Medical Sciences, University College Dublin, Dublin 4, Republic of Ireland

Anton Krumm
Institute for Stem Cell and Regenerative Medicine, University of Washington School of Medicine, 815 Mercer St., Seattle, Washington, USA

Niklas Krumm
Institute for Stem Cell and Regenerative Medicine, University of Washington School of Medicine, 815 Mercer St., Seattle, Washington, USA

Marie-Christine Labarthe
Pro-Cure Therapeutics Ltd, The Biocentre, Innovation Way, York Science Park, Heslington, York, UK

Herbert H. Lindner
Division of Cell Biology, Department of Clinical and Experimental Medicine, Linköping University, SE-58185 Linköping, Sweden

Anita Lönn
Division of Cell Biology, Department of Clinical and Experimental Medicine, Linköping University, SE-58185 Linköping, Sweden

Matthew C. Lorincz
Department of Medical Genetics, Life Sciences Institute, University of British Columbia, 2350 Health Sciences Mall, Vancouver, BC, Canada

Karen M. Lower
MRC Molecular Haematology Unit, Weatherall Institute of Molecular Medicine, University of Oxford, Oxford, UK

Sidinh Luc
Haemopoietic Stem Cell Laboratory, Weatherall Institute of Molecular Medicine, University of Oxford, Oxford, UK

Magnus Lynch
MRC Molecular Haematology Unit, Weatherall Institute of Molecular Medicine, University of Oxford, Oxford, UK

Dixie L. Mager
Department of Medical Genetics, Life Sciences Institute, University of British Columbia, 2350 Health Sciences Mall, Vancouver, BC, Canada

Norman J. Maitland
YCR Cancer Research Unit, Department of Biology, University of York, Wentworth Way, York, UK

Irina A. Maksakova
Department of Medical Genetics, Life Sciences Institute, University of British Columbia, 2350 Health Sciences Mall, Vancouver, BC, Canada

Deborah E. McFadden
Department of Medical Genetics, University of British Columbia, 2329 West Mall, Vancouver, BC, Canada

Emma E. Oldridge
YCR Cancer Research Unit, Department of Biology, University of York, Wentworth Way, York, UK

Richard J. Packer
YCR Cancer Research Unit, Department of Biology, University of York, Wentworth Way, York, UK

Davide Pellacani
YCR Cancer Research Unit, Department of Biology, University of York, Wentworth Way, York, UK

Maria S. Peñaherrera
Department of Medical Genetics, University of British Columbia, 2329 West Mall, Vancouver, BC, Canada

Arturas Petronis
The Centre for Addiction and Mental Health, Departments of Psychiatry and Pharmacology, and the Institute of Medical Science at the University of Toronto, Ontario, Canada

Cristina Pina
MRC Molecular Haematology Unit, Weatherall Institute of Molecular Medicine, University of Oxford, Oxford, UK

Raffaele Renella
MRC Molecular Haematology Unit, Weatherall Institute of Molecular Medicine, University of Oxford, Oxford, UK

Wendy P. Robinson
Department of Medical Genetics, University of British Columbia, 2329 West Mall, Vancouver, BC, Canada

Eric D. Rubio
Institute for Stem Cell and Regenerative Medicine, University of Washington School of Medicine, 815 Mercer St, Seattle, Washington, USA

Ingemar Rundquist
Division of Cell Biology, Department of Clinical and Experimental Medicine, Linköping University, SE-58185 Linköping, Sweden

Bettina Sarg
Division of Cell Biology, Department of Clinical and Experimental Medicine, Linköping University, SE-58185 Linköping, Sweden

Denis C. Shields
UCD Conway Institute of Biomolecular and Biomedical Research, School of Medicine and Medical Sciences, University College Dublin, Dublin 4, Republic of Ireland

Matthew S. Simms
Castle Hill Hospital, Castle Rd, Cottingham, East Yorkshire, UK

Prim B. Singh
Department of Medical Genetics, Life Sciences Institute, University of British Columbia, 2350 Health Sciences Mall, Vancouver, BC, Canada

Jacqueline A. Sloane-Stanley
MRC Molecular Haematology Unit, Weatherall Institute of Molecular Medicine, University of Oxford, Oxford, UK

Shamit Soneji
MRC Molecular Haematology Unit, Weatherall Institute of Molecular Medicine, University of Oxford, Oxford, UK

Leah Spontaneo
Department of Computer Science and Engineering, York University Toronto, Canada

Michael J. Stower
York District Hospital, Wigginton Road, City Centre, York, UK

Alan J. Tackett
University of Arkansas for Medical Sciences, 4301 West Markham Street, Little Rock, Arkansas, USA

Sean D. Taverna
Johns Hopkins School of Medicine, 855 North Wolfe Street, Baltimore, Maryland, USA

Stephen Taylor
Computational Biology Research Group (CBRG), University of Oxford, Oxford, UK

Brandon J. Thomas
Institute for Stem Cell and Regenerative Medicine, University of Washington School of Medicine, 815 Mercer St., Seattle, Washington, USA

Douglas Vernimmen
MRC Molecular Haematology Unit, Weatherall Institute of Molecular Medicine, University of Oxford, Oxford, UK

Paresh Vyas
MRC Molecular Haematology Unit, Weatherall Institute of Molecular Medicine, University of Oxford, Oxford, UK; Department of Haematology, University of Oxford, John Radcliffe Hospital, Oxford, UK

Piri Welcsh
Institute for Stem Cell and Regenerative Medicine, University of Washington School of Medicine, 815 Mercer St,, Seattle, Washington, USA

Albert H. C. Wong
The Centre for Addiction and Mental Health, Departments of Psychiatry and Pharmacology, and the Institute of Medical Science at the University of Toronto, Ontario, Canada

Ryan K. C. Yuen
Department of Medical Genetics, University of British Columbia, 2329 West Mall, Vancouver, BC, Canada

LIST OF ABBREVIATIONS

ADP	Adenosine diphosphate
APC	Allophycocyanin
BPH	*Benign prostatic hyperplasia*
BSA	Bovine serum albumin
BWS	Beckwith–Wiedemann syndrome
CB	Committed basal cells
CEPH	Center d'Etude du Polymorphisme Humaine
ChIP	Chromatin immunoprecipitation
CHM	Complete hydatidiform mole
CNV	Copy number variation
CSCs	Cancer stem cells
DAPI	6-diamidine-2-phynelindole
DAVID	Database for Annotation, Visualization and Integrated Discovery
DMEM	Dulbecco's modified Eagle's medium
DML	Differentially methylated loci
DMR	Differentially methylated region
DNA	Deoxyribonucleic acid
DZ	Dizygotic
EBV	Epstein-Barr virus
EDR	Electrodermal response
ERVs	Endogenous retroviral sequences
ES cells	Embryonic stem cells
FA	Formaldehyde
FACS	Fluorescence-activated cell sorting
FBS	Fetal bovine serum
FCS	Fetal calf serum
FDR	False discovery rate
FITC	Fluorescein isothiocyanate
FPR	False-positive rate

GFP	Green fluorescent protein
GO	Gene ontology
hESC	human embryonic stem cell
HIV	*Human immunodeficiency virus*
IAP	Intracisternal-A particle
ICR	Imprinting control region
I-DIRT	Isotopic differentiation of interactions as random or targeted
IGF2R	Insulin-like growth factor 2 receptor
IMDM	Iscove's modified Dulbecco's medium
IP	Immunoprecipitation
KD	Knockdown
LCL	Lymphoblastoid cell line
LTRs	Long terminal repeats
MACS	Magnetic-activated cell sorting
MaLR	Mouse apparent LTR retrotransposons
MERV-L	Mouse endogenous retrovirus type L
mESCs	mouse Embryonic. Stem Cells
MLVs	Murine leukaemia viruses
MPQ	Multidimensional Personality Questionnaire
mRNA	messenger Ribonucleic acid
MSCV	Murine stem cell virus
MSP	*Methylation specific PCR*
MZ	Monozygotic
MZA	Morozygotic twins reared apart
MZT	Monozygotic twins
NTC	No-template control
ORF	Open reading frame
PBS	Phosphate-buffered saline
PCR	Polymerase chain reaction
PCSCs	*Prostate Cancer* Stem Cells
PE	Phycoerythrin
PHD	Plant homeodomain
piRNA	piwi-interacting Ribonucleic acid
PRC2	PcG repressive complex 2
PTMs	Post-translational modifications
qPCR	quantitative polymerase chain reaction

RT	Reverse transcriptase
SAM	Significance analysis of microarrays
SC	Stem *cells*
siRNA	Small interfering RNA
SLR	Sitewise likelihood ratio
SNP	Single-nucleotide polymorphism
SOM	Self-organizing maps
TA	Transit-amplifying
TSSs	Transcription start sites
UPDs	Uniparental disomies

INTRODUCTION

Epigenetics refers to DNA and chromatin modifications that play an important role in regulation of various genomic functions. The different cellular phenotypes that comprise multicellular organisms are developed by the expression of housekeeping and cell-type-specific genes and by the suppression of inappropriate ones. The pattern of gene expression that designates a cell type is termed the "epigenotype," which is created and maintained by "epigenetic" mechanisms that are able to execute gene expression regardless of the underlying genetic code. Genomic imprinting, where genes are expressed from only one of the inherited parental alleles, shows a typical example of epigenetic gene regulation. Although the genotype of most cells of a given organism is the same (with the exception of gametes and the cells of the immune system), cellular phenotypes and functions differ thoroughly, and this can be (at least to some extent) governed by differential epigenetic regulation that is established during cell differentiation and embryonic morphogenesis.

In chapter 1, Wong and his colleagues introduce us to the epigenetic perspective. Human monozygotic twins and other genetically identical organisms are almost always strikingly similar in appearance, yet they are often discordant for important phenotypes including complex diseases. Such variation among organisms with virtually identical chromosomal DNA sequences has largely been attributed to the effects of environment. Environmental factors can have a strong effect on some phenotypes, but evidence from both animal and human experiments suggests that the impact of environment has been overstated and that our views on the causes of phenotypic differences in genetically identical organisms require revision. New theoretical and experimental opportunities arise if epigenetic factors are considered as part of the molecular control of phenotype. Epigenetic mechanisms may explain paradoxical findings in twin and inbred animal studies when phenotypic differences occur in the absence of observable environmental differences and also when environmental differences do not significantly increase the degree of phenotypic variation.

CpG islands are important regions in DNA. They usually appear at the 5' end of genes containing GC-rich dinucleotides. When DNA methylation occurs, gene regulation is affected and it sometimes leads to carcinogenesis. In chapter 2, Spontaneo and Cercone propose a new detection program using a hidden-markov model alongside the Viterbi algorithm. Their solution provides a graphical user interface not seen in many of the other CGI detection programs. They unified the detection and analysis under one program to allow researchers to scan a genetic sequence, detect the significant CGIs, and analyze the sequence once the scan is complete for any noteworthy findings. Using human chromosome 21, they showed that their algorithm found a significant number of CGIs. Running an analysis on a dataset of promoters revealed that the characteristics of methylated and unmethylated CGIs are significantly different. Finally, they detected significantly different motifs between methylated and unmethylated CGI promoters using MEME and MAST. Spontaneo and Cercone conclude that developing this new tool for the community using powerful algorithms showed that combining analysis with CGI detection will improve the continued research within the field of epigenetics.

Epigenetic control is essential for maintenance of tissue hierarchy and correct differentiation. In cancer, this hierarchical structure is altered and epigenetic control deregulated, but the relationship between these two phenomena is still unclear. In chapter 3, Pellacani et al. examine CD133 as a marker for adult stem cells in various tissues and tumor types. They note that stem cell specificity is maintained by tight regulation of CD133 expression at both transcriptional and post-translational levels, and in their study they investigated the role of epigenetic regulation of CD133 in epithelial differentiation and cancer. DNA methylation analysis of the CD133 promoter was done by pyrosequencing and methylation specific PCR; qRT-PCR was used to measure CD133 expression and chromatin structure was determined by ChIP. Cells were treated with DNA demethylating agents and HDAC inhibitors. All the experiments were carried out in both cell lines and primary samples. They found that CD133 expression is repressed by DNA methylation in the majority of prostate epithelial cell lines examined, where the promoter is heavily CpG hypermethylated, whereas in primary prostate cancer and benign prostatic hyperplasia, low levels of DNA methylation, accompanied by low levels of mRNA, were found. Moreover, differential methylation of CD133 was absent from both benign or malignant CD133+/α2β1integrinhi prostate (stem) cells, when compared to CD133-/α2β1integrinhi (transit amplifying) cells or CD133-/α2β1integrinlow (basal committed) cells, selected from primary epithelial cultures. Condensed chromatin was associated with CD133 downregulation in all of the cell lines, and treatment with HDAC inhibitors resulted in CD133 re-expression in both cell lines and primary samples. Pellacani and colleagues concluded that CD133 is tightly regulated by DNA methylation only in cell lines, where promoter methylation and gene expression inversely correlate. This highlights the crucial choice of cell model systems when studying epigenetic control in cancer biology and stem cell biology. Significantly, in both benign and malignant prostate primary tissues, regulation of CD133 is independent of DNA methylation, but is under the dynamic control of chromatin condensation. This indicates that CD133 expression is not altered in prostate cancer and it is consistent with an important role for CD133 in the maintenance of the hierarchical cell differentiation patterns in cancer.

Meanwhile, genomic imprinting is an important epigenetic process involved in regulating placental and fetal growth. Imprinted genes are typically associated with differentially methylated regions (DMRs) whereby one of the two alleles is DNA methylated depending on the parent of origin. Identifying imprinted DMRs in humans is complicated by species- and tissue-specific differences in imprinting status and the presence of multiple regulatory regions associated with a particular gene, only some of which may be imprinted. In chapter 4, Yuen and colleagues have taken advantage of the unbalanced parental genomic constitutions in triploidies to further characterize human DMRs associated with known imprinted genes and identify novel imprinted DMRs. By comparing the promoter methylation status of over 14,000 genes in human placentas from ten diandries (extra paternal haploid set) and ten digynies (extra maternal haploid set) and using 6 complete hydatidiform moles (paternal origin) and ten chromosomally normal placentas for comparison, they identified 62 genes with apparently imprinted DMRs (false discovery rate <0.1%). Of these 62 genes, 11 have been reported previously as DMRs that act as imprinting control regions, and the observed parental methylation patterns were concordant with those previously reported. They demonstrated that novel imprinted genes, such as FAM50B, as well as novel imprinted DMRs associated with known imprinted genes (for example, CDKN1C and RASGRF1), can be identified by using this approach. Furthermore, they demonstrated how comparison of DNA methylation for known imprinted genes (for example, GNAS and CDKN1C) between placentas of different gestations and other somatic tissues (brain, kidney, muscle and blood) provides a detailed analysis of specific CpG sites associated with tissue-specific imprinting and gestational age-specific methylation. Yuen et al. concluded that DNA methylation profiling of triploidies in different tissues and developmental ages can be a powerful and effective way to map and characterize imprinted regions in the genome.

Endogenous retroviruses (ERVs) are parasitic sequences whose derepression is associated with cancer and genomic instability. Many ERV families are silenced in mouse embryonic stem cells (mESCs) via SETDB1-deposited trimethylated lysine 9 of histone 3 (H3K9me3), but the mechanism of H3K9me3-dependent repression remains unknown. Multiple proteins, including members of the heterochromatin protein 1 (HP1) family, bind H3K9me2/3 and are involved in transcriptional silencing in model organisms. In chapter 5, Maksakova and colleagues address the role of such H3K9me2/3 "readers" in the silencing of ERVs in mESCs. They demonstrate that despite the reported function of HP1 proteins in H3K9me-dependent gene repression and the critical role of H3K9me3 in transcriptional silencing of class I and class II ERVs, the depletion of HP1α, HP1β and HP1γ, alone or in combination, is not sufficient for derepression of these elements in mESCs. While loss of HP1α or HP1β leads to modest defects in DNA methylation of ERVs or spreading of H4K20me3 into flanking genomic sequence, respectively, neither protein affects H3K9me3 or H4K20me3 in ERV bodies. Furthermore, using novel ERV reporter constructs targeted to a specific genomic site, they demonstrate that, relative to Setdb1, knockdown of the remaining known H3K9me3 readers expressed in mESCs, including Cdyl, Cdyl2, Cbx2, Cbx7, Mpp8, Uhrf1 and Jarid1a-c, leads to only modest proviral reactivation. Taken together, these results reveal that each of the known H3K9me3-binding proteins is dispensable for SETDB1-mediated ERV silencing. Maksakove and her colleagues speculate that H3K9me3 might maintain ERVs in a silent state in mESCs by directly inhibiting deposition of active covalent histone marks.

In chapter 6, Gréen et al. focus on histone H1 interphase phosphorylation. Histone H1 is an important constituent of chromatin, and is involved in regulation of its structure. During the cell cycle, chromatin becomes locally decondensed in S phase, highly condensed during metaphase, and again decondensed before re-entry into G1. This has been connected to increasing phosphorylation of H1 histones through the cell cycle. However, many of these experiments have been performed using cell-synchronization techniques and cell cycle-arresting drugs. Gréen and her colleagues investigated the H1 subtype composition and phosphorylation pattern in the cell cycle of normal human activated T cells and Jurkat T-lymphoblastoid cells by capillary electrophoresis after sorting of exponentially growing cells into G1, S and G2/M populations. They found that the relative amount of H1.5 protein increased significantly after T-cell activation. Serine phosphorylation of H1 subtypes occurred to a large extent in late G1 or early S phase in both activated T cells and Jurkat cells. Furthermore, their data confirmed that the H1 molecules newly synthesized during S phase achieve a similar phosphorylation pattern to the previous ones. Jurkat cells had more extended H1.5 phosphorylation in G1 compared with T cells, a difference that can be explained by faster cell growth and/or the presence of enhanced H1 kinase activity in G1 in Jurkat cells. The data was consistent with a model in which a major part of interphase H1 phosphorylation takes place in G1 or early S phase. This implies that H1 serine phosphorylation may be coupled to changes in chromatin structure necessary for DNA replication. In addition, the increased H1 phosphorylation of malignant cells in G1 may be affecting the G1/S transition control and enabling facilitated S-phase entry as a result of relaxed chromatin condensation. Furthermore, increased H1.5 expression may be coupled to the proliferative capacity of growth-stimulated T cells.

Genome-wide studies use techniques, like chromatin immunoprecipitation, to purify small chromatin sections so that protein-protein and protein-DNA interactions can be analyzed for their roles in modulating gene transcription. Histone post-translational modifications (PTMs) are key regulators of gene transcription and are therefore prime targets for these types of studies. Chromatin purification protocols vary in the amount of chemical cross-linking used to preserve in vivo interactions. A balanced level of chemical cross-linking is required to preserve the native chromatin state during purification, while still allowing for solubility and interaction with affinity reagents. Byrum and her colleagues previously used an isotopic labeling technique combining affinity purification and mass spectrometry called transient isotopic differentiation of interactions as random or targeted (transient I-DIRT) to identify the amounts of chemical cross-linking required to prevent histone exchange during chromatin purification. New bioinformatic analyzes reported in chapter 7 reveal that histones containing transcription activating PTMs exchange more rapidly relative to bulk histones and therefore require a higher level of cross-linking to preserve the in vivo chromatin structure. Byrum et al. conclude that the bioinformatic approach described here is widely applicable to other studies requiring the analysis and purification of cognate histones and their modifications. Histones containing PTMs correlated to active gene transcription exchange more readily than bulk histones; therefore, it is necessary to use more rigorous in vivo chemical cross-linking to stabilize these marks during chromatin purification.

In chapter 8, Thomas et al. acknowledge the ways in which random monoallelic expression contributes to the phenotypic variation of cells and organisms. However, the epigenetic mechanisms by which individual alleles are randomly selected for expression are not known. Taking cues from chromatin signatures at imprinted gene loci such as the insulin-like growth factor 2 gene 2 (IGF2), they evaluated the contribution of CTCF, a zinc finger protein required for parent-of-origin-specific expression of the IGF2 gene, as well as a role for allele-specific association with DNA methylation, histone modification and RNA polymerase II. Using array-based chromatin immunoprecipitation, Thomas and his colleagues identified 293 genomic loci that are associated with both CTCF and histone H3 trimethylated at lysine 9 (H3K9me3). A comparison of their genomic positions with those of previously published monoallelically expressed genes revealed no significant overlap between allele-specifically expressed genes and colocalized CTCF/H3K9me3. To analyze the contributions of CTCF and H3K9me3 to gene regulation in more detail, Thomas et al. focused on the monoallelically expressed IGF2BP1 gene. In vitro binding assays using the CTCF target motif at the IGF2BP1 gene, as well as allele-specific analysis of cytosine methylation and CTCF binding, revealed that CTCF does not regulate mono- or biallelic IGF2BP1 expression. Surprisingly, they found that RNA polymerase II is detected on both the maternal and paternal alleles in B lymphoblasts that express IGF2BP1 primarily from one allele. Thus, allele-specific control of RNA polymerase II elongation regulates the allelic bias of IGF2BP1 gene expression. Thomas and his colleagues concluded that colocalization of CTCF and H3K9me3 does not represent a reliable chromatin signature indicative of monoallelic expression. Moreover, association of individual alleles with both active (H3K-4me3) and silent (H3K27me3) chromatin modifications (allelic bivalent chromatin) or with RNA polymerase II also fails to identify monoallelically expressed gene loci. The selection of individual alleles for expression occurs in part during transcription elongation.

In self-renewing, pluripotent cells, bivalent chromatin modification is thought to silence (H3K27me3) lineage control genes while "poising" (H3K4me3) them for subsequent activation during differentiation, implying an important role for epigenetic modification in directing cell fate decisions. However, rather than representing an equivalently balanced epigenetic mark, the patterns and levels of histone modifications at bivalent genes can vary widely and the criteria for identifying this chromatin signature are poorly defined. In chapter 9, De Gobbi et al. initially show how chromatin status alters during lineage commitment and differentiation at a single well-characterized bivalent locus. In addition they have determined how chromatin modifications at this locus change with gene expression in both ensemble and single cell analyzes. They also show, on a global scale, how mRNA expression may be reflected in the ratio of H3K4me3/ H3K27me3. While truly "poised" bivalently modified genes may exist, the original hypothesis that all bivalent genes are epigenetically premarked for subsequent expression might be oversimplistic. In fact, from the data presented in chapter 9, it is equally possible that many genes that appear to be bivalent in pluripotent and multipotent cells may simply be stochastically expressed at low levels in the process of multilineage priming. Although both situations could be considered to be forms of "poising," the underlying mechanisms and the associated implications are clearly different.

Milk proteins are required to proceed through a variety of conditions of radically varying pH, which are not identical across mammalian digestive systems. In chapter 10, Khaldi and Shields investigate the shifts in these requirements to determine if they have resulted in marked changes in the isoelectric point and charge of milk proteins during evolution. They investigated nine major milk proteins in 13 mammals, and in comparison with a group of orthologous non-milk proteins, they found that 3 proteins, κ-casein, lactadherin, and muc1, had undergone the highest change in isoelectric point during evolution. The pattern of non-synonymous substitutions indicated that selection had played a role in the isoelectric point shift, since residues that showed significant evidence of positive selection are much more likely to be charged ($p = 0.03$ for κ-casein; $p < 10-8$ for muc1). However, this selection did not appear to be solely due to adaptation to the diversity of mammalian digestive systems, since striking changes are seen among species that resemble each other in terms of their digestion. Khaldi and Shields determined that the changes in charge are most likely due to changes of other protein functions, rather than an adaptation to the different mammalian digestive systems. These functions may include differences in bioactive peptide releases in the gut between different mammals, which are known to be a major contributing factor in the functional and nutritional value of mammalian milk. This raises the pertinent question of whether bovine milk is optimal in terms of particular protein functions, for human nutrition and possibly disease resistance.

— Sun Woo Kang, MD, PhD

PART I

FROM GENES TO EPIGENOMES

CHAPTER 1

THE EPIGENETIC PERSPECTIVE

ALBERT H. C. WONG, IRVING I. GOTTESMAN, and ARTURAS PETRONIS

CONTENTS

1.1 INTRODUCTION

Identical human twins have been a source of superstition and fascination throughout human history, from Romulus and Remus, the mythical founders of Rome, to movies such as Cronenberg's "Dead Ringers," to the twin paradox in the theory of special relativity [1]. For biologists and psychologists, twins have been an important resource for exploring the etiology of disease and for understanding the role of genetic and environmental factors in determining phenotype, and this fundamental question was first enunciated in its alliterated form by Galton in the 19th century [2]. The relative contribution of nature versus nurture can be estimated by comparing the degree of phenotypic similarities in monozygotic (MZ) versus dizygotic (DZ) twins. MZ twins arise from the same zygote, whereas dizygotic twins arise from a pair of separate eggs, fertilized by two different sperm. As a result, MZ twins have the same chromosomal DNA sequence, except for very small errors of DNA replication after the four to eight cell zygote stage. MZ twins share all of their nuclear DNA, whereas DZ twins share only 50% of DNA sequence variation, on average. Therefore, the degree of genetic contribution to a given phenotype can be estimated from the comparison of MZ to DZ concordance rates or intra-class correlation coefficients. Traits that show higher MZ versus DZ similarity are assumed to have a genetic component because the degree of genetic sharing and the degree of phenotypic similarity are correlated. The amount of genetic contribution can be expressed as the 'heritability' (h^2), which is calculated in various ways (e.g. as twice the difference between the MZ and DZ concordance rates) [3, 4].

Most (if not all) common human diseases show a significant heritability by this definition; however, while MZ twins appear virtually identical, they are often discordant for disease. For example, the heritability of schizophrenia is variously reported to be in the range of 0.70–0.84, on the basis of an MZ twin concordance of 50% and a DZ twin concordance of 10–15% [5-7]. A heritability figure of 0.8 (or 80%) seems to imply that the genetic contribution to disease susceptibility is the main component of risk. Yet, half of MZ twin pairs, in the case of schizophrenia, do not share the disease. The situation is similar for virtually all complex non-Mendelian diseases in which there is clearly some appreciable degree of heritable risk, yet a significant proportion of MZ twin pairs is discordant for the disease. Figure 1 shows the MZ and DZ concordance rates for some common behavioral and medical disorders.

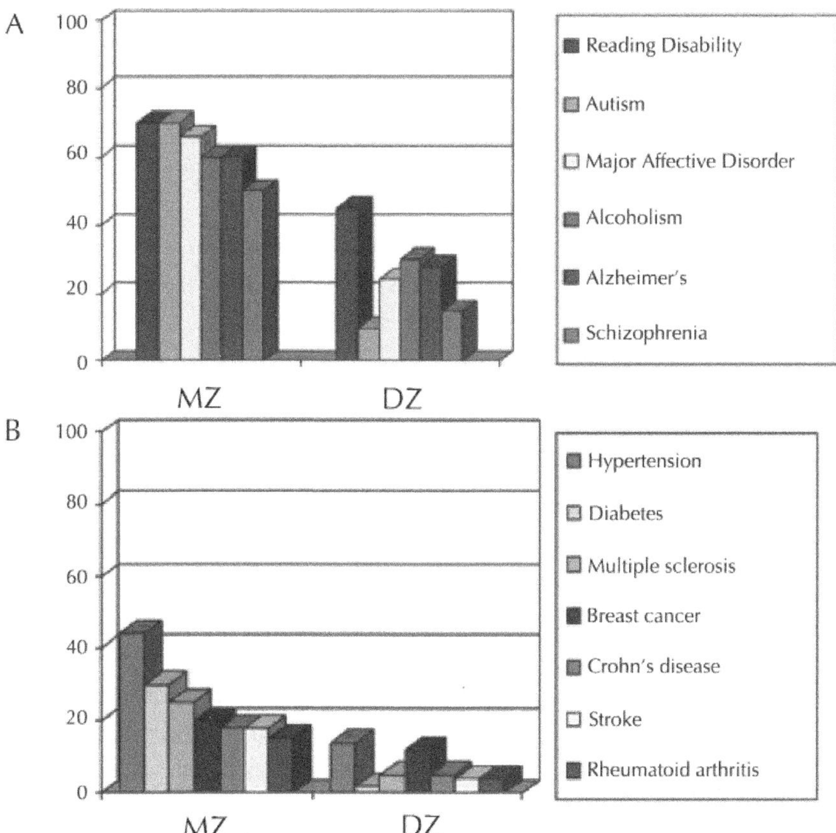

FIGURE 1 The MZ and DZ concordance rates for (A) some common behavioral [5-11] and (B) medical disorders [12-18]. The concordance rates (%) shown are an approximate mid-range value derived from multiple reported figures.

Although heredity clearly influences disease risk, the substantial discordance between MZ twins indicates that chromosomal DNA sequence alone cannot completely determine susceptibility [19]. The imperfect disease concordance in MZ twins is an example of a more general phenomenon: (i.e. phenotypic differences between or among genetically identical organisms). These differences have usually been attributed to the effects of environment (the "non-shared environment" in the case of MZ twins) [20], as a default explanation for variation that remains after genetic effects are accounted for. It is not easy to measure empirically the amount of non-genetic variation that is due to environmental factors. There are examples of significant environmental effects on disease risk, such as smoking and lung cancer [21], but direct evidence of other measured environmental effects on phenotype is rare. In addition, there is an increasing body of experimental evidence suggesting that the generally accepted assumption—variation not attributable to genetic factors must therefore be environmental—may require revision. This chapter will review human twin and animal data that highlight paradoxical findings regarding the contribution of heredity and environment to phenotype, followed by a reinterpretation of these experiments that incorporates epigenetic factors.

1.2 STUDIES OF MZ TWINS RAISED APART OR TOGETHER: THE EPHEMERAL ROLE OF ENVIRONMENT

One of the landmark studies in human twin research that challenges the received importance of environment is the Minnesota Study of Twins Reared Apart, in which detailed physical and psychological assessments were conducted longitudinally in over 100 MZ and DZ pairs of twins who had been reared apart since early childhood [22, 23]. In a variation of the traditional twin study comparing MZ twins reared together (MZT) with DZ twins reared together, the Minnesota study compares MZT with MZ twins raised apart (MZA). This study design allows comparisons between genetically identical MZ twin pairs who have been raised in a shared environment, at least as similar as for any two siblings, and those who have been raised in different homes, cities and states. Thus, the degree of dissimilarity between the MZT and the MZA pairs can be assumed to be the result of different environments [24]. A series of tests were administered simultaneously to each pair of MZA and MZT twins, and the correlations of their scores on each scale were calculated and compared with test–retest correlations as a measure of the reliability of each scale. The intra-class correlation (R) within pairs of MZA (R_{MZA}) and MZT (R_{MZT}) was then expressed as a ratio (R_{MZA}/R_{MZT}). Surprisingly, the correlations within MZT and MZA twin pairs on personality measurements were almost identical (e.g. R_{MZA}=0.50 and R_{MZT}=0.49 on the Multidimensional Personality Questionnaire—MPQ). The R_{MZA}/R_{MZT} ratio for the MPQ was 1.02, compared with 1.01 for fingerprint ridge counts. Out of 22 measurements for which the R_{MZA}/R_{MZT} ratio was reported, 15 had a value over 0.9.

In addition to the traits mentioned previously, these included: electroencephalographic patterns; systolic blood pressure; heart rate; electrodermal response (EDR) amplitude in males and number of EDR trials to reach habituation; the performance scale on the WAIS-IQ; the Raven Mill–Hill IQ test; the California Psychological Inventory; social attitudes on religious and non-religious scales and various scales of MPQ [22, 23].

The findings of the Minnesota study are generally consistent with other studies of MZA twins. For example, a recent effort to look at the etiology of migraine headaches gathered data from the Swedish Twin Registry and found that susceptibility to migraine was mostly inherited and that the twins separated earlier had even greater similarity in migraine status. For women in particular, migraine profiles were very similar in MZA compared to the MZT group, R_{MZA}=0.58 and R_{MZT}=0.46 (R_{MZA}/R_{MZT}=1.26), whereas for men, the migraine incidence was too low and the confidence intervals for correlations were too wide to be able to draw firm conclusions [25]. The Swedish Twin sample was also utilized to explore the factors affecting regular tobacco smoking, and again the results were similar for the two groups of MZ male twins, R_{MZA}=0.84 and R_{MZT}=0.83 (R_{MZA}/R_{MZT}=1.01). For women, the data were more difficult to interpret, R_{MZA}=0.44 and R_{MZT}=0.68 (R_{MZA}/R_{MZT}=0.65), but when more recent cohorts were examined, the rates of smoking in women were similar to men and the heritability was much the same as for men. Smoking rates in women in the early 1900s (the oldest cohort in this sample) were very low and social factors inhibiting smoking in women could result in there being strong local environmental effects, whereas more modern cohorts, with less restrictions on acceptable female behavior, were free to smoke for the same reasons (genetic) as men [26].

Pepticulcer, a disease that contains an evident environmental component (i.e. exposure to *Helicobacter pylori*), has also been subjected to an MZT/MZA study design. The results are another example of paradoxical findings in which the contribution of environmental and genetic factors is unclear. Again, using the Swedish Twin Registry, MZ and DZ concordance rates in twins raised together for peptic ulcer were reported to be 0.65 and 0.35, respectively, suggesting that genetic factors are indeed important in vulnerability (heritability 0.62). However, comparisons between MZA or MZT showed R_{MZA}=0.67 and R_{MZT}=0.65 (R_{MZA}/R_{MZT}=1.03), suggesting that the common home environment has little effect on susceptibility to peptic ulcer [27]. So, the question that remains after analyzing these data is: what can account for the discordance in MZ twin pairs? If environment were a significant factor, then why is the R_{MZA}/R_{MZT} almost 1 (1.03)?

All the aforementioned examples of the MZT/MZA comparisons, from anthropometric data to psychological traits to pathogen susceptibility, lead to a paradoxical conclusion regarding the role of environmental factors. The fact that the MZ twin concordance (correlation) is well below 100% argues that environment is important. However, different environments in MZA do not result in a higher degree of phenotypic discordance when compared with MZT. These MZT/MZA studies clearly illustrate the core problem that would be explored in this chapter; though some studies implicate non-genetic factors in susceptibility to disease or other phenotypes, the available data do not support the interpretation that the remaining variation in phenotype is due to environmental factors. Similar inconsistencies regarding the impact of environmental effects have also been detected in the studies of experimental animals.

1.3 PHENOTYPIC VARIATION IN GENETICALLY IDENTICAL ANIMALS: "THE THIRD COMPONENT"

Some of the questions raised by human twin studies can be re-examined by experimental manipulations of laboratory animals. Animal strains that have been inbred for many generations have almost identical genomes, that is, they are virtually isogenic. True MZ twins can also be generated through in vitro embryo manipulations that provide an opportunity to directly separate the effects of genes from pre-natal environment. Although environmental "twins" do not exist in humans, a close approximation can be created by strictly controlling the environment of laboratory animals in a way that is impossible with humans. At the very least, the effects of constrained versus diverse environments can be quantified to determine the relative contribution of specific environmental factors to phenotypic variation.

In an elegant series of experiments designed to explore the relative contributions of genes, environment and other factors to laboratory animal phenotype, Gartner [28] was able to demonstrate that the majority of random non-genetic variability was not due to the environment. Genetic sources of variation were minimized by using inbred animals, but reduction of genetic variation did not substantially reduce the amount of observed variation in phenotypes such as body weight or kidney size. Strict standardization of the environment within a laboratory did not have a major effect on inter-individual variability when compared with tremendous environmental variability in a natural setting. Only 20–30% of the variability could be attributed to environmental factors, with the remaining 70–80% of non-genetic variation due to a "third component…effective at or before fertilization" [28].

To directly segregate genetic from pre- and post-natal effects, in vitro embryo manipulations in isogenic animals can be performed. In two mouse strains and in Friesian cattle, Gartner artificially created MZ and DZ twins by transplanting divided and non-divided eight-cell embryos into pseudopregnant surrogates. The effect of different uterine environments was tested by transplanting pairs of MZ or DZ embryos into the same or into two different surrogate dams. Pre-natal and post-natal environments were tightly controlled and, most importantly, were equally variable between the isogenic DZ and MZ twin pairs. The variance (s^2) of mean body weights and time to reach certain developmental milestones like eye opening, between twin pairs (s_b^2) and within twin pairs (s_w^2) was calculated for both MZ and DZ groups. The F-test comparisons for variation between the MZ and DZ groups overall were not significantly different. However, variance within twin pairs (s_w^2) was significantly much greater for DZ twins than for MZ twins (ranging from $P<0.01$ to $P<0.001$ for individual weight measurements). Therefore, despite the fact that all mice were isogenic, and developed in identical pre- and post-natal environments, the MZ twin pairs showed a greater degree of phenotypic similarity among co-twins than did the DZ twin pairs, thus implicating non-DNA sequence—and non-environment—based influences on the zygote at or before the eight-cell stage as the main source of phenotypic variation. Gartner and Baunack

[29] referred to this non-genetic influence as the "third component," after genes and environment, the molecular basis of which remains unknown.

The cloning of mammals has recently been accomplished in a variety of species, and these experiments, technical feats in themselves, also present an opportunity to differentiate the effects of chromosomal DNA sequence from other factors that can influence phenotype. Although the offspring of these cloning experiments have the same genome as the donor animals, they exhibit a variety of phenotypic abnormalities that obviously cannot be attributed to genetic causes [30]. In some cases, the phenotypes are pathological and represent disease states, whereas other abnormalities are more subtle, suggesting that these observations are relevant to understanding both susceptibility to human complex diseases and variation within a normal functional spectrum [31]. The most famous of these cloning experiments was performed with sheep, but along with seemingly healthy lambs, many clone siblings died perinatally as a result of overgrowth, pulmonary hypertension and renal, hepatobiliary and body-wall defects [32]. Some cloned mice are susceptible to obesity [33]. In addition to higher overall weight, the cloned mice have the same inter-individual variability in weight as non-cloned control mice of the same genetic background [34]. Clones in other species also show considerable variation in lifespan and disease phenotypes between genetically identical clones and non-cloned members of that species. This has been reported in pigs [35] and in cattle, where the main post-natal abnormality is musculoskeletal in origin [36].

This list is far from comprehensive and is meant to illustrate the point that significant phenotypic variation, including crossing a threshold to fatal disease, can emerge from animals that have an identical, cloned genetic background, and frequently-occurring differences in mitochondrial DNA cannot be a universal mechanism for a wide spectrum of phenotypic differences. These early examples of cloned animals were subjected to intense scrutiny in highly supervised and controlled environments, yet they still exhibit disease in an inconsistent fashion. If environment were the source of this phenotypic variation, then one would expect the same emergence of disease among non-cloned members of this species, in an even greater extent, because their environment is not usually so tightly constrained. More likely, there are other potential explanations for this variation.

The general conclusion drawn from the previously described experiments is that substantial phenotypic variation may occur in the absence of either genetic background differences or identifiable environmental variation. When genetic sources of variation are excluded, environmental factors are usually considered to be the source of the remaining variation. However, the previously described data do not support this hypothesis. It is easy to see how the environment is often blamed for this non-genetic variation in phenotype. It is difficult to prove that environmental factors are not affecting phenotype. Environmental sources of phenotypic variation can only be excluded by showing that variation persists in a zero-variation environment. Obviously it is difficult to design such an experiment in which environmental variation can be shown to be near zero, but the studies described previously circumvent this problem. They did so by either directly controlling the degree of environmental variation (as in Gartner's experiments) or by using naturally occurring (human twins) or artificially induced (through in vitro embryo manipulations) controls as comparison groups. In all of these examples, there exists a component of phenotypic variation whose source remains unexplained.

1.4 THE EPIGENETIC PERSPECTIVE

Epigenetics refers to DNA and chromatin modifications that play a critical role in regulation of various genomic functions. Although the genotype of most cells of a given organism is the same (with the exception of gametes and the cells of the immune system), cellular phenotypes and functions differ radically, and this can be (at least to some extent) controlled by differential epigenetic regulation that is set up during cell differentiation and embryonic morphogenesis [37-39]. Once the cellular phenotype is established, genomes of somatic cells are "locked" in tissue-specific patterns of gene expression, generation after generation. This heritability of epigenetic information in somatic cells has been called an "epigenetic inheritance system" [40].

Even after the epigenomic profiles are established, a substantial degree of epigenetic variation can be generated during the mitotic divisions of a cell in the absence of any specific environmental factors. Such variation is most likely to be the outcome of stochastic events in the somatic inheritance of epigenetic profiles. One example of stochastic epigenetic event is a failure of DNA methyltransferase to identify a post-replicative hemimethylated DNA sequence, which would result in loss of methylation signal in the next round of DNA replication (reviewed in [41]). In tissue culture experiments, the fidelity of maintenance DNA methylation in mammalian cells was detected to be between 97 and 99.9% [42]. In addition, there was also de novo methylation activity, which reached 3–5% per mitosis [42]. Thus, the epigenetic status of genes and genomes varies quite dynamically when compared with the relatively static DNA sequence. This partial epigenetic stability and the role of epigenetic regulation in orchestrating various genomic activities make epigenetics an attractive candidate molecular mechanism for phenotypic variation in genetically identical organisms.

From the epigenetic point of view, phenotypic differences in MZ twins could result, in part, from their epigenetic differences. Because of the partial stability of epigenetic regulation, a substantial degree of epigenetic dissimilarity can be accumulated over millions of mitotic divisions of cells in genetically identical organisms. This is consistent with experimental findings in MZ twins discordant for Beckwith–Wiedemann syndrome (BWS) [43]. In skin fibroblasts from five MZ twin pairs discordant for BWS, the affected co-twins had an imprinting defect at the *KCNQ1OT1* gene. The epigenetic defect is thought to arise from the unequal splitting of the inner cell mass (containing the DNA methylation enzymes) during twinning, which results in differential maintenance of imprinting at KCNQ1OT1. In another twin study, the bisulfite DNA modification-based mapping of methylated cytosines revealed numerous subtle inter-individual epigenetic differences, which are likely to be a genome-wide phenomenon [44]. The finding that differences in MZT are similar to MZA, for a large number of traits, suggests that in such twins stochastic events may be a more important cause of phenotypic differences than specific environmental effects. If the emphasis is shifted from environment to stochasticity, it may become clear why MZ twins reared apart are not more different from each other than MZ twins reared together. It is possible that MZ twins are different for some traits, not because they are exposed to different environments but because those traits are determined by meta-stable epigenetic regulation on which environmental factors have only a modest impact.

The intention is not to argue that environment has no effect in generating phenotypic differences in genetically identical organisms. Rather, it is suggested that epigenetic studies of disease may help to understand the pathophysiology of, and susceptibility to, etiologically complex, common illnesses. The current method of studying most diseases includes molecular genetic approaches to identify gene-sequence variants that affect susceptibility and epidemiological efforts to identify environmental factors affecting either susceptibility or outcomes. However, epidemiological studies in humans are limited by a number of methodological issues. Obviously, it is unethical to deliberately expose people to putative disease-causing agents in a prospective randomized controlled trial and it is impossible to control human environments in a way that eliminates most sources of bias in epidemiological studies [45]. Such designs may be possible in animal studies, but adequate animal models are available for only a small proportion of human conditions. In this situation, epigenetic studies may help identification of the molecular effects of the environmental factors. There is an increasing list of environmental events that result in epigenetic changes [46-52], including the recent finding of maternal behavior-induced epigenetic modification at the gene for the glucocorticoid receptor in animals [53]. The advantage of the epigenetic perspective is that, especially in humans, identification of molecular epigenetic effects of environmental factors might be easier and more efficient than direct (but methodologically limited) epidemiological studies.

Epigenetic mechanisms can easily be integrated into a model of phenotypic variation in multicellular organisms, which can explain some of the phenotypic differences among genetically identical organisms. MZ twin discordance for complex, chronic, non-Mendelian disorders such as schizophrenia, multiple sclerosis or asthma could arise as a result of a chain of unfavorable epigenetic events in the affected twin. During embryogenesis, childhood and adolescence there is ample opportunity for multidirectional effects of tissue differentiation, stochastic factors, hormones and probably some external environmental factors (nutrition, medications, addictions, etc.) [50, 54] to accumulate in only one of the two identical twins (Figure 2) [41, 44].

Variation in phenotype among isogenic animals can also be attributed to meta-stable epigenetic regulation. Gartner's experiments with controlled versus chaotic environmental conditions showed that non-environmental factors were responsible for the majority of phenotypic variation among inbred animals. Similar observations among cloned animals can be accounted for by epigenetic differences among these animals. The role of dysregulated epigenetic mechanisms in disease is also consistent with the experimental observations in cloned animals. The derivation of embryos from somatic cells, which contain quite different epigenetic profiles when compared with the germline, generates abnormalities of development that can arise from inadequate or inappropriate nuclear programming [33, 55, 56].

Evidence of epigenetic, non-environmentally mediated sources of variation in genetically identical organisms can be found in the examples of the mouse agouti and AxinFu loci [57, 58]. The *agouti* gene (A) is responsible for the coat color of wild-type mice and isogenic heterozygous c57BL/6 Avy/a mice have a range of coat colors from yellow to black (pseudoagouti). The darkness of the Avy/a mice was proportional to the amount of DNA methylation in the agouti locus, with complete methylation in black psuedoagouti mice and reduced methylation in yellow ones [57]. Transplantation experiments of fertilized oocytes to surrogate dams demonstrated that color was influenced by the phenotype of the genetic dam, not the foster dam [57]. Thus, an obvious phenotype of this isogenic mouse strain is controlled by epigenetic factors that are partially heritable.

Another example of the role of epigenetic mechanisms on phenotype is the murine axin-fused (AxinFu) allele, which in some cases produces a characteristically kinked tail. Like the agouti gene locus, the Axin gene contains an intracisternal-A particle (IAP) retrotransposon that is subjected to epigenetic modification. The methylation status of the long terminal repeat of the IAP in the AxinFu allele correlates with the degree of tail deformity. Furthermore, the presence of the deformity and associated methylation pattern in either sires or dams increases the probability of the same deformity in the offspring [58]. These experiments demonstrate both stochastic and heritable features of epigenetic mechanisms on variability in isogenic animals.

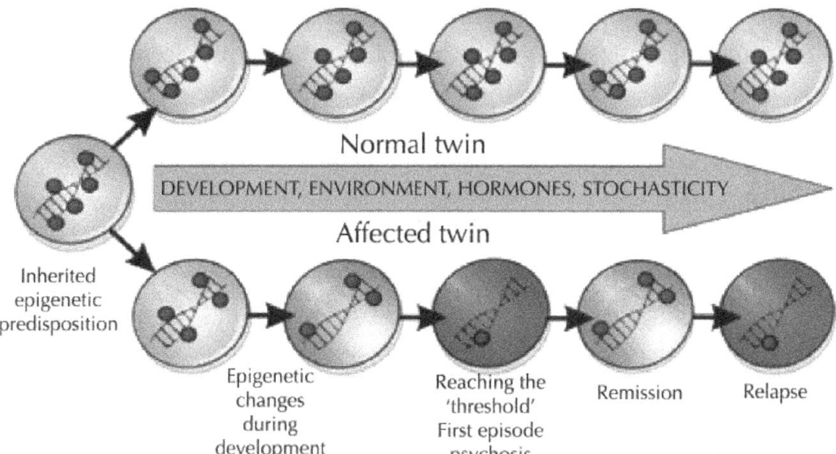

FIGURE 2 Epigenetic model of MZ twin discordance in complex disease, e.g. schizophrenia. Red circles represent methylated cytosines.

From the epigenetic point of view, phenotypic disease differences in MZ twins result from their epigenetic differences. Due to the partial stability of epigenetic signals, a substantial degree of epigenetic dissimilarity can be accumulated over millions of mitotic divisions of cells of MZ co-twins. Although the figure shows that disease is caused by gene hypomethylation, scenarios where pathological condition is associated with gene hypermethylation are equally possible.

The two epigenetic mouse studies described previously as well as experimental data from other species [59] suggest epigenetic signals can exhibit meiotic stability, i.e. epigenetic information can be transmitted from one generation to another. Traditionally, it has been thought that during the maturation of the germline, gametes re-program their epigenetic status by erasing the old and re-establishing a new epigenetic profile. Although the extent of meiotic epigenetic stability remains unknown, the implications are potentially dramatic, blurring the distinction between epigenetic and DNA sequence-based inheritance. The inheritance of epigenetic information and the potential for this to affect disease susceptibility also challenge the dominant paradigm of human morbid genetics, which is almost exclusively concentrated on DNA sequence variation [60]. This partial heritability of epigenetic status of the germline may explain the molecular origin of Gartner's "third component" (Figure 3). The variance within twin pairs (s_w^2) was significantly lower for isogenic MZ twins than for isogenic DZ twins because MZ twins derived from the same zygote shared the same epigenomic background. DZ animals, however, originated from different zygotes that had different epigenetic backgrounds. This interpretation suggests that epigenetic meta-stability is not only limited to somatic cells but also applies to the germline and that germ cells of the same individuals may be carriers of different epigenomes despite their DNA sequence identity. Additionally, the inherited epigenetic signals have a significant impact on the phenotype despite numerous epigenetic changes that take place during embryogenesis [38, 61].

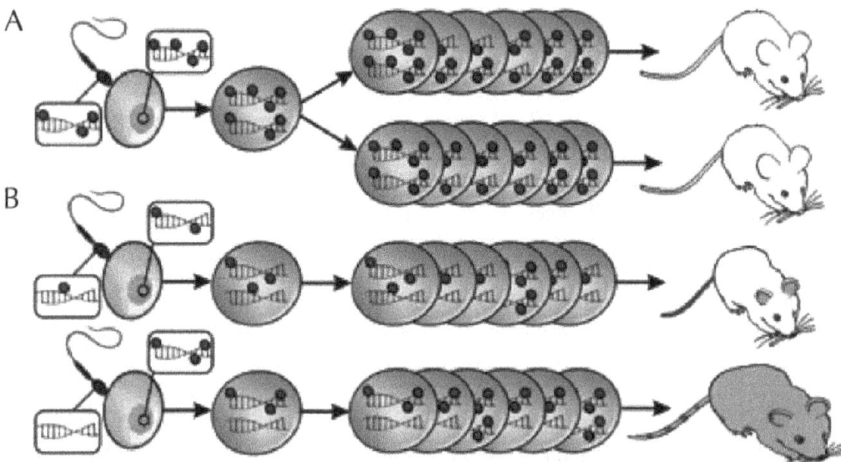

FIGURE 3 Epigenetic interpretation of Gartner's 'third component'.
Phenotypic differences in MZ isogenic animals (A) and DZ isogenic animals (B) can be explained by epigenetic variation in the germline. MZ animals derive from the same zygote and therefore their epigenetic 'starting point' is identical, whereas DZ animals originate from different sperm and oocytes that may carry quite different epigenomic profiles. As in Figure 2, red circles represent methyl groups attached to cytosines.

Until recently, it has not been feasible to test these epigenetic interpretations of phenotypic differences directly among genetically identical organisms. Technologies for high-throughput, large scale epigenomic profiling have been developed [62-66], which along with more well-established techniques, such as focused fine-mapping of methylated cytosines using bisulfite modification or identification of histone modification status using chromatin immunoprecipitation, can evaluate epigenetic profiles in a target tissue and permit comparisons of epigenetic profile among different phenotypes. Methods such as these could be applied to genetically identical organisms to determine whether phenotypic differences are indeed correlated with differences in epigenetic profiles and where in the genome the crucial epigenetic signals may be located. The classical genetic phenomena of incomplete penetrance and variable expressivity may in part be explained by differences in epigenetic regulation of certain genes and their expression levels. We now have the experimental tools to test these hypotheses directly and characterize the extent to which epigenetic factors may influence the traditional dyad of genes and environment.

Vogel and Motulsky [67] wisely said that "human genetics is by no way a completed and closed complex of theory… [with] results that only need to be supplemented in a straightforward way and without major changes in conceptualization…anomalies and discrepancies may exist, but we often do not identify them because we share the "blind spots" with other members of our paradigm group" [67]. The source of phenotypic differences in genetically identical organisms may be one such blind spot among geneticists. Apart from human diseases, various concerns have been raised regarding the limitations of the DNA sequence-based paradigm, and the importance of epigenetic factors has been emphasized. Strohman [68] concluded that the Watson–Crick genetic code "which began as a narrowly defined and proper theory and paradigm of the gene, has mistakenly evolved into a theory and a paradigm of life." In a similar way, Fedoroff et al. [69] stated that "our traditional genetic picture…which is concerned almost exclusively with the effect of nucleotide sequence changes on gene expression and function is substantially incomplete," and that "epigenetic factors are significantly more important than it is generally thought" [69]. As seen in other fields of science [70], identification of the areas where inconsistencies or controversies lie may provide new opportunities for rethinking fundamental laws, lead to new experimental designs, and may even result in major paradigmatic shifts.

KEYWORDS

- **Cloning**
- **Cytosines**
- **MZ twins**
- **Non-Mendelian disease**
- **Zygote**

ACKNOWLEDGMENT

We thank Dr Axel Schumacher for his help with drawing figures for this chapter. This research has been supported by the Special Initiative grant from the Ontario Mental Health Foundation and also by NARSAD, the Canadian Psychiatric Research Foundation, the Stanley Foundation, the Juvenile Diabetes Foundation International and the Crohn's and Colitis Foundation of Canada to A.P.

REFERENCES

1. Resnick, R. and Halliday, D. *Basic concepts in relativity*. New York, Macmillan (1992).
2. Galton, F. English men of science, their nature and nurture. In Proceedings of the Meetings of members of the Royal Institution, Vol. 7, pp. 227–236. London, Macmillan (1874).
3. Boomsma, D., Busjahn, A., and Peltonen, L. Classical twin studies and beyond. *Nat Rev Genet* **3,** 872–882 (2002).
4. Carey, G. Human genetics for the social sciences. Thousand Oaks, Calif.: Sage Publications (2003).
5. Gottesman, I. I. Schizophrenia genesis: the origins of madness. New York, W.H. Freeman (1991).
6. Cardno, A. G. and Gottesman, I. I., II. Twin studies of schizophrenia: from bow-and-arrow concordances to Star Wars Mx and functional genomics. *Am J Med Genet* **97,** 12–17 (2000).
7. Kendler, K. S. Twin studies of psychiatric illness: an update. *Arch Gen Psychiatry* **58,** 1005–1014 (2001).

8. Kieseppa, T., Partonen, T., Haukka, J., Kaprio, J., and Lonnqvist, J. High concordance of bipolar I disorder in a nationwide sample of twins. *Am J Psychiatry* **161,** 1814–1821 (2004).
9. Kendler, K. S. and Prescott, C. A. A population-based twin study of lifetime major depression in men and women. *Arch Gen Psychiatry* **56,** 39–44 (1999).
10. Pedersen, N. L., Gatz, M., Berg, S., and Johansson, B. How heritable is Alzheimer's disease late in life? Findings from Swedish twins. *Ann Neurol* **55,** 180–185 (2004).
11. Muhle, R., Trentacoste, S. V., and Rapin, I. The genetics of autism. *Pediatrics* **113,** e472–e486 (2004).
12. Hamilton, A. S. and Mack, T. M. Puberty and genetic susceptibility to breast cancer in a case-control study in twins. *N Engl J Med* **348,** 2313–2322 (2003).
13. Orholm, M., Binder, V., Sorensen, T. I., Rasmussen, L. P., and Kyvik, K. O. Concordance of inflammatory bowel disease among Danish twins. Results of a nationwide study. *Scand J Gastroenterol* **35,** 1075–1081 (2000).
14. Willer, C. J., Dyment, D. A., Risch, N. J., Sadovnick, A. D., and Ebers, G. C. Twin concordance and sibling recurrence rates in multiple sclerosis. *Proc Natl Acad Sci USA* **100,** 12877–12882 (2000).
15. Selby, J. V., Newman, B., Quiroga, J., Christian, J. C., Austin, M. A., and Fabsitz, R. R. Concordance for dyslipidemic hypertension in male twins. *JAMA* **265,** 2079–2084 (1991).
16. Hyttinen, V., Kaprio, J., Kinnunen, L., Koskenvuo, M., and Tuomilehto, J. Genetic liability of type 1 diabetes and the onset age among 22,650 young Finnish twin pairs: a nationwide follow-up study. *Diabetes* **52,** 1052–1055 (2003).
17. Brass, L. M., Isaacsohn, J. L., Merikangas, K. R., and Robinette, C. D. A study of twins and stroke. *Stroke* **23,** 221–223 (1992).
18. Grant, S. F., Thorleifsson, G., Frigge, M. L., Thorsteinsson, J., Gunnlaugsdottir, B., Geirsson, A. J., Gudmundsson, M., Vikingsson, A., Erlendsson, K., Valsson, J. et al. The inheritance of rheumatoid arthritis in Iceland. *Arthritis Rheum* **44,** 2247–2254 (2001).
19. Chakravarti, A. and Little, P. Nature, nurture and human disease. *Nature* **421,** 412–414 (2003).
20. Plomin, R. and Daniels, D. Why are children in the same family so different from one another? *Behav Brain Sci* **10,** 1–60 (1987).
21. Alberg, A.J. and Samet, J.M. Epidemiology of lung cancer. *Chest* 123, 21S–49S (2003).
22. Bouchard, T. J. Jr., Lykken, D. T., McGue, M., Segal, N. L. and Tellegen, A. Sources of human psychological differences: the Minnesota Study of Twins Reared Apart. *Science* **250,** 223–228 (1990).
23. Bouchard, T. J. Jr. and McGue, M. Genetic and environmental influences on human psychological differences. *J Neurobiol* **54,** 4–45 (2003).
24. Bouchard, T. J. Jr., Heston, L., Eckert, E., Keyes, M., and Resnick, S. The Minnesota study of twins reared apart: project description and sample results in the developmental domain. *Prog Clin Biol Res* **69** (Pt B), 227–233 (1981).
25. Svensson, D. A., Larsson, B. Waldenlind, E., and Pedersen, N. L. Shared rearing environment in migraine: Results from twins reared apart and twins reared together. *Headache* **43,** 235–244 (2003).

26. Kendler, K. S., Thornton, L. M., and Pedersen, N. L. Tobacco consumption in Swedish twins reared apart and reared together. *Arch Gen Psychiatry* **57,** 886–892 (2000).
27. Malaty, H. M., Graham, D. Y., Isaksson, I., Engstrand, L., and Pedersen, N. L. Are genetic influences on peptic ulcer dependent or independent of genetic influences for Helicobacter pylori infection? *Arch Intern Med* **160,** 105–109 (2000).
28. Gartner, K. A third component causing random variability beside environment and genotype. A reason for the limited success of a 30 year long effort to standardize laboratory animals? *Lab Anim* **24,** 71–77 (1990).
29. Gartner, K. and Baunack, E. Is the similarity of monozygotic twins due to genetic factors alone? *Nature* **292,** 646–647 (1981).
30. Edwards, J. L., Schrick, F. N., McCracken, M. D., van Amstel, S. R., Hopkins, F. M., Welborn, M. G., and Davies, C. J. Cloning adult farm animals: a review of the possibilities and problems associated with somatic cell nuclear transfer. *Am J Reprod Immunol* **50,** 113–123 (2003).
31. Cibelli, J. B., Campbell, K. H., Seidel, G. E. West, M. D., and Lanza, R. P. The health profile of cloned animals. *Nat Biotechnol* **20,** 13–14 (2002).
32. Rhind, S. M., King, T. J., Harkness, L. M., Bellamy, C., Wallace, W., DeSousa, P., and Wilmut, I. Cloned lambs--lessons from pathology. *Nat Biotechnol* **21,** 744–745 (2003).
33. Yanagimachi, R. Cloning: experience from the mouse and other animals. *Mol Cell Endocrinol* **187,** 241–248 (2002).
34. Tamashiro, K. L., Wakayama, T., Yamazaki, Y., Akutsu, H., Woods, S. C., Kondo, S., Yanagimachi, R. and Sakai, R. R. Phenotype of cloned mice: development, behavior, and physiology. *Exp Biol Med (Maywood)* **228,** 1193–1200 (2003).
35. Carter, D. B., Lai, L., Park, K. W., Samuel, M., Lattimer, J. C., Jordan, K. R., Estes, D. M., Besch-Williford, C. and Prather, R. S. Phenotyping of transgenic cloned piglets. *Cloning Stem Cells* **4,** 131–145 (2002).
36. Wells, D. N., Forsyth, J. T., McMillan, V., and Oback, B. The health of somatic cell cloned cattle and their offspring. *Cloning Stem Cells* **6,** 101–110 (2004).
37. Monk, M., Boubelik, M., and Lehnert, S. Temporal and regional changes in DNA methylation in the embryonic, extraembryonic and germ cell lineages during mouse embryo development. *Development* **99,** 371–382 (1987).
38. Oswald, J., Engemann, S., Lane, N., Mayer, W., Olek, A., Fundele, R., Dean, W., Reik, W., and Walter, J. Active demethylation of the paternal genome in the mouse zygote. *Curr Biol* **10,** 475–478 (2000).
39. Mayer, W., Niveleau, A., Walter, J., Fundele, R., and Haaf, T. Demethylation of the zygotic paternal genome. *Nature* **403,** 501–502 (2000).
40. Maynard Smith, J. Models of a dual inheritance system. *J Theor Biol* **143,** 41–53 (1990).
41. Petronis, A. The origin of schizophrenia: genetic thesis, epigenetic antithesis, and resolving synthesis. *Biol Psychiatry* **55,** 965–970 (2004).
42. Riggs, A. D., Xiong, Z., Wang, L. and LeBon, J. M. Methylation dynamics, epigenetic fidelity and X chromosome structure. In *Epigenetics* Wolffe, A. (Ed.). Novartis Foundation Symposium. Chistester, UK, John Wiley & Sons Vol. **214,** pp. 214–225 (1998).
43. Weksberg, R., Shuman, C., Caluseriu, O., Smith, A.C., Fei, Y.L., Nishikawa, J., Stockley, T. L., Best, L., Chitayat, D., and Olney, A. Discordant KCNQ1OT1 imprinting in sets of

monozygotic twins discordant for Beckwith-Wiedemann syndrome. *Hum Mol Genet* **11,** 1317–1325 (2002).

44. Petronis, A., Gottesman, II, Kan, P., Kennedy, J. L., Basile, V. S., Paterson, A. D., and Popendikyte, V. Monozygotic twins exhibit numerous epigenetic differences: clues to twin discordance? *Schizophr Bull* **29,** 169–178 (2003).
45. Taubes, G. Epidemiology faces its limits. *Science* **269,** 164–169 (1995).
46. Jablonka, E. and Lamb, M. *Epigenetic Inheritance and Evolution.* London, Oxford University Press (1995).
47. Li, S., Hursting, S. D., Davis, B. J., McLachlan, J. A., and Barrett, J. C. Environmental exposure, DNA methylation, and gene regulation: lessons from diethylstilbesterol-induced cancers. *Ann NY Acad Sci* **983,** 161–169 (2003).
48. Moore, L. E., Huang, W. Y., Chung, J., and Hayes, R. B. Epidemiologic considerations to assess altered DNA methylation from environmental exposures in cancer. *Ann NY Acad Sci* **983,** 181–196 (2003).
49. Ross, S. A. Diet and DNA methylation interactions in cancer prevention. *Ann NY Acad Sci* **983,** 197–207 (2003).
50. Sutherland, J. E. and Costa, M. Epigenetics and the environment. *Ann NY Acad Sci* **983,** 151–160 (2003).
51. Waterland, R. A. and Jirtle, R. L. Transposable elements: targets for early nutritional effects on epigenetic gene regulation. *Mol Cell Biol* **23,** 5293–5300 (2003).
52. Ruden, D. M., Garfinkel, M. D., Sollars, V. E., and Lu, X. Waddington's widget: Hsp90 and the inheritance of acquired characters. *Semin Cell Dev Biol* **14,** 301–310 (2003).
53. Weaver, I. C., Cervoni, N., Champagne, F. A., D'Alessio, A. C., Sharma, S., Seckl, J. R., Dymov, S., Szyf, M., and Meaney, M. J. Epigenetic programming by maternal behavior. *Nat Neurosci* **7,** 847–854 (2004).
54. Jaenisch, R. and Bird, A. Epigenetic regulation of gene expression: how the genome integrates intrinsic and environmental signals. *Nat Genet* **33** (suppl.), 245–254 (2003).
55. Inui, A. Obesity: a chronic health problem in cloned mice? *Trends Pharmacol Sci* **24,** 77–80 (2003).
56. McEvoy, T. G., Ashworth, C. J., Rooke, J. A., and Sinclair, K. D. Consequences of manipulating gametes and embryos of ruminant species. *Reprod Suppl* **61,** 167–182 (2003).
57. Morgan, H. D., Sutherland, H. G., Martin, D. I., and Whitelaw, E. Epigenetic inheritance at the agouti locus in the mouse. *Nat Genet* **23,** 314–318 (1999).
58. Rakyan, V. K., Chong, S., Champ, M. E., Cuthbert, P. C., Morgan, H. D., Luu, K. V., and Whitelaw, E. Transgenerational inheritance of epigenetic states at the murine Axin(Fu) allele occurs after maternal and paternal transmission. *Proc Natl Acad Sci USA* **100,** 2538–2543 (2003).
59. Klar, A. J. Propagating epigenetic states through meiosis: where Mendel's gene is more than a DNA moiety. *Trends Genet* **14,** 299–301 (1998).
60. Petronis, A. Human morbid genetics revisited: relevance of epigenetics. *Trends Genet* **17,** 142–146 (2001).
61. Mayer, W., Niveleau, A., Walter, J., Fundele, R., and Haaf, T. Demethylation of the zygotic paternal genome. *Nature* **403,** 501–502 (2000).

62. Gitan, R. S., Shi, H., Chen, C. M., Yan, P. S., and Huang, T. H. Methylation-specific oligonucleotide microarray: a new potential for high-throughput methylation analysis. *Genome Res* **12,** 158–164 (2002).
63. Tompa, R., McCallum, C. M., Delrow, J., Henikoff, J. G., van Steensel, B., and Henikoff, S. Genome-wide profiling of dna methylation reveals transposon targets of chromomethylase3. *Curr Biol* **12,** 65–68 (2002).
64. Adorjan, P., Distler, J., Lipscher, E., Model, F., Muller, J., Pelet, C., Braun, A., Florl, A. R., Gutig, D., and Grabs, G. Tumor class prediction and discovery by microarray-based DNA methylation analysis. *Nucleic Acids Res* **30,** e21 (2002).
65. Van Steensel, B. and Henikoff, S. Epigenomic profiling using microarrays. *Biotechniques* **35,** 346–350, 352–344, 356–347 (2003).
66. Novik, K. L., Nimmrich, I., Genc, B., Maier, S., Piepenbrock, C., Olek, A., and Beck, S. Epigenomics: genome-wide study of methylation phenomena. *Curr Issues Mol Biol* **4,** 111–128 (2002).
67. Vogel, F. and Motulsky, A. Human genetics: problems and approaches, 3rd edn. Berlin, New York, Springer-Verlag (1997).
68. Strohman, R. C. The coming Kuhnian revolution in biology. *Nat Biotechnol* **15,** 194–200 (1997).
69. Fedoroff, N., Masson, P., and Banks, J. A. Mutations, epimutations, and the developmental programming of the maize suppressor–mutator transposable element. *Bioessays* **10,** 139–144 (1989).
70. Weinberg, S. Scientist: four golden lessons. *Nature* **426,** 389 (2003).

CHAPTER 2

CORRELATING CpG ISLANDS, MOTIFS, AND SEQUENCE VARIANTS

LEAH SPONTANEO and NICK CERCONE

CONTENTS

2.1 INTRODUCTION

Epigenetic studies the changes in gene function and gene expression that are not discernable by mutations in the DNA sequence. The area of biology devoted to epigenetic is a recent development and has a large amount of room for growth with new research on cancer, mammalian gene expression, and technological advances constantly being brought forth from the community. Epigenetic inheritance focuses on both mitotic and meiotic cellular changes and the processes involved. Looking at cell differentiation and genetic imprinting through epigenetic has created new leads for cancer research in terms of tumor growth. The chromatin that controls DNA processes is an epigenetic mechanism in either an active or repressive state. There are three main mechanisms in epigenetic: DNA methylation, histone modifications, and the binding of non-histone proteins [1].

The CpG islands (CGIs) usually appear at the 5' end of genes containing GC-rich dinucleotides. Normally, these regions are unmethylated; however, when methylation occurs, gene regulation is affected and methylation sometimes leads to carcinogenesis. The importance of CGIs has produced numerous algorithms throughout the community dedicated to locating and understanding these regions in DNA [2]. Many of the traditional algorithms use the measures of length, GC content, and the number of observed over expected CpGs when determining if a section of DNA is a CGI. However, some newer algorithms employ a distance based detection method to identify CpG clusters [3]. Some of the features of unmethylated CpGs are their affinity to bind to a protein domain (*CXXC3*), their low mutation rate, and their association with open chromatin. Low methylation frequency has been correlated with high CpG density and *vice versa.*

The DNA methylation refers to the replacement of the attached hydrogen with a methyl group on a cytosine base, which causes the chromatin to become more compact affecting transcription factor binding [1]. Methylated cytosines will produce thymine when deaminated unlike unmethylated cytosines which produce uracil. The CpG density is lost due to the thymine mutation and this often inhibits the promoter regions of genes [4]. The affect of methylation is particularly important in the case of cancer and the epigenetic modifications that silence tumor suppressor genes. It has been observed that genes silenced epigenetically share sequence motifs in their promoter regions. This is one of the possible ways to detect cancer hypermethylation [5].

In genetics, a sequence motif is a short pattern of nucleotides that is deemed to have biological significance. If a motif appears in the exonic region of a gene, it may encode the structural motif of a specific protein. Regulatory sequence motifs are located in areas where regulatory proteins such as transcription factors bind to the DNA. In order to find the most significant motifs in DNA, many algorithms have been designed and applied to different organisms [6]. When describing a motif, pattern notation using regular expressions is what depicts the sequence. The *de novo* computational methods to discover important motifs take multiple input sequences and try to generate candidate motifs. Two of the most well-known algorithms often used by many researchers are BLAST and multiple EM for motif elicitation (MEME) [7, 8]. The CGIs are found in 40% of promoter and exonic regions of mammalian genes.

Other areas of the genome contain very few CpG dinucleotides and these areas are normally methylated [9]. Methylation of promoter CpGs is known to cause gene silencing and is heavily implicated in carcinogenesis. Gardiner–Garden and Frommer were the first researchers to use computational methods to detect and analyze CGIs using specific criteria: 200 bp (base pair) length DNA region, GC content greater than 50%, and observed CpG/expected CpG ratio (Obs_{CpG}/Exp_{CpG}) greater than 0.6 [10]. Although the original criteria provided a good starting point for CpG island detection, it did not take into account repeating regions of DNA. Alu repeats are short interspersed elements repeating within the genome that are approximately 280 bp in length often containing a high GC content and Obs_{CpG}/Exp_{CpG} ratio [11].

Takai and Jones analyzed human chromosomes 21 and 22 for CGIs using their algorithm that built upon the original criteria coined by Gardiner-Garden and Frommer. The new algorithm reduced the number of detected CGIs from 14,062 to 1,101, which is closer to the number of genes located on the two chromosomes (~750 genes). Using the same criteria, the new algorithm modifies the constraints of each criterion to produce better detection results. Now repeating elements such as Alu are not considered as often as before (from 7,651 to 122 Alus detected as CpG islands). The constraints for the new algorithm are as follows: length ≥500 bp, GC content ≥55%, and Obs_{CpG}/Exp_{CpG} ≥0.65.

The CpGcluster [3] discovers clusters of CpGs by looking at the distance between other CpGs on the same chromosome and applying statistical significance. The two algorithms were compared using many different factors (length, GC content, Obs_{CpG}/Exp_{CpG}) to determine which is better at finding CGIs [2]. The CpGcluster locates a much larger number of CGIs; however, after further analysis only 14.7 and 16.2% mapped to promoter regions of the human and mouse genomes, respectively. The study found that often multiple clusters of CpGs from CpGcluster were substrings of one large CGI detected by the Takai and Jones algorithm. Since there is no length limitation when CpGcluster detects CGIs, a much larger number of clusters are discovered.

Gene expression is one of the most important functions in all forms of life. Transcriptions factors are encoded in about 3–5% of genes in eukaryotes allow the repression or activation of specific genes within DNA. Chromatin plays another significant role in gene regulation and the network of interactions between transcription factors and chromatin structure is becoming increasingly important in epigenetic research. The DNA methylation is an epigenetic memory mechanism involved in the silencing of genes within eukaryotic organisms. Histone modifications are another instrument in the epigenetic inheritance that passes information from parent to daughter cells. Post-translational modifications of core histone proteins have been linked to transcription repression and activation [12].

Aberrant methylation of promoter regions of genes has been linked to gene silencing and loss of expression in diseases like cancer [1] and it is known that cancer mutation can cause alterations to protein signaling genes [13]. Single nucleotide polymorphisms (SNPs) are the most common variation in the genetic sequence of the human genome. The Studies continue to provide an increasing amount of evidence that SNPs are correlated with cancer and can be used as indicators of the disease [14]. Analyzing sequence variants in motifs within promoter regions of methylated genes could provide significant disease markers and possible sites for therapeutic study.

In this chapter, we examine several problems related to CGIs and DNA methylation of the promoter regions of genes:

(a) Design a method of CGI detection using powerful algorithms that improves performance while incorporating the ability to correlate the methylation status of the DNA with the location of the CGI.
(b) Incorporate the power of analysis alongside CGI detection for an all-in-one program that covers the needs of the community.
(c) Integrate motif finding into the detection algorithm, determine if the motifs are within CGIs, and verify if they are within a transcriptional start site (TSS).

The rest of the chapter is organized as follows. We first present the conceptual framework on which the program was designed. Then the methodology of the finalized program is discussed. Finally, the first set of experiments is presented.

2.2 METHODS

The detection of CGIs has evolved since the first detection algorithm was proposed by Gardiner–Garden and Frommer. Recently, studies have shown that incorporating the use of a hidden Markov model (HMM) in a detection algorithm can improve results [15]. When establishing new techniques, it is still important to consider the traditional methods and integrate the best features of both into a new algorithm. We combined a HMM, the Baum–Welch algorithm, and the Viterbi algorithm along with the traditional sliding window criteria to lower the detection of repeating elements in this work.

2.2.1 HMM

The HMM consists of a Markov process in which the state is unobservable. A Markov process is a random phenomenon where future probabilities are determined based on the most recent values. The HMM requires the knowledge of a few probabilities before it can be run on any data. These probabilities are split into three separate groups:

(a) Initial probabilities—the probabilities that determine which state the system will be in during the start of the algorithm. Often, the initial probabilities are equal among the different states (i.e. 2 states: P(i) = 0.5, P(j) = 0.5).

(b) Transition probabilities—the probabilities that provide the occurrence of a change from state i to state j (i.e. P(i|j) = 0.25, P(i|i) = 0.75).

(c) Emission probabilities—the probabilities distinguishing each state based on the observations of the system (i.e. P(x|i) = 0.22, P(x|j) = 0.36).

2.2.2 ESTIMATING PARAMETERS

The HMMs have three distinct sets of parameters or probabilities: the initial state probabilities that determine which state the system starts in; the transition probabilities that decide if the state will switch after a certain period of time; the emission probabilities showing whether the current symbol output belongs to one state or another. For an HMM to decode a sequence into a path of states, the parameters need to be trained on a sequence of symbols to detect the final probabilities the system will use when finding the Viterbi path. One of the best and most efficient methods of estimating probabilities is through the use of the Baum–Welch algorithm, which is a special case of expectation-maximization.

The Baum–Welch algorithm is often used in HMMs to estimate the unknown parameters or probabilities. It is also known as the forward-backwards algorithm and is a special case of the generalized expectation-maximization algorithm. It can produce maximum likelihood and posterior mode estimates for model parameters when given only the emission probabilities to work with. The algorithm starts by assigning initial probabilities to all of the model parameters. Then it continues until convergence happens by adjusting the probabilities of each model parameter to increase the probability of the model in accordance with the training set being scanned.

2.2.3 VITERBI ALGORITHM

The Viterbi algorithm uses dynamic programming to find the most likely sequence that the hidden states would take based on the observations in a parameterized model. This sequence is called the Viterbi path and it is usually related to HMMs. The Viterbi algorithm is very similar to the forward algorithm, which computes the probability that a set of observed events was generated by the model. The algorithm was designed in 1967 by Andrew Viterbi to decode convolutional codes within the noise of digital communication links.

The algorithm takes a HMM with possible Q states, initial probabilities π_i where i is the current state of the model, and transition probabilities $a_{i,j}$ where i, j is the change from state i to state j. Given a sequence of observable data $x_0,\ldots,x_L$, the algorithm will generate a state sequence $q_0, \ldots,q_L$ for each observable value. The algorithm produces the final output using recurrence relations.

$$V_{0,k} = P(X_0 \mid K)\cdot n_k$$
$$V_{l,k} = P(X_t \mid k)\cdot \max_{q\in Q}(a_{q,k} V_{l-1,q})$$

$V_{l,k}$ is the probability of the most likely state sequence based on the current $l + 1$ observations. The state sequence can be recovered by saving in memory the state q is in during the run through the 2nd equation. Then say there is a function, $St(k,l)$ that returns the value of q which produced $V_{l,k}$ when $l > 0$ and k when $l = 0$. The Viterbi path can be discovered using the following:

$$q_L = \arg\max_{q\in Q} (V_{L,q})$$
$$q_{l-1} = St(q_p l)$$

2.2.4 DNA METHYLATION ANALYSIS

Once the CGI detection algorithm runs and scans the genetic sequence, the researcher can use the detected island locations to create primer sequences to determine the methylation status of the CGI. Often, a separate statistics program is used to calculate significance. In work, the analysis of the data is available using the p-value derived from the Kolmogorov–Smirnov two-sample test and the distribution of methylated to unmethylated islands is tabulated through the calculation of the z-score. The Kolmogorov-Smirnov test uses minimum distance estimation to compare sample datasets with reference probability distributions equating them with a one-dimensional probability distribution. The test can be performed with one sample dataset (one sample K-S test) or with two sample datasets (two sample K-S test). The test either defines the mathematical distance between the empirical distribution function of a set of data and the cumulative distribution function of the reference distribution (one sample) or the distance between the empirical distributions of two separate sets of data (two sample). The samples calculated under the null hypothesis are taken from the reference distribution (one sample) or the same distribution (two samples) and form the null distribution for the test. When the Kolmogorov-Smirnov test is used as a goodness of fit test, the data is normalized and compared to a standard normal distribution.

The Kolmogorov-Smirnov statistic uses the empirical distribution function where $x_1,\ldots\ldots,x_n$ are a set of ordered data points,

$$E_n(X) = \frac{1}{n}\sum_{i=1}^{n} 1\{y_i \leq X\}$$

where $1\{y_i \leq x\}$ is the indicator function. The Kolmogorov statistic for a cumulative distribution function $E(x)$ can be calculated using:

$$K_n = \sup_x \left|E_n(x) - E(x)\right|$$

which calculates the supremum of the distances in the set, sup x. The Kolmogorov-Smirnov statistic often requires a large set of data to give an accurate acceptance or rejection of the null hypothesis; however, since we are working with the human genome and chromosomes seem to contain a large amount of CGIs [16, 17] and thus datasets should be large enough to produce accurate p-values. In work, we use the two sample K-S test to determine if two datasets (unmethylated and methylated CGIs) differ in their probability distributions in regards to length, GC content, and Obs/Exp ratio. The two sample K-S test uses the Kolmogorov statistic:

$$K_{n,\hat{n}} = \sup_x \left| E_{1,n}(x) - E_{2,\hat{n}}(x) \right|$$

where $E_{1,n}$ and $E_{2,\hat{n}}$ are the empirical distribution functions of each sample.

2.2.5 PROGRAM ARCHITECTURE

There are a multitude of detection programs that use many different algorithms to accurately detect CGIs within a genetic sequence. Some of the most popular CGI detection algorithms are Gardiner–Garden and Frommer [10], CpG Island Searcher [9], and CpGProd [18]. The Gardiner–Garden and Frommer algorithm is the original CGI detection algorithm which uses a sliding window of 200 bp along with a GC content greater than 50% and an observed-to-expected CpG ratio greater than 0.6. The CpG Island Searcher built upon the original foundation using a window of 500 bp, GC content ≥55%, and Obs_{CpG}/Exp_{CpG} ≥0.65. The CpGProd algorithm searches a genome for CGIs using two steps: (1) search for all CGIs in a submitted sequence based of the traditional criteria and (2) predict the orientation of promoters once the start CGI is discovered [18].

We decided to design a graphical user interface (GUI) for CGI detection program to provide accessibility for researchers that are not well versed in scripting or working with programs depending on the command prompt. While both the CpG Island Searcher and CpGProd have a web server that provides a user interface, it is not feasible to run a dataset as large as a human chromosome through the web. Both programs must be run using typed commands and parameters if not, running them through the web and do not provide the same interface as on the web for the user.

2.2.6 LAYOUT AND DESIGN

The original CGI detection layout and code was released by Tanner Helland (http://www.tannerhelland.com/ webcite) under the BSD license. It was originally written and designed in Microsoft Visual Basic 6.0, but we updated and modified the code using Microsoft Visual Basic .NET and Microsoft Visual Studio 2008. The GUI provides an intuitive method for loading the FASTA sequence file, setting up the HMM parameters, estimating the parameters based on the file, and running the algorithm. Once the Viterbi algorithm has defined when the sequence is in an island state ("I") or a normal state ("B"), the sliding window can be run to detect where the islands are located within the genomic sequence which show inside the graphs for Obs_{CpG}/Exp_{CpG} ratio and I/B ratio. Once the sliding window has scanned for CGIs, the results are shown in the textbox to the right (Figure 1).

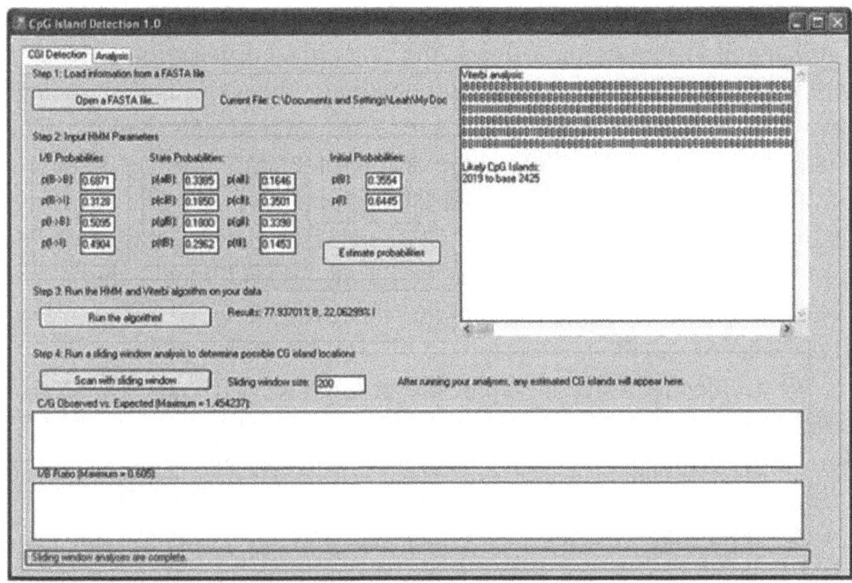

FIGURE 1 The CGI detection 1.0 layout.

We modified the layout to use a tab control in order to switch between the CGI detection and the analysis of the results. The analysis tab contains a small spreadsheet for the detected islands where the user can change the methylation status (methlyated/ unmethylated) and can enter a methylation score for each island. Methylation scores have been used in many different studies [16, 19, 20] as a measure of the strength of the methylation of a specific CGI to determine whether to classify it as methylated or unmethylated. Once the user has filled in the values for methylation status and score for each CGI, the analysis of the data can be completed.

The table next to the spreadsheet displays the separate mean values for CGI length, GC content, and Obs_{CpG}/Exp_{CpG} ratio for the unmethylated and methylated CGIs. The p-values are calculated using the Kolmogorov-Smirnov two sample test. The chart underneath the spreadsheet displays the z-score distribution for the methylated and unmethylated CGIs determined by using the methylation scores entereds by the user. The final chart shows the distribution of lengths of the CGIs across the two sets of data (Figure 2).

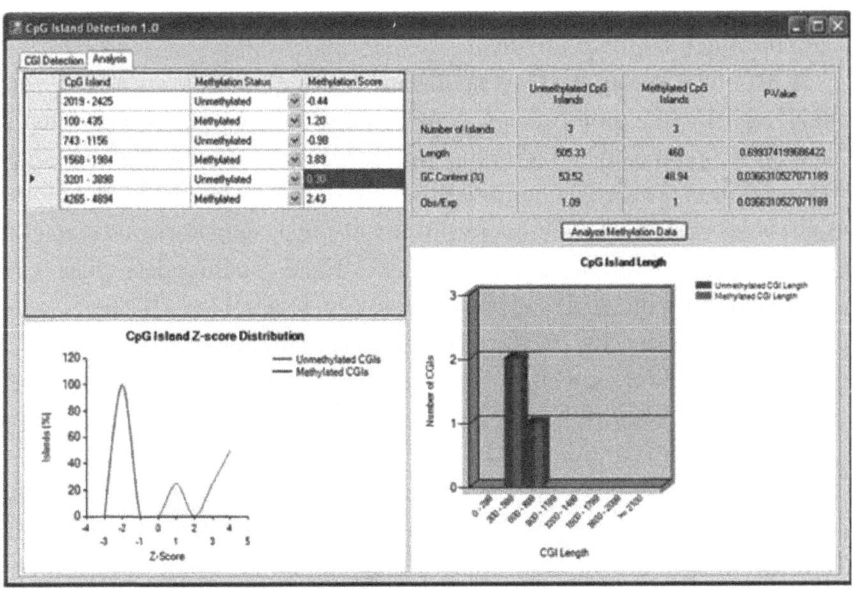

FIGURE 2 The CGI analysis layout and design.

We created this layout and design to provide practicality for users that are not computational experts and to offer an all-in-one solution for detecting and analyzing CGI data.

2.3 DISCUSSION

In this work, we designed a brand new CGI detection tool, CpG Island Detection 1.0. It uses a HMM and the Viterbi algorithm to detect CGIs within mammalian genomes. The parameters of the model are estimated using the Baum-Welch algorithm and the final method of locating islands uses a sliding window of a size specified by the program user. The tool provides a GUI for users, allowing textbox entry and one-click results. Even with the limitations in computational power, CpG Island Detection 1.0 stood up well against the Takai and Jones CpG Island Searcher. It was able to detect 347 CGIs within human chromosome 21 and the average lengths and Obs_{CpG}/Exp_{CpG} ratios were in line with that of the Takai and Jones method, 1206 bp ,and 0.87, respectively.

A list of 163 promoters within chromosome 21 from Weber et al. was analyzed using the tool's analysis tab. The methylation status was entered from the original data and after the tool's analysis some significant results were obtained. There is a definite difference between methylated and unmethylated islands and the compositions of their characteristics (Length, GC content, Obs_{CpG}/Exp_{CpG} ratio) which all had p-values less than 0.05 (as indicated by the Kolmogorov-Smirnov two-sample test). The z-score graph produced by the dataset showed that unmethylated CGIs show a normal distribution as well. A larger dataset of methylated islands is required in order to see if there is anything interesting in regards to how they are distributed.

Finally, the 13 methylated promoter regions and another 13 unmethylated regions were run through MEME and Motif Alignment and Search Tool (MAST) to determine if there are any significant motifs shared between CGIs. The methylated islands shared the 3 top motifs with percentages of 76.9, 69.2, and 61.5%, respectively; whereas, those motifs were only found in less than 50% of the unmethylated CGIs. The motifs found within the unmethylated islands were equally distributed throughout both the methylated and unmethylated regions. This indicates a possible correlation between motifs and methylation in regards to gene silencing. Those genes more likely to become methylated may contain motifs prone to methylation and mutation.

2.4 RESULTS

2.4.1 METHYLATION ANALYSIS

Looking at the composition of CGIs when methylated and unmethylated is important in the study of the epigenetic mechanism of methylation. The list of promoters from Weber et al. [21] was evaluated by taking the promoters found within chromosome 21 and analyzing them using the CGI Detection 1.0 program analysis tab. The 163 promoters were selected based on the promoter class given to each in the work. Those with a class of HCP or ICP were considered to contain CGIs within or covering the promoter region when looking at methylation. The CGI was considered methylated if the 5 mC log2 ratio 0.4 and unmethylated otherwise. Using these criteria, the promoter regions were run through the program, using the 5 mC log2 ratio for the methylation score of each island (Figure 3).

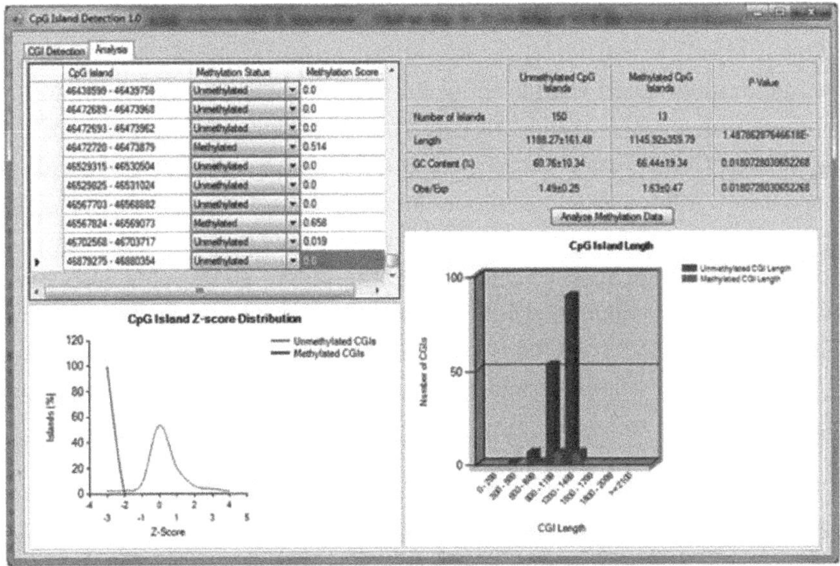

FIGURE 3 Analysis of promoters found on chromosome 21.

Of the 163 promoters, 13 were considered methylated and 150 were unmethylated. This distribution is unsurprising considering that promoters which are methylated tend to silence the function associated with the gene. The p-values calculated by the Kolmogorov-Smirnov test clearly show the significance between methylated and unmethylated CGIs with all categories being notably less than 0.05 as shown in Table 1.

TABLE 1 Comparison of methylated and unmethylated promoter CGIs.

	Unmethylated CpG Islands	**Methylated CpG Islands**	**P-Value**
Number of Islands	150	13	
Length	1188.27±161.48	1145.92±359.79	$<1.48\times10^{-11}$
GC Contents (%)	60.76±10.34	66.44±19.34	$<1.81\times10^{-2}$
ObsCpG/ExpCpG	1.49±0.25	1.63±0.47	$<1.81\times10^{-2}$

Methylated and unmethylated promoter CGIs were compared by sequence for average length, GC content (%), and ObsCpG/ExpCpG ratio. The p-values were determined using the Kolmogorov-Smirnov two sample test.

Spontaneo and Cercone *BMC Genomics* 2011 **12**(Suppl 2):S10 doi:10.1186/1471-2164-12-S2-S10

The unmethylated islands show a normal distribution according to the z-scores; however, it is hard to discern any important findings from the methylated CGIs. This is possibly due to the lack of methylated islands in this dataset. A larger count of methylated CGIs might have provided some deeper insight into how they tend to be distributed throughout the genome. Looking at the allocation of island length, it is clear that most islands within promoters are somewhere in the range of 900–1500 bp long. With a larger set of methylated CGIs, there might have been a more informative conclusion as to whether longer islands, such as those in promoters, are more susceptible to methylation.

2.4.2 CGI MOTIFS

It is still widely unknown as to why certain CGIs are more susceptible to DNA methylation as compared to others. The possibility that CGIs, especially those located within promoter regions and covering TSSs, contain motifs that are more likely to mutate or allow methylation has not yet been thoroughly researched. To determine if any of the CGIs contained similar motifs, MEME was run on 5 of the total 13 methylated promoters to discover the top 3 most significant methylated sequence motifs. The motifs were then checked against the entire 13 CGIs by using the MAST tool for alignment. A similar experiment was performed for 13 randomly chosen unmethylated promoter CGIs. It is interesting to note that even from this small sample of CGIs, a large percentage (76.9%) of methylated islands contain the top motif; whereas, less than half of the unmethylated islands contain that same motif (38.4%). The 3rd motif for methylated CGIs contains the most surprising result showing that it was almost non-existant within the scanned unmethylated CGIs as compared to the methylated promoters (7.7 and 61.5%, respectively).

However, when looking at the unmethylated island motifs, both the methylated and unmethylated CGIs have almost equal occurrences for the top 3 hits. With a larger database of methylated CGIs, a more significant result could be obtained in terms of the motifs found within the islands, especially the CGIs that cover promoter regions and TSSs.

2.5 CONCLUSION

Using an entire genome analysis will provide better insight into the analysis of methylated and unmethylated islands. A more recognizable distribution for methylated islands might be possible with a larger set of data points to consider. Acquiring more data will better showcase the power of the analysis tool and the assistance it provides to users scanning genomic regions for CGIs. The tool will be improved to allow a user to save their results and export the located CGIs to a file for other experiments and analyzes. It will also be upgraded to allow a user to import located CGIs from a file, rather than requiring a rescan of a sequence every time an analysis needs to be completed.

Obtaining a larger number of methylated islands for analysis with MEME and MAST will provide more significant results in terms of motifs within CGIs. Looking across the entire genome and the promoter regions found within a multitude of chromosomes could show whether the methylation-prone motifs are consistent within susceptible CGIs across the genome and provide an insight into why certain genes become silenced within diseases such as cancer. Developing this new tool for the community using powerful algorithms has shown that combining analysis with CGI detection will improve the continued research within the field of epigenetic.

KEYWORDS

- **Baum-Welch algorithm**
- **Gardiner-Garden and Frommer algorithm**
- **Hidden Markov model**
- **Kolmogorov-Smirnov test**
- **Viterbi algorithm**

ACKNOWLEDGMENT

Thank you to IBM for the use of the Blade Center computer system and to Steven Chen for the assistance in getting the experiments up and running.

AUTHORS' CONTRIBUTIONS

Leah Spontaneo created the program and carried out the experiments. Nick Cercone provided advice and resources to perform all of the required tests and design the algorithm and helped draft the manuscript. All authors read and approved the final manuscript.

REFERENCES

1. Bock, C. and Lengauer, T. Computational epigenetics. *Bioinformatics*, **24**(1), 110 (2008).

2. Zhao, Z. and Han, L. Cpg islands: algorithms and applications in methylation studies. *Biochemical and biophysical research communications*, **382**(4), 643–645 (2009).
3. Hackenberg, M., Previti, C., Luque-Escamilla, P. L., Carpena, P., Martinez-Aroza, J., and Oliver, J. L. Cpg cluster: A distance-based algorithm for cpg-island detection. *BMC Bioinformatics*, **7**, 446 (2006).
4. Siegfried, Z. and Simon, I. DNA methylation and gene expression. Wiley Interdisciplinary Reviews, *Systems Biology and Medicine,* **2**, 362–371 (2009).
5. Goh, L., Murphy, S. K., Muhkerjee, S., and Furey, T. S. Genomic sweeping for hypermethylated genes. *Bioinformatics*, **23**(3), 281–288 (2007).
6. Das, M. K. and Dai, H. K. K. A survey of dna motif finding algorithms. *BMC bioinformatics*, **8**(Suppl 7), S21+ (2007).
7. Li, N. and Tompa, M. Analysis of computational approaches for motif discovery. *Algorithms for molecular biology*, **1**(1), 8+ (2006).
8. McGinnis, S. and Madden, T. L. Blast: at the core of a powerful and diverse set of sequence analysis tools. *Nucleic Acids Res*, **32** (2004)
9. Takai, D. and Jones, P. A. Comprehensive analysis of Cpg islands in human chromosomes 21 and 22. *Proc Natl Acad Sci USA*, **99**(6), 3740–3745 (2002).
10. Gardiner-Garden, M. and Frommer, M. Cpg islands in vertebrate genomes. *Journal of molecular biology*, **196**(2), 261–282 (1987).
11. Schmid, C. W. Does sine evolution preclude alu function? *Nucl. Acids Res*, **26**(20), 4541–4550 (1998).
12. Van Steensel, B. Mapping of genetic and epigenetic regulatory networks using microarrays. *Nature Genetics*, **37**(Suppl), S18S24 (2005).
13. Bianco, R., Melisi, D., Ciardiello, F., and Tortora, G. Key cancer cell signal transduction pathways as therapeutic targets. *European journal of cancer*, **42**(3), 290–294 (2006).
14. Bond, G. L., Hu, W., and Levine, A. A single nucleotide polymorphism in the MDM2 gene: from a molecular and cellular explanation to clinical effect. *Cancer Res*, **65**, 5481–5484 (2005).
15. Wu, H., Caffo, B., Jaffee, H. A., Irizarry, R. A., and Feinberg, A. P. Redefining Cpg islands using hidden markov models. *Biostatistics*, **11**(3), 499–514 (2010).
16. Straussman, R., Nejman, D., Roberts, D., Steinfeld, I., Blum, B., Benvenisty, N., Simon, I., Yakhini, Z., and Cedar, H. Developmental programming of cpg island methylation profiles in the human genome. *Nature structural and molecular biology*, **16**(5), 564–571 (2009).
17. Yamada, Y., Shirakawa, T., Taylor, T. D., Okamura, K., Soejima, H., Uchiyama, M., Iwasaka, T., Mukai, T., Muramoto, K., Sakaki, Y.,and Ito, T. A comprehensive analysis of allelic methylation status of Cpg islands on human chromosome 11q: comparison with chromosome 21q. *DNA sequence, the journal of DNA sequencing and mapping*, **17**(4), 300–306 (2006).
18. Ponger, L. and Mouchiroud, D. CpGProd: identifying CpG islands associated with transcription start sites in large genomic mammalian sequences. *Bioinformatics*, **18**, 631–633 (2002).
19. Dai, W., Teodoridis, J. M., Graham, J., Zeller, C., Huang, T. H. M., Yan, P., Vass, K. J., Brown, R., and Paul, J. Methylation linear discriminant analysis (mlda) for identifying differentially methylated cpg islands. *BMC Bioinformatics*, **9**, 337+ (2008).

20. Lapidus, R. G., Nass, S. J., Butash, K. A., Parl, F. F., Weitzman, S. A., Graff, J. G., Herman, J. G., and Davidson, N. E. Mapping of ER Gene CpG Island Methylation by Methylation-specific Polymerase Chain Reaction. *Cancer Res*, **58**, 2515–2519 (1998).
21. Bird, A. P. Cpg islands as gene markers in the vertebrate nucleus. *Trends in Genetics*, **3**, 342–347 (1987).
22. Weber, M., Hellmann, I., Stadler, M. B., Ramos, L., Pääbo, S., Rebhan, M., and Schübeler, D. Distribution, silencing potential and evolutionary impact of promoter DNA methylation in the human genome. *Nature genetics*, **39**(4), 457–466 (2007).

PART II

DNA METHYLATION

CHAPTER 3

PROMOTER HYPERMETHYLATION

DAVIDE PELLACANI, RICHARD J. PACKER, FIONA M. FRAME, EMMA E. OLDRIDGE, PAUL A. BERRY, MARIE-CHRISTINE LABARTHE, MICHAEL J. STOWER, MATTHEW S. SIMMS, ANNE T. COLLINS, and NORMAN J. MAITLAND

CONTENTS

3.1 INTRODUCTION

Epigenetic regulation of gene expression is a dynamic mechanism, which permits precise regulation throughout differentiation [1]. It plays a crucial role in preserving the hierarchical structure of tissues and is involved in maintaining stemness and fate determination of adult stem cells [2, 3]. Indeed, DNA methylation varies throughout cell differentiation [4] and epigenetic control is required for the multipotency of hematopoietic stem cells [5].

There is mounting evidence to support the hypothesis that cancers can retain the hierarchical structure present in normal tissues, but that homeostasis is disrupted, leading to aberrant replication and differentiation. As in normal tissues, a small percentage of cancer cells can persist and initiate new tumor growth (tumor-initiating cells or cancer stem cells [CSCs]), while most cells proceed to terminal differentiation [6].

The CD133 is a pentaspan membrane glycoprotein first identified in humans as a hematopoietic stem cell marker [7] and is currently used for the identification of stem cells from several tissues and cancer types [8]. In non-malignant human prostate, CD133 and $\alpha_2\beta_1$integrin are two markers that, when used in combination, have been demonstrated to enrich for a cell population with stem cell features [9]. Indeed, CD133$^+$ cells from human primary tissues represent a very small subpopulation (0.1–3.0% of the total prostatic epithelium) located in the basal compartment. Tissue stem cells are restricted to the $\alpha_2\beta_1$integrin^{hi} population, have a high colony-forming ability and proliferative potential, and are able to regenerate fully differentiated prostate epithelium in vivo. The stem cells express basal cell markers, such as CD44 and CK5/14, but do not express luminal markers AR, PSA or depend on androgens for their survival [9–11]. Other markers have been used to identify prostate stem cells, for example, in murine models, where selection markers include Sca-1 [12] and CD117 [13].

The CD133$^+$ cells from prostate cancer biopsies (PCSCs) are similar in phenotype to normal prostate stem cells. They are rare (0.1–0.5% of the total cell population) and represent the clonogenic population with highest proliferative potential. Moreover, they express basal cell cytokeratins, but do not express AR [14].

Disruption of epigenetic mechanisms is found in all cancers and, together with genetic changes, plays a key role in cancer initiation and progression [15]. Epigenetic studies in prostate cancer (CaP) have resulted in the identification of hundreds of hypermethylated genes, of which GSTP1 is the most studied [16], as well as changes to chromatin structure and histone-modifying enzymes [15].

However, these studies do not take into account the hierarchical structure of cancer, since they describe epigenetic alterations that occur in the bulk population of cancer cells. It has been proposed that disruption of epigenetic control may result in formation of aberrant self-renewing cells [17], culminating in a complete deregulation of the hierarchical system, ultimately leading to cancer [18]. Thus, understanding how epigenetic regulation of gene expression controls the differentiation process and its deregulation in cancer is of great importance in order to develop new therapeutic strategies for cancer, directed to the therapy resistant cancer stem cell population.

The CD133 expression is controlled at multiple steps including transcriptional regulation, alternative transcription initiation sites, alternative splicing and post-translational modifications [8, 19]. This fine regulation results in the maintenance of stem cell-specific CD133 expression patterns. Twelve different mRNA isoforms, generated by alternative splicing, have been described in various mammals (CD133.s1–s12), of which at least seven are expressed in human cells [19]. Moreover, five different alternative first exons, regulated by five TATA-less promoters, have been described [20] (Figure 1A). Transcription is initiated from different first exons in a tissue specific manner. In particular, only exon 1A is expressed in prostate, indicative of promoter P1 activity. The presence of a large CpG island in the CD133 promoter area and the silencing of promoters P1 and P2 transcriptional activity by in vitro DNA methylation, suggest that this gene can be regulated by epigenetic mechanisms [20].

Here, it is shown that CD133 expression is efficiently repressed by promoter methylation in prostate cell lines. However, in primary cultured cells from prostate epithelium and tumor xenografts, regulation of CD133 expression is independent of DNA methylation, indicating that, for this gene, cell lines are not indicative of DNA methylation status in primary tissues. In addition, the CD133 promoter is not hypermethylated in prostate cancer tissues, highlighting the important role for CD133 in the maintenance of the hierarchical structure of cancer.

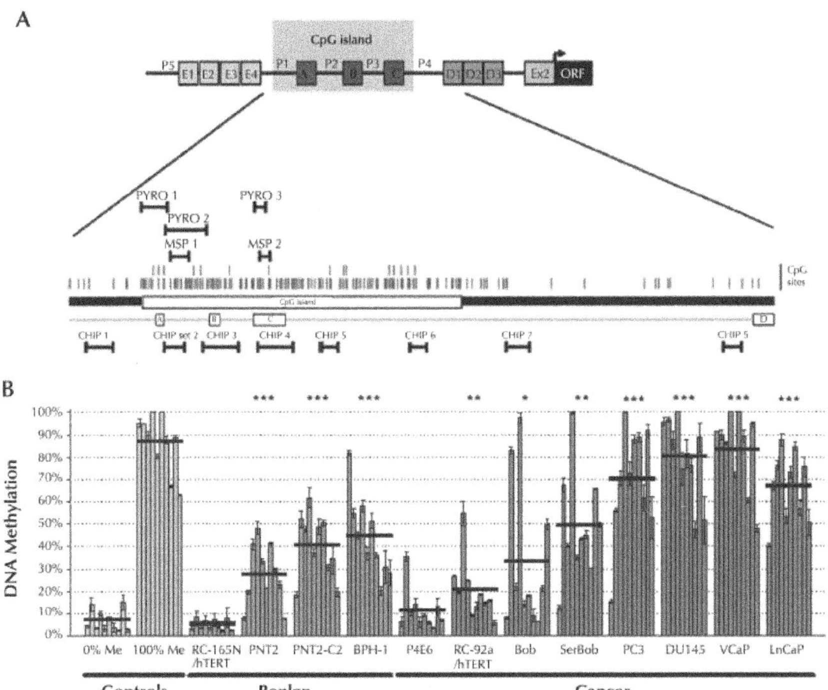

FIGURE 1 CD133 promoter is hypermethylated in prostate cancer cell lines. (A) The CD133 promoter (adapted from Shmelkov et al. [20]) (A,B,C,D1–3, E1–4 = alternative first exons; P1–P5 = promoters; Ex2 = Exon 2; ORF = open reading frame; PYRO 1–3: pyrosequencing assays; MSP1–2: MSP assays; CHIP 1–8: ChIP primers). (B) Pyrosequencing analysis of the CD133 promoter performed in prostate cell lines (PYRO 1 assay; bars = single CpG sites; n = 3 technical replicas; x ± SD; black line = average of 10 CpG sites; * = $p < 0.05$; ** = $p < 0.01$; *** = $p < 0.001$ in an unpaired t-test).

3.2 MATERIALS AND METHODS

3.2.1 CELL LINES, TISSUE PROCESSING, PRIMARY EPITHELIAL CELL CULTURE, AND XENOGRAFTS

Human prostatic tissue was obtained from patients undergoing transurethral and retropubic prostatectomy for BPH or undergoing radical prostatectomy for CaP (with patient consent and full ethical permission from York District Hospital and Castlehill Hospital, Hull). BPH or CaP diagnosis was confirmed by histological examination of representative adjacent fragments.

Tissues were disaggregated and cultured as described previously [9, 10]. Basal cells were then cultured and further fractionated on the basis of adhesion to type I collagen [10]. $CD133^+$ cells were selected from cells that adhered within 20 min using MACS microbeads linked to anti-human CD133, according to the manufacturer's instruction (Miltenyi Biotec) [14].

Xenografts were generated by subcutaneous grafting of CaP tissue in $RAG2^{-/-}$ gamma $C^{-/-}$ mice. Tumors generated were serially passaged in vivo and routinely genotyped to confirm the original patient's genotype. Early passages (1–3) were used for DNA methylation studies.

PC3 xenografts were generated by injecting subcutaneously 106 PC3 cells embedded in 100 µl of Matrigel (BD Biosciences) in Balb/c Nude mice. Tumors were harvested after 29 days from the injection.

3.2.2 DNA PURIFICATION AND SODIUM BISULFITE CONVERSION

The DNA was extracted using the DNeasy Blood and Tissue Kit (Qiagen) and the QIAamp DNA micro Kit (Qiagen) for small samples. 0% and 100% methylated controls were purchased from Qiagen (EpiTect PCR Control DNA). 50 ng/1 µg of DNA was bisulfite converted using the EpiTect Bisulfite Kit (Qiagen). Converted DNA from SCs was amplified using the EpiTect Whole Bisulfitome Amplification Kit (Qiagen).

3.2.3 PYROSEQUENCING ASSAY

The CD133 promoter sequences were amplified by PCR using specific primers for 3 regions of the CpG island and sequenced using the PyroMark Q24 System (Qiagen). Data were analyzed with PyroMark Q24 software. In primary epithelial cultures and xenografts, a mixture of human and mouse DNA was analyzed. PYRO 1 and PYRO 2 (not shown) assays were human specific. PYRO 3 gave non-specific PCR products with mouse DNA (not shown) so it was not used for mixed (human/mouse) samples.

3.2.4 METHYLATION-SPECIFIC PCR

Methylation-specific PCR was carried out as previously described [21] in a final 25 μL reaction mixture using 10 ng of bisulfite converted DNA as template and Platinum Taq DNA Polymerase (Invitrogen). Primer sequences are listed in Supplementary Table 2. The PCR program was: 94°C for 2 min, then 35 cycles of 94°C for 20 seconds, 55°C for 20 seconds, and 72°C for 20 seconds, with a final extension of 5 min at 72°C. PCR products were separated by electrophoresis through a 2% agarose GelRed (Invitrogen) stained gel for 1 h at 80 V and then visualized using a Gene Genius bio-imaging system.

3.2.5 RNA EXTRACTION, CDNA SYNTHESIS, AND QRT- PCR

The RNA was extracted using the RNeasy kit (Qiagen) and reverse transcribed using random hexamers and reverse transcriptase (Superscript III, Invitrogen). Real time PCR was carried out using either Power mix SYBR Green and specific primers or TaqMan gene expression pre-synthesized reagents and master mix (CD133: Hs01009257_m1 and RPLP0: Hs99999902_m1, Applied Biosystems). Reaction volumes were reduced to 25 μl for SYBR Green and 10 μl for TaqMan, and were carried out as previously described [22]. Gene expression was considered undetectable if fewer than 2/3 reactions were positive after 40 cycles of PCR.

3.2.6 FLOW CYTOMETRY

Live cells were stained with CD133/2(293C)-APC antibody (1:11) (Miltenyi Biotec) and analyzed on a CyAn ADP flow cytometer (Dako Cytomation,

Denmark). Doublet cells were gated out with pulse width. Dead cells were gated out by DAPI exclusion. More than 250,000 events were analyzed.

3.2.7 CHROMATIN IMMUNOPRECIPITATION

The ChIP assays were performed as previously described [23]. The antibodies histone H3 (Abcam), rabbit IgG, H3K4me2, and H3K27me3 (Millipore) were used at a 1:100 dilution to immunoprecipitate an equivalent 20 μg of DNA in ChIP assay. To standardize between experiments, the percentage of immunoprecipitation (IP) was calculated by dividing the value of the IP by the value of the corresponding input (both values normalized for dilution factors).

3.3 DISCUSSION

The CD133 is widely used as a marker for CSCs in many different solid tumors including: colon [24, 25], brain [26, 27], skin [28], pancreatic [29], liver [30, 31], and prostate [14]. In both normal and cancerous prostate, the expression of CD133 is restricted to a very small subpopulation of cells with stem-like features, suggesting a tight regulation for the expression of this gene [9, 14, 32].

The data presented here indicated that DNA methylation mediated suppression of CD133 expression in prostate epithelial cell lines, where an inverse correlation between expression and DNA methylation was clearly seen, together with re-expression of both the mRNA and the glycosylated protein after gene demethylation by 5-Aza-2'-deoxycytidine treatment.

However, data obtained from prostate tissue and primary epithelial cultures displayed a stark contrast to that obtained in cell lines. A huge variation (spanning nearly 100,000 fold) in CD133 expression was found across the cell lines analyzed, while CD133 was undetectable or expressed at very low levels in prostate primary epithelial cultures. These results provide strong evidence that regulation of CD133 expression can be disrupted during long-term culture in vitro.

Moreover, the low levels of CD133 mRNA detected in primary epithelial cultures suggested that the CD133 gene was repressed in the vast majority of prostate basal epithelial cells (with the exception of the stem cell population). The very low levels of promoter methylation found in these samples indicated that repression of CD133 expression was independent of promoter methylation and implied that other mechanisms were required to control CD133 expression in prostate tissues and primary cultures. Interestingly, CD133 expression was also not deregulated in primary prostate cancer, suggesting that a tight regulation of CD133 expression was important in the hierarchical structure of both normal and cancerous prostate.

Prostate epithelium is heterogeneous and composed of various cell populations at different differentiation stages. When SC ($CD133^{+}$), TA and CB ($CD133^{-}$) cell populations from primary epithelial cultures were analyzed separately, the results obtained showed no differential methylation of the CD133 promoter in individual populations. Therefore mechanisms other than DNA methylation must regulate CD133 expression in the prostate epithelial cell hierarchy.

Our results indicated that one such mechanism is chromatin condensation. In prostate cell lines, a condensed status correlated with low levels of mRNA (PC3 and P4E6 cells). This is in line with the concept of crosstalk between different epigenetic mechanisms [33] and with previous findings in glioblastoma and colon cancer cell lines [34] where CD133 was regulated by both DNA methylation and histone modifications. Interestingly, chromatin structure seemed not only to parallel DNA methylation in repressing CD133 expression, but also to have an active role in repressing transcription even when hypermethylation was not present (P4E6 cells). The same mechanism was also present in primary epithelial cultures from both BPH and CaP, clearly indicating an important role for chromatin structure in repressing CD133 expression in primary prostate.

In prostate tissues, the glyscosylated form of CD133 was expressed only in a very small subpopulation of basal epithelial cells [9, 35]. However, it has recently been shown that an isoform of CD133 protein that does not have the same glycosylation pattern is present in terminally differentiated prostate luminal cells [35]. So, it is clear that the CD133 gene needs to be dynamically regulated throughout differentiation of prostate epithelia, clearly supporting the hypothesis that long-term transcriptional silencing caused by DNA methylation is unlikely to affect CD133 expression in prostate epithelia. This also supports our findings that more dynamic mechanisms, such as changes in histone modifications and chromatin structure are the more likely control mechanisms.

Although CD133 is widely used as a stem cell marker in various types of cancer, very little is known about its molecular function and its functional involvement in tumor and metastasis formation. Even if there is evidence that CD133-positive cells from various cancers are more resistant to anti-cancer therapies [36-38], it is not known whether CD133 has a primary role in this resistance or it just happens to mark resistant cells. Indeed CD133 has been shown to be involved in maintaining neuroblastoma cells in an undifferentiated state, and downregulation of CD133 led to inhibition of tumor formation [39]. However, the wide expression of this surface marker in various human tissues [40] and the sparse knowledge of its molecular function pose great difficulties in using CD133 as a target for cancer stem cell therapy.

The results obtained show two different mechanisms for regulation of CD133 expression in cell lines (DNA methylation dependent) and primary tissues and recently established prostate cancer xenografts (DNA methylation independent). Although discrepancies between cell lines and primary samples have been discussed in the literature for at least 20 years [41], cell lines remain the most frequently used model for epigenetics and cancer epigenetics studies; in many cases with weak correlations to the original tissue/cancer. It is known that *de novo* methylation is a common event during cell line establishment [42, 43] as part of the adaptation process that cells undergo during long-term culture. This process results in downregulation of genes that are nonessential in culture, including many tissue-specific genes [41]. The results presented here for the expression of CD133, a common stem cell marker in multiple tissues, emphasise that cell lines do not represent a valid model for DNA methylation studies, since culture conditions influence promoter methylation.

Thus, the choice of the correct cell model system is of paramount importance when studying epigenetic regulation of gene expression in cancer biology and stem cell biology. The results with CD133 highlight the need for verification (using primary tissues) of the results obtained with cell lines. For example, in the last few years, several high-throughput epigenetic studies compared commercially available normal epithelial cultures (non immortalized and cultured for a limited number of passages) with established cancer cell lines in prostate [44, 45] and other tissues. In these comparisons, epigenetic adaptation of cell lines to culture conditions was not taken into account and the results obtained might be biased by *in vitro* adaptation.

Finally, in recent studies designed to isolate cells with stem cell features from prostate cancer cell lines a great discordance was reported regarding the expression of CD133 on the surface of several CaP cell lines. The data presented here, reveal the limitations of the use of cell lines in such studies, and indicate that prolonged *in vitro* culture affects the fine gene regulation that is essential for the maintenance of prostate epithelial hierarchy. Moreover, the data presented by Pfeiffer and Schalken [46], is in accordance with our data, confirming the lack of cell surface expression of CD133 in many established prostate cancer cell lines, and when expressed, CD133 did not appear to select for cells with stem cell characteristics. Our results now provide a mechanistic explanation for the apparently contrasting results presented in that study.

3.4 RESULTS

3.4.1 THE CD133 PROMOTER REGION IS HYPERMETHYLATED IN PROSTATE CANCER CELL LINES

The CD133 promoter methylation was quantified by pyrosequencing in prostate cell lines using 3 assays (PYRO 1–3 assays) located on promoter P1/exon 1A, promoter P2 and exon 1C respectively (Figure 1A). Using the PYRO 1 assay, significant hypermethylation was found in the majority of cell lines analyzed, relative to the 0% methylation control (0% Me) ($p < 0.05$), apart from the benign cell line RC-165N/hTERT and the cancer cell line P4E6 (Figure 1B). However, when each individual CpG site was analyzed separately, significant hypermethylation ($p < 0.05$) was found at CpG sites 2-3-4 in P4E6 cells. Very high levels of CD133 promoter methylation were found in the CaP cell lines PC3 (70%), DU145 (80%), VCaP (83%) and LnCaP (67%) ($p < 0.001$). The benign cell lines BPH-1, PNT2 and PNT2-C2, also showed significant hypermethylation ($p < 0.001$, average methylation between 30% and 50%). The CaP-derived Bob and SerBob cell lines showed a very heterogeneous pattern of methylation throughout the sequence analyzed, with higher levels of methylation in SerBob ($p < 0.01$) than Bob ($p < 0.05$). Significant, but low methylation levels ($p < 0.01$) were found in the RC-92a/hTERT cancer cell line.

To assess whether the levels of methylation were consistent along the entire CpG island, PYRO 2 and 3 assays were carried out, and both showed comparable patterns of methylation with the PYRO 1 assay. Finally, to confirm our findings, a standard methylation-specific PCR (MSP) was also carried out and gave results that matched with those obtained by pyrosequencing.

Both malignancy and culture conditions (presence of fetal calf serum - FCS) influences the methylation status of the CD133 CpG island. The cell lines analyzed could be divided into four groups, depending on the type of tissue from which they were derived (benign or CaP) and the amount of FCS present in the culture medium (≤ 2% or >2%). A significant difference ($p < 0.01$) in average CD133 promoter methylation was found between benign and cancer derived cell lines cultured in high levels of FCS, but also between CaP cell lines cultured in low or high levels of FCS

3.4.2 CD133 EXPRESSION IS REGULATED BY DNA METHYLATION IN PROSTATE CELL LINES

In order to test whether CD133 expression was directly regulated by DNA methylation, prostate cell lines were treated with the demethylating agent 5-Aza-2'-deoxycytidine (1 μM for 96 h). CD133 expression (measured by qRT-PCR) was induced in 2 out of 4 benign cell lines (BPH-1 and PNT1A) and all of the CaP cell lines analyzed (P4E6, RC-92a, Bob, SerBob, PC3, DU145 and LnCaP) with the exception of VCaP (Figure 2A). The lack of induction in VCaP, could be explained by the fact that this cell line has a doubling time of 5–6 days [47], which is insufficient time for the

5-Aza-2'-deoxycytidine to be incorporated into the genome during the 96 h treatment and to exploit its demethylating function. Importantly, DNA demethylation did not induce CD133 expression in RC-165N/hTERT, confirming that CD133 is not repressed by DNA methylation in this cell line.

A

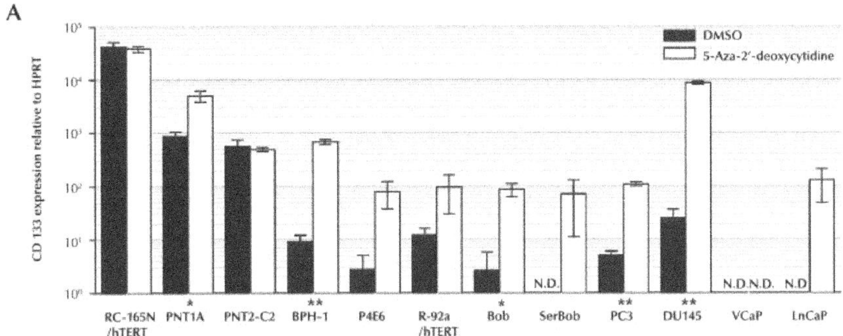

B

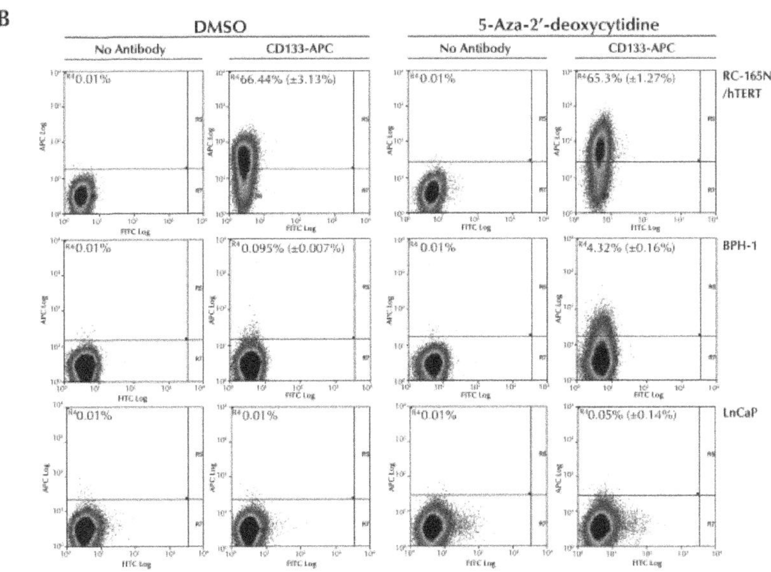

FIGURE 2 *(Continued)*

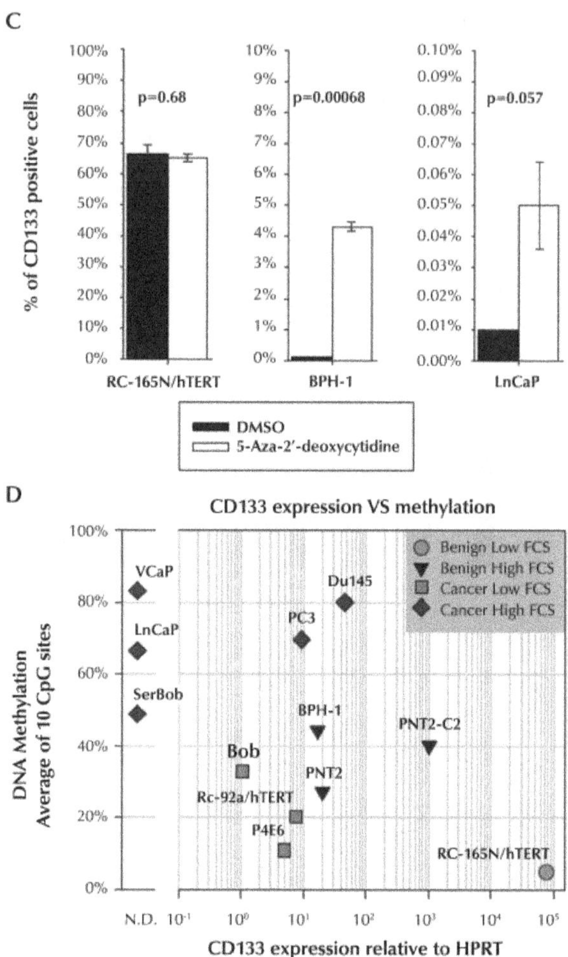

FIGURE 2 CD133 expression is repressed by DNA methylation in prostate cancer cell lines. (A) qRT-PCR analysis (relative to HPRT, calibrator = P4E6) of CD133 expression in prostate cell lines treated with 1 μM 5-Aza-2'-deoxycytidine for 96 hr. (N.D. = expression undetectable after 40 cycles, n = 2; x ± SD, * = p <0.05; ** = p <0.01 in a paired t-test, p-value not calculated in cell lines with non-detectable levels of CD133). (B, C) FACs analysis of CD133 expression in prostate cell lines treated with 1 μM 5-Aza-2'-deoxycytidine for 96 hr. Live cells were stained with CD133/2(293C)-APC antibody (Miltenyi Biotec) (CD133-APC) or without any antibody (No Antibody) and analyzed by FACs. For each dot plot, X axes: FITC channel fluorescence (not stained); Y axes: APC channel fluorescence (CD133 or No Antibody control); Percent of CD133$^+$ cells is indicated and the results are summarised in C. (D) Dot-plot showing expression (qRT-PCR, relative to HPRT, calibrator = Bob) and methylation of CD133 in a panel of prostate cell lines grouped on the basis of their origin (benign or cancer) and their culture conditions (low FCS = 0–2%; high FCS = 5–10%).

Next, 3 cell lines representative of the panel used were treated with 1 μM 5-Aza-2'-deoxycytidine for 96 h and analyzed by FACs for the expression of the glycosylated form of CD133 (Figure 2B and 2C). As expected, no significant change in CD133 expression was seen in RC-165n/hTERT after demethylation, while a significant increase was seen in BPH-1 and marked increase (although not significant) in LnCaP.

Taken together, these results show that DNA demethylation induces CD133 upregulation, indicating that promoter methylation suppresses CD133 expression in prostate cell lines.

A direct comparison between CD133 expression, measured by qRT-PCR, and DNA methylation (PYRO 1 assay), confirmed that hypermethylation of the CD133 promoter results in downregulation of gene expression (Figure 2D). Cell lines expressing high levels of mRNA had the lowest levels of promoter methylation (e.g. RC-165N/hTERT), while CD133 was strongly downregulated when high levels of methylation were present (e.g. VCaP and LnCaP). However, in cancer cells grown in low levels of FCS, the CD133 promoter showed low DNA methylation (less than 20%) associated with low, but detectable expression (P4E6 and RC-92a/hTERT). This confirmed that culture conditions influence DNA methylation of the CD133 promoter and suggests that other mechanisms, in addition to DNA methylation, resulted in downregulation of CD133 expression.

Moreover, mRNA abundance and expression of the glycosylated isoform of CD133 (detected by the 293C3 antibody) directly correlated in prostate cell lines. All the benign cell lines (RC-165N/hTERT, PNT1A, PNT2-C2, BPH-1) and 3 of the cancer cell lines (Bob, P4E6 and RC-92a/hTERT), analyzed for CD133 expression (by flow cytometry), contained $CD133^+$ cells and had detectable levels of mRNA. In contrast, $CD133^+$ cells were not found in PC3 and DU145 cells, although detectable levels of mRNA were found; while in LnCaP and VCaP cells, CD133 was not detectable at either the mRNA or protein level. Finally, in SerBob cells, although CD133 was undetectable at the mRNA level, a very small subpopulation of cells expressing very low levels of CD133 protein was detected.

3.4.3 LOW LEVELS OF CD133 PROMOTER METHYLATION ARE FOUND IN PROSTATE TISSUES AND PRIMARY EPITHELIAL CULTURES

The CD133 expression (qRT-PCR - Figure 3A) and methylation of its promoter (PYRO 1 assay - Figure 3B-D) were then measured in primary epithelial cultures derived from clinical samples of benign prostatic hyperplasia (BPH), CaP and castration resistant prostate cancer (CR-CaP). Expression of CD133 was undetectable in 7 out of 10 samples, and detectable, but at low levels, in the remaining 3. No differences in CD133 mRNA expression were seen between BPH and CaP. These results are in line with previous results showing that CD133 is expressed (at the protein level) only in a small subpopulation of the prostate primary epithelial cultures [9, 14, 22, 48]. DNA methylation levels were unexpectedly low, with the average methylation less than 20%, in all the samples analyzed. No significant distinction was seen between BPH, CaP or CR-CaP. In fact, only 2/6 BPH, 1/8 CaP and none of the CR-CaP samples contained significantly hypermethylated DNA compared to the 0% Me control. CaP 17 (Figure 3C) and CaP 28 (Figure 3D) clearly showed hypermethylation of a

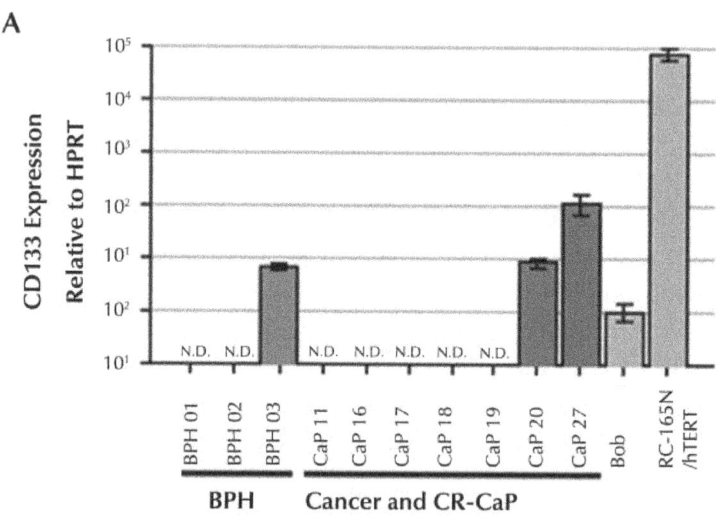

FIGURE 3 *(Continued)*

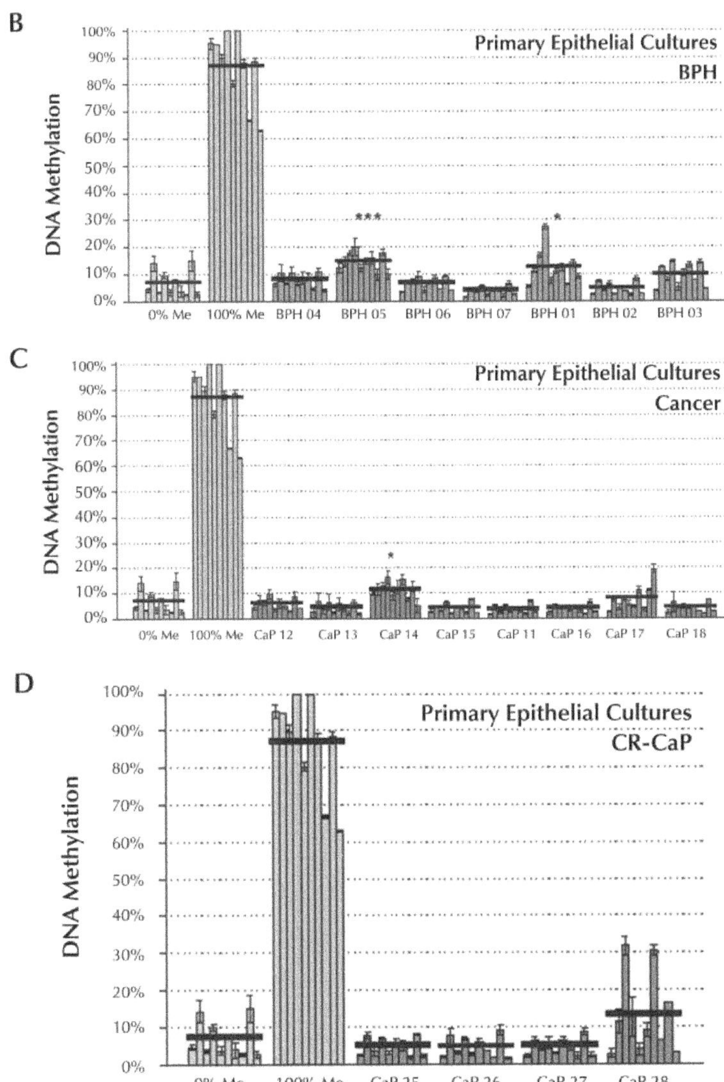

FIGURE 3 CD133 expression and methylation in prostate primary epithelial cultures. (A) qRT-PCR analysis (relative to HPRT, calibrator = Bob; N.D. = expression undetectable after 40 cycles) of CD133 expression in prostate primary epithelial cultures (RC-165N/hTERT and Bob were used as controls). Methylation analysis of the CD133 promoter in prostate primary epithelial cultures derived from BPH (B), CaP (C) and CR-CaP (D) by pyrosequencing (PYRO 1 assay; bars = single CpG sites; n = 3 technical replicas; x ± SD; black line = average of 10 CpG sites; * = p <0.05; ** = p <0.01; *** = p <0.001 in an unpaired t-test).

few CpG sites, and separate analysis of each individual CpG site revealed significant hypermethylation ($p < 0.05$) in CpG site 10 for CaP 17 and sites 3 and 7 for CaP 28. The results obtained were then confirmed by pyrosequencing using the PYRO 2 assay and by MSP. The results indicate that CD133 is not tightly regulated by DNA methylation in prostate primary epithelial cells, which was confirmed by a lack of a strong and consistent increase in CD133 expression after treatment with 5-Aza-2'-deoxycytidine (1 μM for 96 h - Not shown).

Lack of hypermethylation of the CD133 promoter in CaP was then confirmed in CaP xenografts recently established in RAG2$^{-/-}$ gamma C$^{-/-}$ immunocompromised mice (Figure 4A). None of the samples showed any significant hypermethylation of the CpG island, although Xeno 30 showed hypermethylation of CpG sites 3 and 4 ($p < 0.05$). Moreover, a small subpopulation of CD133 positive cells, reminiscent of the stem-like cells present in prostate cancer primary epithelial cultures [14], was present in these xenografts (Figure 4B).

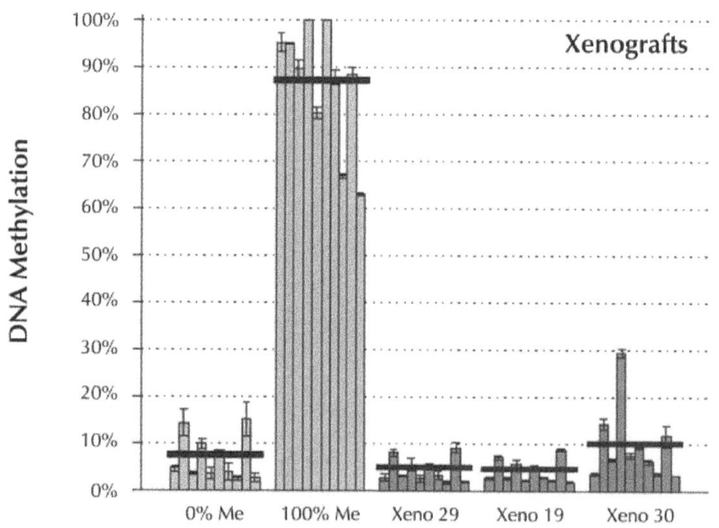

FIGURE 4 *(Continued)*

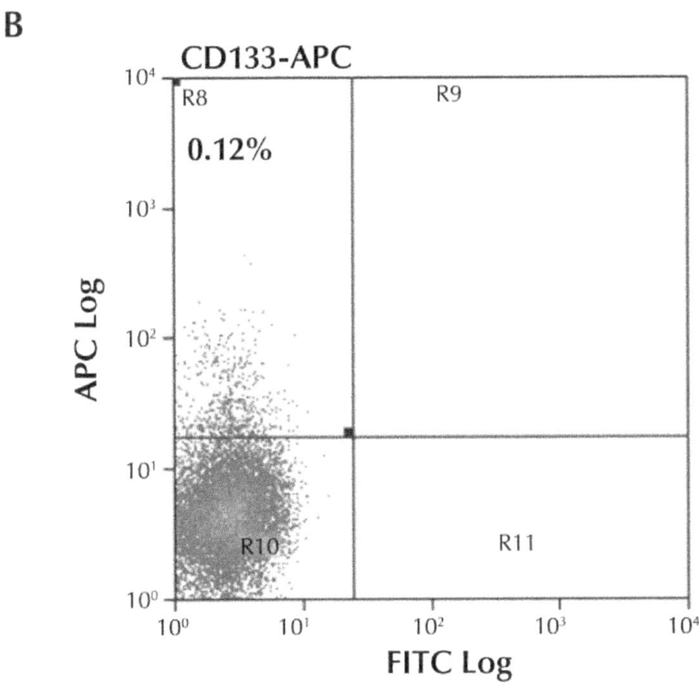

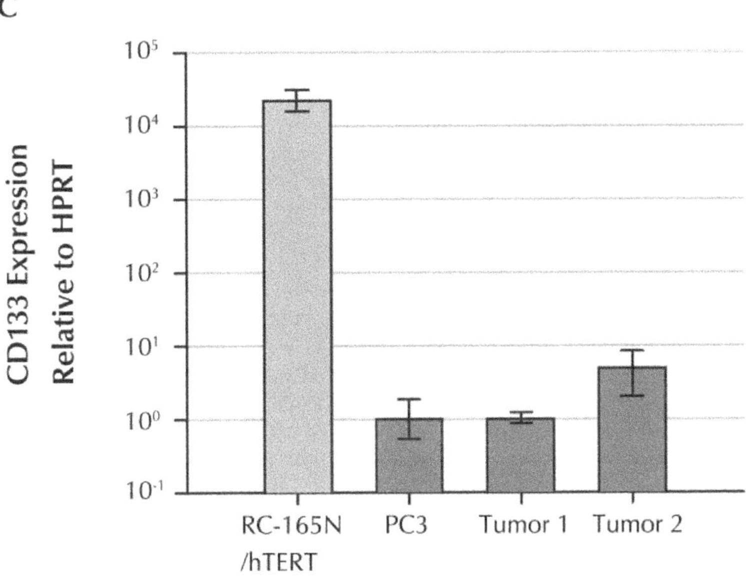

FIGURE 4 *(Continued)*

D

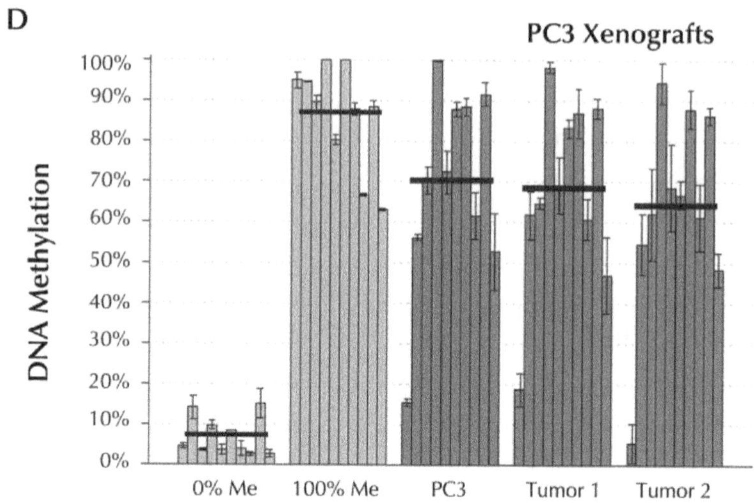

E

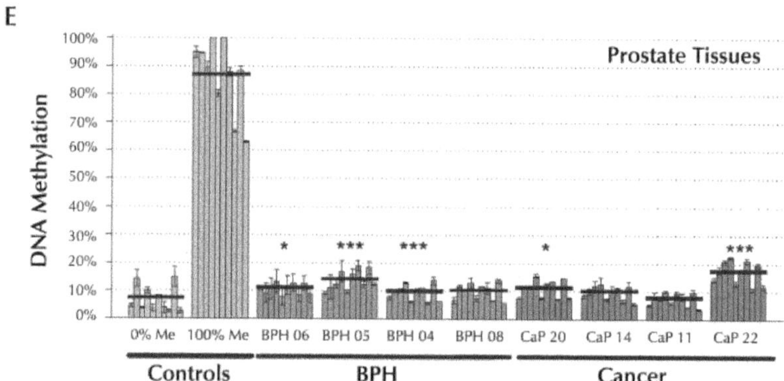

FIGURE 4 Methylation of the CD133 promoter in prostate primary tissues and xenografts. Methylation analysis of the CD133 promoter in primary prostate cancer xenografts generated in RAG2$^{-/-}$ gamma C$^{-/-}$ mice (A), in PC3 xenografts (D) and in prostate tissues (DNA extracted from glass slides of pooled snap frozen tissue sections) (E) and by pyrosequencing (PYRO 1 assay; bars = single CpG sites; n = 3 technical replicas; x ± SD; black line = average of 10 CpG sites; * = $p < 0.05$; ** = $p < 0.01$; *** = $p < 0.001$ in an unpaired t-test). (B) Representative image of CD133 expression measured by FACs in disaggregated primary prostate cancer xenografts. X axes: FITC channel fluorescence (not stained); Y axes: APC channel fluorescence (CD133-APC); Percent of CD133^{+} cells is indicated. (D) qRT-PCR analysis (relative to RPLP0, calibrator = PC3; N.D. = expression undetectable after 40 cycles) of CD133 expression in PC3 xenografts compared to PC3 (RC-165N/hTERT was used as a positive control).

To test whether the methylation of CD133 promoter is a reversible event in vivo, CD133 mRNA expression and promoter methylation was measured in xenografts generated by subcutaneous injection of PC3 cells (Figure 4C and 4D). In both the samples tested, no significant changes in expression or methylation of CD133 were found when comparing PC3 cells *in vitro* and *in vivo*. These results suggest that, once established, methylation patterns within CD133 promoter are stable and not easily reversible.

Lastly, these results were also confirmed by methylation analysis of the CD133 promoter in snap frozen prostate tissues from both benign and CaP samples Figure 4E. 2/4 BPH samples and 2/4 CaP samples showed significantly elevated, but still very low levels (7%–16%) of hypermethylation ($p < 0.05$).

3.4.4 CD133 EXPRESSION IS NOT REGULATED BY DNA METHYLATION IN THE PROSTATE EPITHELIAL HIERARCHY

The CD133 expression was next analyzed by qRT-PCR in SC ($\alpha_2\beta_1^{hi}$/CD133$^+$), TA ($\alpha_2\beta_1^{hi}$/CD133$^-$) and CB ($\alpha_2\beta_1^{low}$/CD133$^-$) cell populations isolated from low passage (<10) primary prostate epithelial cultures (1 BPH and 2 CaP) (Figure 5A). In BPH 09 and CaP 17, CD133 expression was undetectable in unselected cultures, CB cells and TA cells, but was detectable (at low levels) in SCs. CD133 expression was not detectable, in any subpopulation, from culture CaP 18. These results provided evidence that CD133 can be regulated at the transcriptional level during differentiation of prostate epithelia from tissues.

Then it would be analyzed whether this differential regulation was sustained by DNA methylation. Pyrosequencing (PYRO 1 assay - Figure 5B) and MSP were carried out on SC, TA and CB cells. Low levels of methylation (<20%) were found in all the populations. No significant differences were found between the 3 populations analyzed.

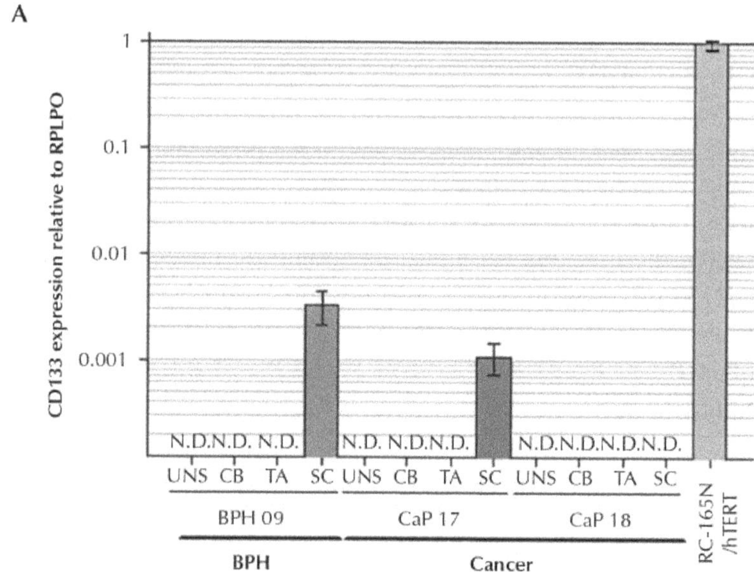

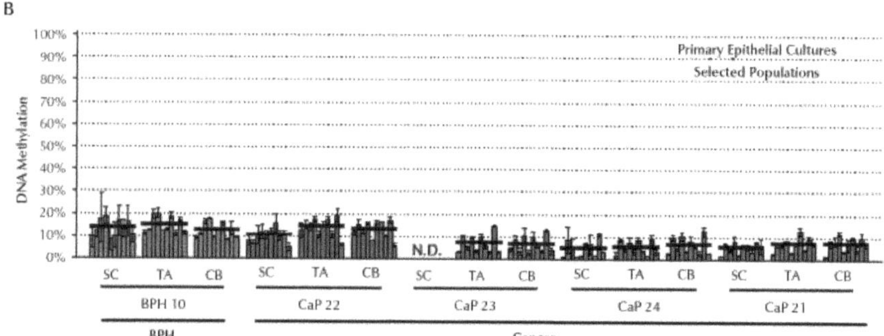

FIGURE 5 CD133 expression and methylation in prostate hierarchy. (A) qRT-PCR analysis of CD133 expression (relative to RPLP0, calibrator cells = RC-165N/hTERT; N.D. = expression undetectable after 40 cycles) and (B) pyrosequencing methylation analysis of the CD133 promoter in selected populations from prostate primary epithelial cultures (SC = stem cells; TA = transit amplifying cells; CB = committed basal cells: PYRO 1 assay; gray bars = single CpG sites; n = 3 technical replicas; x ± SD; the black line = average of 10 CpG sites).

3.4.5 CHROMATIN STRUCTURE, TOGETHER WITH DNA HYPERMETHYLATION, LEAD TO CD133 DOWNREGULATION IN PROSTATE EPITHELIAL CELLS

The presence of active (as detected by dimethylation of lysine 4 of histone H3 - H3K4me2) or inactive (trimethylation of lysine 27 of histone H3–H3K27me3) chromatin around the CD133 promoter was determined by chromatin immunoprecipitation (ChIP) in RC-165N/hTERT, PNT2-C2, P4E6 and PC3 cells (Figure 6A). H3K4 dimethylation was detected in the CD133 promoter of RC-165N/hTERT cells, in accordance with the transcriptional activity. The inactive chromatin mark H3K27me3 was over-represented in P4E6 cells, indicating that chromatin structure, rather than DNA methylation, played a more crucial role in repressing CD133 expression in this cell line. PC3 cells also showed an inactive state of chromatin, with tri-methylation of H3K27 and no enrichment for H3K4me2. PNT2-C2 showed an intermediate state of chromatin condensation where both markers for active and inactive chromatin were present, matching both the intermediate levels of methylation (around 40%) and gene expression (Figure 2A). Taken together, these data indicated that changes in chromatin structure alone, or in co-operation with DNA methylation, could result in the repression of CD133 expression.

The P4E6 cells showed a very similar expression and DNA methylation pattern to primary epithelial cultures and primary tissues, containing low levels of CD133 mRNA and promoter methylation. In this cell line CD133 was maintained in a repressed state by a highly condensed chromatin structure. In line with these results, CD133 mRNA was highly overexpressed after treatment of P4E6 cells with trichostatin A (TSA, 0.6 μM for 24 and 48 hr) (Figure 6B), a well characterised histone deacetylase (HDAC) inhibitor that relaxes chromatin by inducing acetylation of histones.

Treatment of BPH and CaP derived primary epithelial cultures with 0.6 μM TSA and 10 mM NaBu (a less potent and specific HDAC inhibitor) also resulted in overexpression of CD133 mRNA (Figure 6C-D) after 48 hr, in agreement with a clear role for condensed chromatin structure in maintaining CD133 repression in both cell lines and primary samples.

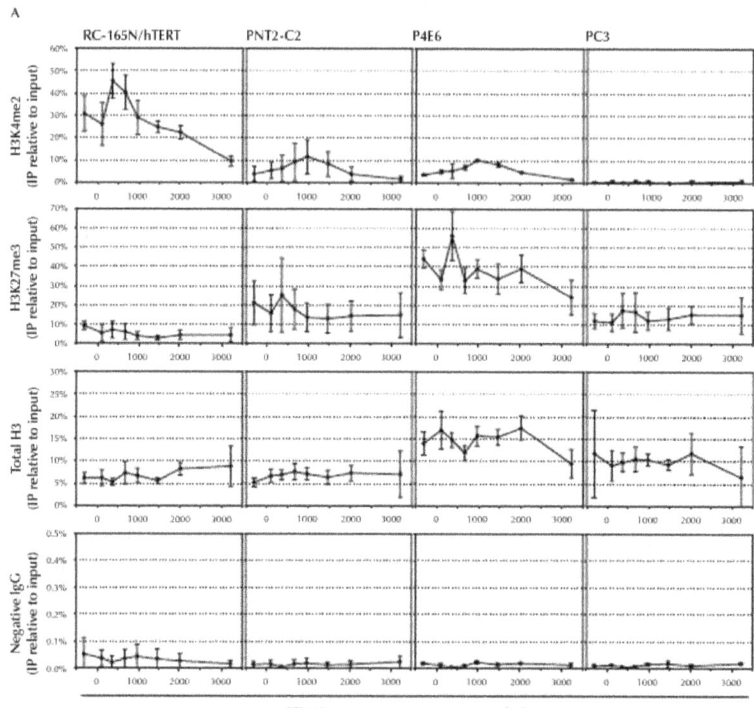

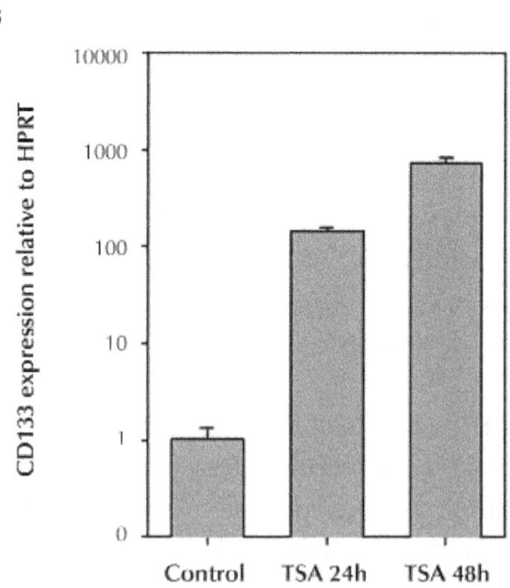

FIGURE 6 *(Continued)*

C

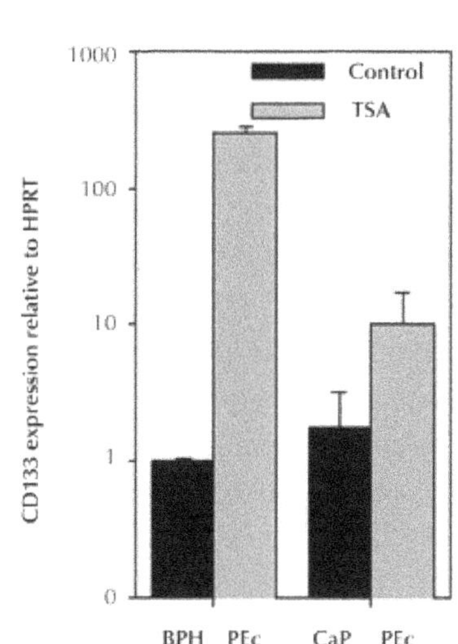

D

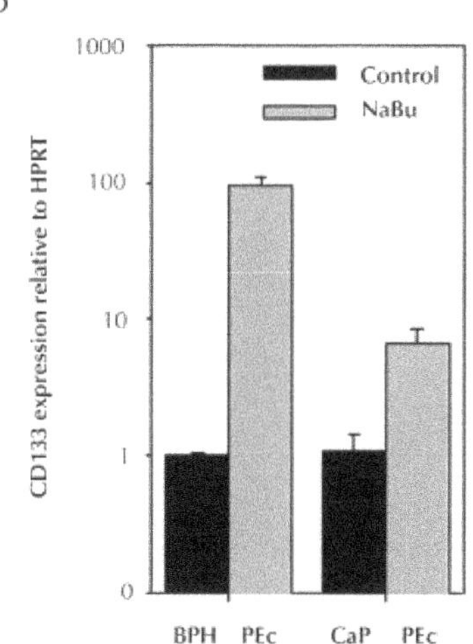

FIGURE 6 CD133 expression is regulated by chromatin structure in prostate cell lines and primary epithelial cultures. (A) ChIP-qPCR analysis carried out in RC-165N/hTERT, PNT2-C2, P4E6, PC3 with rabbit IgG, anti-histone H3, anti-H3K4me2, and anti-H3K27me3. qPCR primer positions are shown in Figure 1A (X axes: CD133 promoter sequence, 0 = 5'end of exon 1A; Y axes: percentage of immunoprecipitated DNA relative to input DNA; n = 3; x± SD, data shown on logarithmic scale). (B) qRT-PCR analysis of CD133 expression in P4E6 cells lines with 0.6 μM TSA for 24 and 48 hr (relative to HPRT, calibrator = Control, x± SD). (C,D) qRT-PCR analysis of CD133 expression in prostate primary epithelial cultures (PEc) treated with 0.6 μM TSA (C) or 10 mM NaBu (D) for 48 hr (relative to HPRT, calibrator = Control, x± SD, data shown on logarithmic scale).

3.5 CONCLUSION

Here a comprehensive study of the epigenetic regulation of CD133 promoter in cell lines is presented, primary epithelial cultures, tissue, and tumor xenografts from the human prostate. It is concluded that CD133 expression is regulated by different mechanisms in cell lines relative to the other samples, and that regulation in primary cultures is independent of methylation, where this gene is maintained in a repressed state by condensed chromatin structure. These results also have implications in the choice of models that are chosen to analyze epigenetic changes in cancer cells, and highlight the complexity of regulation of this common stem cell marker.

KEYWORDS

- **Chromatin Structure**
- **DNA Methylation**
- **Epithelial Cultures**
- **RNA Extraction**
- **Xenografts**

ACKNOWLEDGMENT

We thank Dr. Simon Hayward for provision of BPH-1 cells, Dr Johng Rhim for provision of RC-165N/hTERT and RC-92a/hTERT cells and Dr. David Hudson for provision of Bob and SerBob cells. We thank Dr. Pablo Navarro for his help with chromatin immunoprecipitation experiments. This work was supported by Yorkshire Cancer Research and The Freemasons' Grand Charity.

COMPETING INTERESTS

The authors declare that they have no competing interests.

AUTHORS' CONTRIBUTIONS

DP: Conception and design, collection and assembly of data, data analysis and interpretation, manuscript writing. RJP: Collection and assembly of data, data analysis and interpretation. FMF: Data analysis and interpretation, manuscript writing, final approval of the manuscript. EEO, PAB, MCL, MJS and MSS: Provision of study material or patient samples, collection and assembly of data. ATC and NJM: Conception and design, manuscript writing, final approval of the manuscript. All authors read and approved the final manuscript.

REFERENCES

1. Li, X. and Zhao, X. Epigenetic regulation of mammalian stem cells. *Stem Cells Dev* **17,** 1043–1052 (2008).
2. Hsieh J. and Gage F. H. Epigenetic control of neural stem cell fate. *Curr Opin Genet Dev* **14,** 461–469 (2004).
3. Xi, R. and Xie, T. Stem cell self-renewal controlled by chromatin remodeling factors. *Science* **310,** 1487–1489 (2005).
4. Meissner, A., Mikkelsen, T. S., Gu, H., Wernig, M., Hanna, J., Sivachenko, A., Zhang, X., Bernstein, B. E., Nusbaum, C., Jaffe, D. B., Gnirke, A., Jaenisch, R., and Lander, E. S. Genome-scale DNA methylation maps of pluripotent and differentiated cells. *Nature* **454,** 766–770 (2008).
5. Bröske, A. M., Vockentanz, L., Kharazi, S., Huska, M. R., Mancini, E., Scheller, M., Kuhl, C., Enns, A., Prinz, M., Jaenisch, R., Nerlov, C., Leutz, A., rade-Navarro, M. A., Jacobsen, S. E. W., and Rosenbauer, F. DNA methylation protects hematopoietic stem cell multipotency from myeloerythroid restriction. *Nat Genet* **41,** 1207–1215 (2009).
6. Miller, S. J., Lavker, R. M., and Sun, T. T. Interpreting epithelial cancer biology in the context of stem cells: tumor properties and therapeutic implications. *Biochim Biophys Acta* **1756,** 25–52 (2005).
7. Yin, A. H., Miraglia, S., Zanjani, E. D., Almeida-Porada, G., Ogawa, M., Leary, A. G., Olweus, J., Kearney, J., and Buck, D. W. AC133, a novel marker for human hematopoietic stem and progenitor cells. *Blood* **90,** 5002–5012 (1997).
8. Shmelkov, S. V., St Clair, R., Lyden, D., and Rafii, S. AC133/CD133/Prominin-1. *Int J Biochem Cell Biol* **37,** 715–719 (2005).
9. Richardson, G. D., Robson, C. N., Lang, S. H., Neal, D. E., Maitland, N. J., and Collins, A. T. CD133, a novel marker for human prostatic epithelial stem cells. *J Cell Sci* **117,** 3539–3545 (2004).
10. Collins, A. T., Habib, F. K., Maitland, N. J., and Neal, D. E. Identification and isolation of human prostate epithelial stem cells based on alpha(2)beta(1)-integrin expression. *J Cell Sci* **114,** 3865–3872 (2001).

11. Hudson, D. L. and Masters, J. R. Prostate epithelial stem cell isolation and culture. *Methods Mol Med* **81,** 59–67 (2003).
12. Lawson, D. A., Xin, L., Lukacs, R. U., Cheng, D., and Witte, O. N. Isolation and functional characterization of murine prostate stem cells. *Proc Natl Acad Sci USA* **104,** 181–186 (2007).
13. Leong, K. G., Wang, B. E., Johnson, L., Gao, W. Q. Generation of a prostate from a single adult stem cell. *Nature* **456,** 804–808 (2008).
14. Collins, A. T., Berry, P. A., Hyde, C., Stower, M. J., Maitland, N. J. Prospective identification of tumorigenic prostate cancer stem cells. *Cancer Res* **65,** 10946–10951 (2005).
15. Schulz, W. A. and Hatina, J. Epigenetics of prostate cancer: beyond DNA methylation. *J Cell Mol Med* **10,** 100–125 (2006).
16. Nakayama, M., Gonzalgo, M. L., Yegnasubramanian, S., Lin, X., De Marzo, A. M., and Nelson, W. G. GSTP1 CpG island hypermethylation as a molecular biomarker for prostate cancer. *J Cell Biochem* **91,** 540–552 (2004).
17. Baylin, S. B. and Ohm, J. E. Epigenetic gene silencing in cancer - a mechanism for early oncogenic pathway addiction? *Nat Rev Cancer* **6,** 107–116 (2006).
18. Feinberg, A. P., Ohlsson, R., Henikoff, S. The epigenetic progenitor origin of human cancer. *Nat Rev Genet* **7,** 21–33 (2006).
19. Fargeas, C. A., Huttner, W. B., and Corbeil, D. Nomenclature of prominin-1 (CD133) splice variants—an update. *Tissue Antigens* **69,** 602–606 (2007).
20. Shmelkov, S. V., Jun, L., St Clair, R., McGarrigle, D., Derderian, C. A., Usenko, J. K., Costa, C., Zhang, F., Guo, X., and Rafii, S. Alternative promoters regulate transcription of the gene that encodes stem cell surface protein AC133. *Blood* **103,** 2055–2061 (2004).
21. Herman, J. G., Graff, J. R., Myohanen, S., Nelkin, B. D., and Baylin, S. B. Methylation-specific PCR: a novel PCR assay for methylation status of CpG islands. *Proc Natl Acad Sci USA* **93,** 9821–9826 (1996).
22. Birnie, R., Bryce, S. D., Roome, C., Dussupt, V., Droop, A., Lang, S. H., Berry, P. A., Hyde, C. F., Lewis, J. L., Stower, M. J., Maitland, N. J., and Collins, A. T. Gene expression profiling of human prostate cancer stem cells reveals a pro-inflammatory phenotype and the importance of extracellular matrix interactions. *Genome Biol* **9,** R83 (2008).
23. Navarro, P., Pichard, S., Ciaudo, C., Avner, P., and Rougeulle, C. Tsix transcription across the Xist gene alters chromatin conformation without affecting Xist transcription: implications for X-chromosome inactivation. *Genes Dev* **19,** 1474–1484 (2005).
24. O'Brien, C. A., Pollett, A., Gallinger, S., and Dick, J. E. A human colon cancer cell capable of initiating tumor growth in immunodeficient mice. *Nature* **445,** 106–110 (2007).
25. Ricci-Vitiani, L., Lombardi, D. G., Pilozzi, E., Biffoni, M., Todaro, M., Peschle, C., and De Maria, R. Identification and expansion of human colon-cancer-initiating cells. *Nature* **445,** 111–115 (2007).
26. Singh, S. K., Hawkins, C., Clarke, I. D., Squire, J. A., Bayani, J., Hide T., Henkelman, R. M., Cusimano, M. D., and Dirks, P. B. Identification of human brain tumor initiating cells. *Nature* **432,** 396–401 (2004).

27. Liu, G., Yuan, X., Zeng, Z., Tunici, P., Ng, H., Abdulkadir, I. R., Lu, L., Irvin, D., Black, K. L., and Yu, J. S. Analysis of gene expression and chemoresistance of CD133+ cancer stem cells in glioblastoma. *Mol Cancer* **5,** 67 (2006).
28. Monzani, E., Facchetti, F., Galmozzi, E., Corsini, E., Benetti, A., Cavazzin, C., Gritti, A., Piccinini, A., Porro, D., Santinami, M., Invernici, G., Parati, E., Alessandri, G., and La Porta, C. A. Melanoma contains CD133 and ABCG2 positive cells with enhanced tumorigenic potential. *Eur J Cancer* **43,** 935–946 (2007).
29. Olempska, M., Eisenach, P. A., Ammerpohl, O., Ungefroren, H., Fandrich, F., and Kalthoff, H. Detection of tumor stem cell markers in pancreatic carcinoma cell lines. *Hepatobiliary Pancreat Dis Int* **6,** 92–97 (2007).
30. Suetsugu, A., Nagaki, M., Aoki, H., Motohashi, T., Kunisada, T., and Moriwaki, H. Characterization of CD133+ hepatocellular carcinoma cells as cancer stem/progenitor cells. *Biochem Biophys Res Commun* **351,** 820–824 (2006).
31. Yin, S., Li, J., Hu, C., Chen, X., Yao, M., Yan, M., Jiang, G., Ge, C., Xie, H., Wan, D., Yang, S., Zheng, S., and Gu, J. CD133 positive hepatocellular carcinoma cells possess high capacity for tumorigenicity. *Int J Cancer* **120,** 1444–1450 (2007).
32. Shepherd, C. J., Rizzo, S., Ledaki, I., Davies, M., Brewer, D., Attard, G., De Bono, J., and Hudson, D. L. Expression profiling of CD133+ and CD133- epithelial cells from human prostate. *Prostate* **68,** 1007–1024 (2008).
33. Vaissière, T., Sawan, C., and Herceg, Z. Epigenetic interplay between histone modifications and DNA methylation in gene silencing. *Mutat Res* **659,** 40–48 (2008).
34. Yi, J. M., Tsai, H. C., Glöckner, S. C., Lin, S., Ohm, J. E., Easwaran, H., James, C. D., Costello, J. F., Riggins, G., Eberhart, C. G., Laterra, J., Vescovi, A. L., Ahuja, N., Herman, J. G., Schuebel, K. E., and Baylin, S. B. Abnormal DNA methylation of CD133 in colorectal and glioblastoma tumors. *Cancer Res* **68,** 8094–8103 (2008).
35. Missol-Kolka, E., Karbanova, J., Janich, P., Haase, M., Fargeas, C. A., Huttner, W. B., and Corbeil, D. Prominin-1 (CD133) is not restricted to stem cells located in the basal compartment of murine and human prostate. *Prostate* **71,** 254–267 (2011).
36. Bao, S., Wu, Q., McLendon, R. E., Hao, Y., Shi, Q., Hjelmeland, A. B., Dewhirst, M. W., Bigner, D. D., and Rich, J. N. Glioma stem cells promote radioresistance by preferential activation of the DNA damage response. *Nature* **444,** 756–760 (2006).
37. Rich, J. N. and Bao, S. Chemotherapy and cancer stem cells. *Cell Stem Cell* **1,** 353–355 (2007).
38. Dean, M., Fojo, T., and Bates, S. Tumor stem cells and drug resistance. *Nat Rev Cancer* **5,** 275–284 (2005).
39. Takenobu, H., Shimozato, O., Nakamura, T., Ochiai, H., Yamaguchi, Y., Ohira, M., Nakagawara, A., and Kamijo, T. CD133 suppresses neuroblastoma cell differentiation via signal pathway modification. *Oncogene* **30,** 97–105 (2011).
40. Florek, M., Haase, M., Marzesco, A. M., Freund, D., Ehninger, G., Huttner, W. B., and Corbeil, D. Prominin-1/CD133, a neural and hematopoietic stem cell marker, is expressed in adult human differentiated cells and certain types of kidney cancer. *Cell Tissue Res* **319,** 15–26 (2005).
41. Antequera, F., Boyes, J., and Bird, A. High levels of de novo methylation and altered chromatin structure at CpG islands in cell lines. *Cell* **62,** 503–514 (1990).

42. Jones, P. A., Wolkowicz, M. J., Rideout, W. M., Gonzales, F. A., Marziasz, C. M., Coetzee, G. A., and Tapscott, S. J. De novo methylation of the MyoD1 CpG island during the establishment of immortal cell lines. *Proc Natl Acad Sci USA* **87,** 6117–6121 (1990).
43. Vertino, P. M. and Issa, J. P., Pereira-Smith, O. M., and Baylin, S. B. Stabilization of DNA methyltransferase levels and CpG island hypermethylation precede SV40-induced immortalization of human fibroblasts. *Cell Growth Differ* **5,** 1395–1402 (1994).
44. Kondo, Y., Shen, L., Cheng, A. S., Ahmed, S., Boumber, Y., Charo, C., Yamochi, T., Urano, T., Furukawa, K., Kwabi-Addo, B., Gold, D. L., Sekido, Y., Huang, T. H-M., and Issa, J-P. J. Gene silencing in cancer by histone H3 lysine 27 trimethylation independent of promoter DNA methylation. *Nat Genet* **40,** 741–750 (2008).
45. Coolen, M. W., Stirzaker, C., Song, J. Z., Statham, A. L., Kassir, Z., Moreno, C. S., Young, A. N., Varma, V., Speed, T. P., Cowley, M., Lacaze, P., Kaplan, W., Robinson, M. D., Clark, S. J. Consolidation of the cancer genome into domains of repressive chromatin by long-range epigenetic silencing (LRES) reduces transcriptional plasticity. *Nat Cell Biol* **12,** 235–246 (2010).
46. Pfeiffer, M. J. and Schalken, J. A. Stem cell characteristics in prostate cancer cell lines. *Eur Urol* **57,** 246–254 (2010).
47. Korenchuk, S., Lehr, J. E., Mclean, L., Lee, Y. G., Whitney, S., Vessella, R., Lin, D. L., Pienta, K. J. VCaP, a cell-based model system of human prostate cancer. *In Vivo* **15,** 163–168 (2001).
48. Lang, S. H., Anderson, E., Fordham, R., and Collins, A. T. Modeling the prostate stem cell niche: an evaluation of stem cell survival and expansion in vitro. *Stem Cells Dev* **19,** 537–546 (2010).

CHAPTER 4

GENOME-WIDE MAPPING

RYAN K. C. YUEN, RUBY JIANG, MARIA S. PECAHERRERA, DEBORAH E. MCFADDEN, and WENDY P. ROBINSON

CONTENTS

4.1 INTRODUCTION

Genomic imprinting is a phenomenon in which one of the two alleles of a gene is expressed in a parent-of-origin manner [1]. Imprinted genes are thought to be particularly important to placental and fetal growth and development and may help regulate growth in response to maternal and fetal signals *in utero* [2]. To date, around 60 imprinted genes have been identified in humans (http://www.geneimprint.com website), largely after first being identified in mice or through characterization of specific imprinting disorders such as Prader–Willi syndrome and Angelman syndrome or Beckwith-Wiedemann syndrome. However, many genes are imprinted in mice but are not known to be in humans, for example, *Impact* [3]. Furthermore, many genes are imprinted only in specific tissues, for example, *Ube3a*, which is maternally expressed in the brain but biparentally expressed in other tissues [4], or may be polymorphically imprinted, for example, *IGF2R* [5]. These issues complicate the discovery and characterization of imprinted genes in humans.

The importance of imprinted genes for placental and fetal development was initially demonstrated in mice by observations that parthenogenetic embryos (maternal origin, digynic diploid) could show embryonic differentiation but failed to form extraembryonic components [6]. In contrast, androgenetic embryos (paternal origin, diandric diploid) had poorly developed embryos, but the trophoblasts showed extensive proliferation [7]. The parallel observations in humans are ovarian teratomas (parthenogenetic), which are a rare form of tumor that consists of a variety of embryonic tissues or organs but no placental tissues, and complete hydatidiform moles (CHMs) (androgenetic), which consist of abnormal placental growth characterized by trophoblast hyperplasia but no (or rare) embryonic structures. The parental conflict theory developed to explain the evolution of imprinted genes [8] suggests that paternally expressed genes tend to promote growth of the offspring at the expense of the mother, while maternally expressed genes act as growth-limiting factors to conserve maternal resources [8].

Most imprinted genes possess differentially methylated regions (DMRs) whereby allelic methylation depends on the parent of origin [1]. DMRs established through the germline are called "gametic" or "primary" DMRs. These often coincide with imprinting control regions (ICRs), which regulate gene expression and further epigenetic modifications [9-11]. Their methylation status is thought to be maintained in all somatic lineages once acquired. Other DMRs, called "somatic" or "secondary" DMRs, are established after fertilization and may be tissue-specific [10, 11].

Since most imprinted genes contain DMRs, comparing DNA methylation profiles between tissues with unbalanced parental constitutions provides an approach to identify and characterize imprinted genes in the genome. One approach is to compare the DNA methylation profile of paternally derived CHMs to that of maternally derived ovarian teratomas [12]. Indeed, several novel imprinted genes have been identified previously by using this strategy [13, 14]. However, such comparisons are limited by the fact that the tissues present in ovarian teratomas and CHMs are highly abnormal and are not of comparable origin, with teratomas being embryonic and CHMs being strictly placental. Many differences may reflect tissue-specific methylated genes, since tissue-specific DMRs are numerous and are established in early pregnancy [15]. CHMs also present with highly proliferative trophoblasts that can lead to increased risk of choriocarcinoma, and hypermethylation of nonimprinted genes has been reported in CHMs [16].

In humans, triploidy (the presence of three complete haploid genomes) occurs spontaneously in 2–3% of pregnancies, and, while such pregnancies frequently end in miscarriage, they can survive into the fetal period and, very rarely, to term [17]. It is proposed that a comparison between diandric and digynic triploidies, in which development is much less severely altered than in CHMs and teratomas, provides a powerful approach for the identification and characterization of imprinted genes in the human genome. The diandric triploid phenotype (two paternal plus one maternal haploid genomes) is characterized by a normal-sized or only moderately growth-restricted fetus with a large and cystic placenta with trophoblast hyperplasia, while the digynic triploid phenotype (two maternal plus one paternal haploid genomes) is characterized by an intrauterine growth-restricted foetus and a very small placenta with no trophoblast hyperplasia [17]. Importantly, embryo and fetal development are largely similar between diandric and digynic triploidy, with growth differences likely arising largely as a consequence of differences in placental function [18]. Furthermore, while small, digynic placentas have a grossly normal structure. Diandric placentas show features similar to a CHM, but their development is much less severely altered than in a CHM, and the placenta can support growth of a foetus at least to some degree.

Although it was previously suggested that DNA methylation may be less important in regulating imprinting in placental tissue as compared to fetal tissue, it was recently demonstrated that the DNA methylation status of many known imprinted DMRs is strictly maintained in triploid placentas and can be used to distinguish diandric from digynic triploidy [19]. Therefore, in the present study, the DNA methylation profiles of placentas from diandric and digynic triploidies were compared using the Infinium HumanMethylation27 BeadChip array (Illumina, Inc., San Diego, CA, USA), which targets over 27,000 CpG loci within the proximal promoter regions of approximately 14,000 genes [20]. Methylation levels in chromosomally normal placentas, CHMs and maternal blood samples were used as reference points for comparison. Using this strategy, the majority of known imprinted ICRs were identified on the array and many novel imprinted DMRs in the genome. For a subset of genes, expressed polymorphisms and informative mother-placenta pairs were identified, which were used to demonstrate parent-of-origin biases in allelic expression. It was also demonstrated that complex DNA methylation domains that regulate imprinted genes can be mapped by comparing the methylation patterns in different tissues and different gestational ages of placentas.

4.2 METHODS

4.2.1 SAMPLE COLLECTION

This study was approved by the ethics committees of the University of British Columbia and the Children's and Women's Health Centre of British Columbia. Early gestation placental samples (ten diandric triploids, ten digynic triploids, six CHMs and ten normal controls) were obtained from spontaneous abortions examined in the Children's and Women's Health Centre of British Columbia pathology laboratory. The parental origin of triploids was determined by using microsatellite polymorphisms as previously described [17-19], and these studies also allowed us to exclude maternal contamination in the placental samples. Midgestation placental samples (n = 10) and fetal tissues (11 muscle samples, 12 kidney samples and 8 brain samples) were obtained from anonymous, chromosomally normal, second-trimester elective terminations for medical reasons. Term placental samples and the corresponding maternal blood samples were collected from Children's and Women's Health Centre of British Columbia with the women's written informed consent. For all placental samples, fragments of about 1 cm^3 were dissected from the fetal side and whole villi were used for investigation. All tissues were karyotyped for chromosomal abnormalities, and genomic DNA was extracted from each tissue sample using standard techniques. Total RNA was extracted from term placentas using an RNeasy kit (Qiagen, Valencia, CA, USA) according to the manufacturer's instructions.

4.2.2 ILLUMINA DNA METHYLATION ARRAY

Genomic DNA was bisulphite-converted using the EZ DNA Methylation Kit (Zymo Research, Orange, CA, USA) according to the manufacturer's instructions. Bisulphite treatment converted unmethylated cytosines to uracils while leaving methylated cytosines unchanged. After DNA purification, bisulphite-converted DNA samples were randomly arrayed and subjected to the Infinium Human Methylation 27 BeadChip panel array-based assay. The array assays methylation levels at 27,578 CpG sites in the human genome. The methylation level for each CpG site was measured by the intensity of fluorescent signals corresponding to the methylated allele (Cy5) and the unmethylated allele (Cy3). The Cy5 and Cy3 fluorescence intensities were corrected independently for background signal and normalized using GenomeStudio software (Illumina, Inc.). Continuous β values that range from 0 (unmethylated) to 1 (methylated) were used to identify the percentage of methylation, from 0 to 100%, for each CpG site. The β value was calculated based on the ratio of methylated/(methylated + unmethylated) signal outputs. The detection P value of each probe was generated by comparison with a series of negative controls embedded in the assay. Probes with detection P values >0.05 in any of the samples were eliminated from the study. The correlation coefficient for technical replicates was >0.98. The microarray data from this study have been submitted to the NCBI Gene Expression Omnibus (http://www.ncbi.nlm.nih.gov/geo webcite) under accession number GSE25966.

4.2.3 DNA METHYLATION ANALYZES FOR TARGETED LOCI

Methylation-unbiased PCR and sequencing primers were designed based on the probe sequences provided by Illumina. All primers were designed in regions free of known SNPs. Pyrosequencing was performed using a PyroMark MD system (Biotage, Uppsala, Sweden). The quantitative levels of methylation for each CpG dinucleotide were evaluated using Pyro Q-CpG software (Biotage). For bisulphite cloning and sequencing, the PCR product from individual samples was generated by using non-biotinylated primers and subsequently TA-cloned into the pGEM-T Easy Vector System (Promega, Madison, WI, USA). Individual clones were picked and PCR-amplified with SP6 and T7 promoter primers. PCR products were sequenced by using Sanger sequencing. The sequencing data were analyzed using BiQ Analyzer Software [21], and sequences with less than an 80% bisulphite conversion rate were eliminated from analysis.

4.2.4 SNP GENOTYPING

Multiplex genotyping of genomic DNA and cDNA was performed by using the iPLEX Gold assay on the MassARRAY platform (Sequenom) at the Génome Québec Innovation Centre (Montréal, PQ, Canada). Primers for SNP genotyping were designed by using primer design software from Sequenom. The primer extended products were analyzed and the genotypes were determined by mass spectrometric detection using the MassARRAY Compact System (Sequenom). Technical replicates showed a correlation of $r = 0.92$. Samples or SNPs with <70% conversion rates (calls) were eliminated. Genotyping by pyrosequencing was performed on a PyroMark MD System, and the relative levels of alleles for SNPs were evaluated by using PSQ 96MA SNP software (Biotage). Genotyping of exonic SNPs was carried out with cDNA prepared using either (1) the Omniscript Reverse Transcriptase Kit (Qiagen) followed by the iPLEX Gold assay or pyrosequencing or (2) the Qiagen OneStep RT-PCR Kit followed by pyrosequencing. Primers for pyrosequencing genotyping were designed by using primer design software from Biotage. PCR without reverse transcriptase was performed on each sample to confirm that there was no genomic DNA contamination.

4.2.5 STATISTICAL ANALYSIS

Unsupervised hierarchical clustering of samples was done using Illumina GenomeStudio software. Differentially methylated probes in the Illumina Infinium HumanMethylation27 BeadChip array from each comparison were identified using the siggenes package from R software with a cutoff of <0.1% FDR. FDRs were generated after comparison of 1,000 random permutations between samples. The Pearson linear correlation coefficient was used to determine the similarity of DNA methylation profiles between samples. The Database for Annotation, Visualization and Integrated Discovery (DAVID) program was used for gene ontology analysis using the total number of genes presented in the array as a background for comparison [22, 23].

4.3 DISCUSSION

Many efforts have been made to identify imprinted genes in the human genome because of their importance in fetal growth and development and their potential for dysregulation [11, 12]. Most known imprinted genes to date were first identified in mice, but many imprinted genes are not conserved across species [5]. In the present study, diandric and digynic triploid placentas were utilized to map imprinted DMRs, sites that are typically associated with imprinted genes, in the human genome. Eleven of the 18 previously reported human ICRs covered by the Illumina Infinium HumanMethylation27 BeadChip panel were identified, with additional ones that showed differences that were insufficient to reach this stringent statistical criteria. Furthermore, the parent-of-origin dependence of methylation and expression in a subset of the candidate novel imprinted genes were confirmed on the basis of independent experiments.

This approach improves upon previous strategies for mapping imprinted DMRs, such as comparing parthenogenotes (ovarian teratomas) and androgenotes (CHMs) [13, 14], which is limited by the grossly abnormal nature of these samples, or comparing maternal and paternal uniparental disomies (UPDs) [24, 25], which is restricted by the rarity of UPDs for many chromosomes and the limited tissues available for analysis. Although triploid placentas do exhibit some abnormal pathology, their cellular composition is comparable and methylation profiles of both types of triploidy were closely correlated with chromosomally normal placentas ($r = 0.99$). In comparison, a previous study showed that mature ovarian teratomas have a methylation profile more similar to that of blood ($r = 0.94$) than to either CHMs ($r = 0.84$) or normal placentas ($r = 0.88$) [14]. Genome-wide transcriptome analysis has also been used to identify imprinted genes [11, 26], but it is gene expression- and SNP-dependent; thus, imprinted genes with tissue-specific expression or lacking a heterozygous exonic SNP would be missed.

As demonstrated, tissue-specific methylation of imprinted DMRs or their flanking regions can readily be assessed by comparing methylation profiles of a variety of tissues, allowing a comprehensive analysis of tissue-specific methylation regulation at complex loci, such as *GNAS* [27]. The regional dependent methylation patterns in the promoters of imprinted genes show the importance of locating the specific CpG sites defining the imprinted DMRs when studying the dynamics of promoter DNA methylation at such genes. While in the present study only loci that demonstrated parent-of-origin-dependent differential DNA methylation in placenta was identified, most known imprinted genes show parent-of-origin-specific expression in this organ [2]. Furthermore, as diandric and digynic triploids can both exist as foetuses, additional comparisons could be used to identify any potential genes that exhibit imprinting specifically in other tissues. This study is limited by the low coverage of CpG sites in the array (about two CpG sites on average for each proximal promoter region of genes), which reduces its power to identify imprinted DMRs as these may be limited to specific regions within the promoter. This analysis could thus be extended further by using microarray or whole-genome sequencing with greater coverage of the genome.

Overall, the number of imprinted DMRs identified in the present study was less than that predicted by bioinformatics approaches [28]. However, the stringent selection criteria (<0.1% FDR and absolute average methylation difference >15%) that was used to pick the top candidate sites caused an underestimation of the number of imprinted loci. Many more candidate imprinted DMRs can be identified with this data set using lower thresholds. In fact, many known imprinted genes that failed to identify on the basis of these criteria did show nominally significant ($P < 0.05$ without correction for multiple comparisons) DNA methylation differences between diandric and digynic triploids. For instance, a recently confirmed imprinted gene, *RB1* [29], was significantly differentially methylated between diandric and digynic triploidies (<0.1% FDR), with a methylation pattern consistent with that of a maternal DMR (data not shown). However, it was excluded because its absolute average methylation difference between diandries and digynies was only 14%. While imprinted DMRs was expected to show a difference of 33% between the triploid groups, smaller differences may be observed owing to a lack of complete methylation at all CpG sites on the inactive allele or to the presence of a mix of cell types, only some of which are imprinted. Similarly, methylation on the inactive X chromosome in females is incomplete (much less than 50%) in the placenta for gene promoter regions that are typically methylated at 50% in somatic tissues, despite still showing a significant increase in methylation relative to male placenta [30].

Only some of the novel putative imprinted DMRs could be confirmed to show monoallelic expression, and others did not show strict parent-of-origin expression in all placentas. There are several possible explanations. First, there may be cell- or tissue-specific imprinting confounding the ability to detect a difference in whole villous samples from term placentas. Many known imprinted genes show imprinted expression only in specific placental cell types, for example, *Mash2* in mice, which is differentially expressed only in diploid trophoblast cells of the postimplantation embryo [31], and *STOX1* in humans, which is maternally expressed in extravillous trophoblast cells [32]. Given the highly heterogeneous cell types present in the placenta [33], nonimprinted expression in some cells may mask allelic expression in others. The possibility that cell heterogeneity exists for the DMRs identified in the present study is supported by the observations that (1) average methylation of some candidate DMRs was not the expected 50% in normal placentas and (2) *DNAJC6* and *RASGRF1* showed differential methylation between trophoblast and mesenchyme. Second, as it is shown, the iPLEX Gold assay may not be sensitive enough to pick up subtle allelic expression biases.

Third, there may be alternative transcripts regulated by alternative promoters that are not imprinted, so the observed expressed allelic ratio at particular SNPs may be complicated by the synergic effect of multiple transcripts. Such complex regulation is observed for known imprinted genes such as *GNAS*, *CDKN1C* and *MEST*. However, allelic expression from either parent in some genes, such as *MOV10L1* and *ST8SIA1*, suggests that some of the identified DMRs may be random monoallelically expressed genes instead of imprinted genes (with DNA methylation differences between diandries and digynies occurring by chance).

The validation of all the putative imprinted DMRs that are identified is limited by the number of samples and common SNPs within regions and by the availability of intact mRNA from the pathological specimens. A proper validation experiment to demonstrate that the DMRs that are identified are associated with imprinted methylation and gene expression requires being able to trace the parental origin of the methylated and expressed alleles in multiple members of the same family, which can be done in mice but is impractical and ethically impossible to do across multiple tissues in humans [34]. The best alternative is to trace the origin of the methylated allele and the expressed allele in multiple individuals. This requires a SNP adjacent to the methylation site that is heterozygous in the test sample but homozygous in one parent. Using this strategy, it was demonstrated for *FAM50B* that (1) a maternal origin of the methylated allele in placenta and blood from multiple individuals and on reciprocal genetic backgrounds and (2) the paternal allele is expressed with either SNP allele in the placenta, thus ruling out the possibility of a genetic effect. Confirming that an imprint represents a primary imprinted DMR requires detailed investigations of postfertilization imprinting dynamics, which is difficult to perform in humans. Nonetheless, it was shown that the methylation level of *FAM50B* is similar in multiple tissues and is unmethylated in sperm, suggesting that it is likely to be a primary maternal DMR. During the revision of this manuscript, the maternal imprint of *FAM50B* was also confirmed by other groups using similar validation methods [35, 36]. The goal of the present study was to demonstrate the ability of the approach to identify imprinted DMRs, not to map and confirm every imprinted DMR on the array. Thus, the putative imprinted DMRs listed in the present study should be considered with caution, and further validation is required.

Two genes identified as potentially being imprinted in the present study, *APC* and *DNMT1*, were excluded as being imprinted in previous studies [37, 38], while *APC* was reported as being imprinted in another study [39]. Of interest, *DNMT1* is a DNA methyltransferase that is important for the maintenance and establishment of DMRs in imprinted genes [40], while *APC* is a negative regulator of the Wnt signalling pathway, which has been implicated in the survival, differentiation and invasion of human trophoblasts [37]. Although *DNMT1* was found to be dispensable for growth of the extraembryonic lineages in mice [41], it is not methylated at the orthologous region in mice [38]. Both the *APC* and *DNMT1* DMRs were reported to be specifically methylated in primate placentas [42], suggesting that the potential imprinting marks of these genes emerged fairly recently in evolution. This is also consistent with the hypothesis that maternal imprints are under selective pressure during early development for methylation-dependent control [43]. This could occur by selecting genes with developmental advantage by gain of imprinting from epipolymorphisms [44].

4.4 RESULTS

4.4.1 DNA METHYLATION PROFILE ANALYSIS IN PLACENTA AND BLOOD SAMPLES

To generate DNA methylation profiles from triploidies, placental DNA was assayed from ten diandric and ten digynic triploidies on the Illumina Infinium HumanMethylation27 BeadChip panel. In addition, ten chromosomally normal placentas, 6 CHMs (diandric diploid, no maternal contribution) and ten maternal whole-blood samples were included for comparison. After background adjustment and normalization, unsupervised hierarchical clustering with all the samples based on a distance measure of 1-r was performed, where r is the Pearson correlation coefficient between different samples. This revealed three distinct groups of clusters: CHMs, triploid and normal placentas, and blood (Figure 1). The blood cluster is more distant from the two other clusters of placentas, confirming that there are many DNA methylation differences between blood and placenta [30, 45, 46]. Although CHMs are trophoblast-derived, they show a distinct methylation profile from the triploid and normal placentas, which probably reflects not only the lack of a maternal genome but also the abnormal development of such tissue. Within triploid and normal placentas, digynic and diandric triploid placentas are clearly separated by their methylation profiles, but, interestingly, they are not separated from the chromosomally normal placentas (Figure 1). This suggests that methylation profiles of triploid placentas closely resemble those of chromosomally normal placentas, but that digynic and diandric triploid placentas have distinguishing DNA methylation differences.

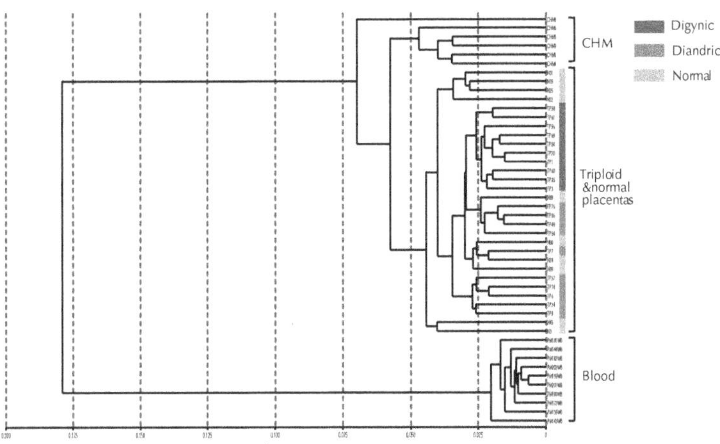

FIGURE 1 Unsupervised clustering of triploid and normal placentas with CHMs and blood samples demonstrates that each tissue type has a distinct methylation profile. Sample names are shown with labelling of corresponding tissue types. Samples were clustered by hierarchical clustering of β values based on 1-r (Illumina GenomeStudio software), where r represents the correlation coefficient between samples. Digynic triploids are indicated by red boxes, diandric triploids by blue boxes and normal placentas by green boxes.

Although clustering can be biased by gender differences resulting from inactivation of an X chromosome in females (that is, higher methylation of the X chromosome CpG islands in female than in male samples) [30, 47], there is no preferential clustering of samples by gender within the triploid and normal placenta cluster. There is a small difference in gestational age (about 3 weeks apart on average) between diandric and digynic placentas ($P < 0.01$), but this also cannot explain the distinct clustering patterns, since the gestational ages of the two groups are largely overlapping.

Further the average DNA methylation of probes was compared between the five sample groups (digynic triploid placentas, diandric triploid placentas, normal placentas, CHMs and blood). As expected, the correlation of average probe methylation values between different sample groups is consistent with that observed in the cluster analysis. In general, blood has the most distinct DNA methylation profile with a greater number of highly methylated probes. Triploid and normal placentas are highly correlated with regard to their methylation profiles ($r = 0.99$), while CHMs are more similar to diandric and normal placentas ($r = 0.98$) than to digynic placentas ($r = 0.96$).

4.4.2 COMPARISON OF DNA METHYLATION PROFILES BETWEEN PLACENTAS FROM DIANDRIC AND DIGYNIC TRIPLOIDIES

After comparing methylation between diandric and digynic placentas by performing Student's *t*-test for all probes, nearly 2,500 probes were identified with a *P* value < 0.01, which is nearly ten times more than expected by chance. To adjust for multiple testing and identify candidates with a very high likelihood of representing true differences, a stringent cutoff of <0.1% false discovery rate (FDR) was used by using the Significance Analysis of Microarrays (SAM) program with 1,000 permutation comparisons for each sample [48]. To further focus on the most meaningful differences, only probes with more than 15% absolute magnitude difference between the mean methylation of diandric and digynic triploidies were also considered. While a theoretical difference of 33.3%

for imprinted sites were expected, a lower cutoff was used because it has been observed that the actual methylation difference may vary for some known imprinted genes [19] and that there may be biases in the Illumina array that result in a nonlinear relationship between the estimated methylation β value and actual methylation. In total, 122 probes were identified with <0.1% FDR and average absolute methylation difference >15% (average absolute Δ β >0.15 from the Illumina array). Probes with higher average methylation in diandric than digynic triploidies were designated putative paternal differentially methylated loci (DML), and probes with higher average methylation in digynic than diandric triploidies were designated putative maternal DML. Plotting DNA methylation of putative DML in all samples from diandric against digynic triploidies showed a clear separation of methylation values of paternal and maternal DML, suggesting that most of the identified differentially methylated probes are consistently methylated within each sample group without much overlap as expected on the basis of the application of stringent statistical criteria.

As some methylation differences between diandric and digynic triploids could theoretically arise as a result of secondary effects, such as altered cell composition, the validity of the identified putative imprinted DML was further evaluated by verifying that the methylation levels of diandric CHMs and chromosomally normal placentas fit the expected pattern (Figure 2). The average methylation in CHMs was closer in value to that of diandric triploidies, while that for normal placentas fell between that for diandric and digynic triploidies for the majority of putative DML as would be expected for imprinted DMRs. The putative maternal DML were more strongly correlated with normal placentas than paternal DML, while putative paternal DML tended to have higher correlation with CHMs than maternal DML. CHMs showed particularly low correlation for maternal DML compared with other placental groups, which was largely due to the low average methylation of putative maternal DML in CHMs as well as more variability in values for CHMs.

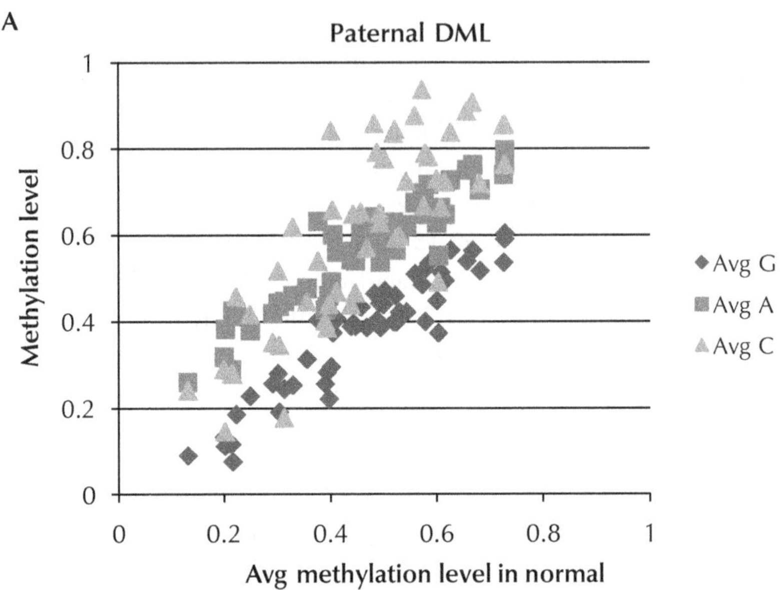

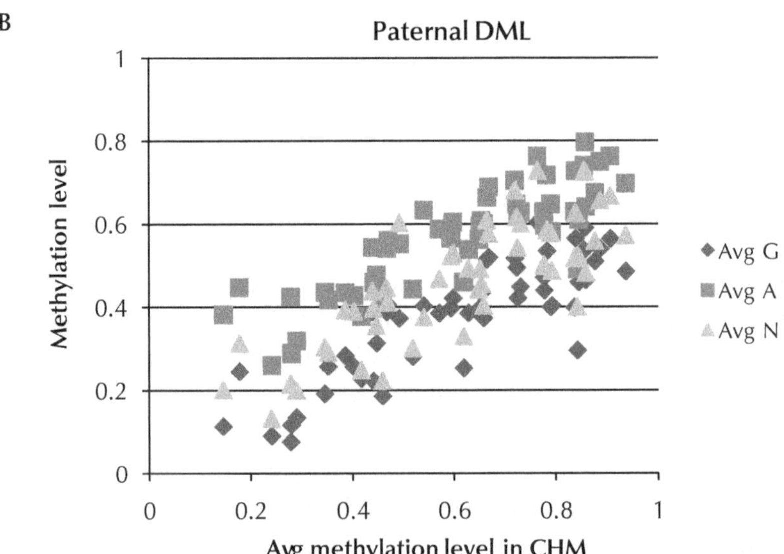

FIGURE 2 *(Continued)*

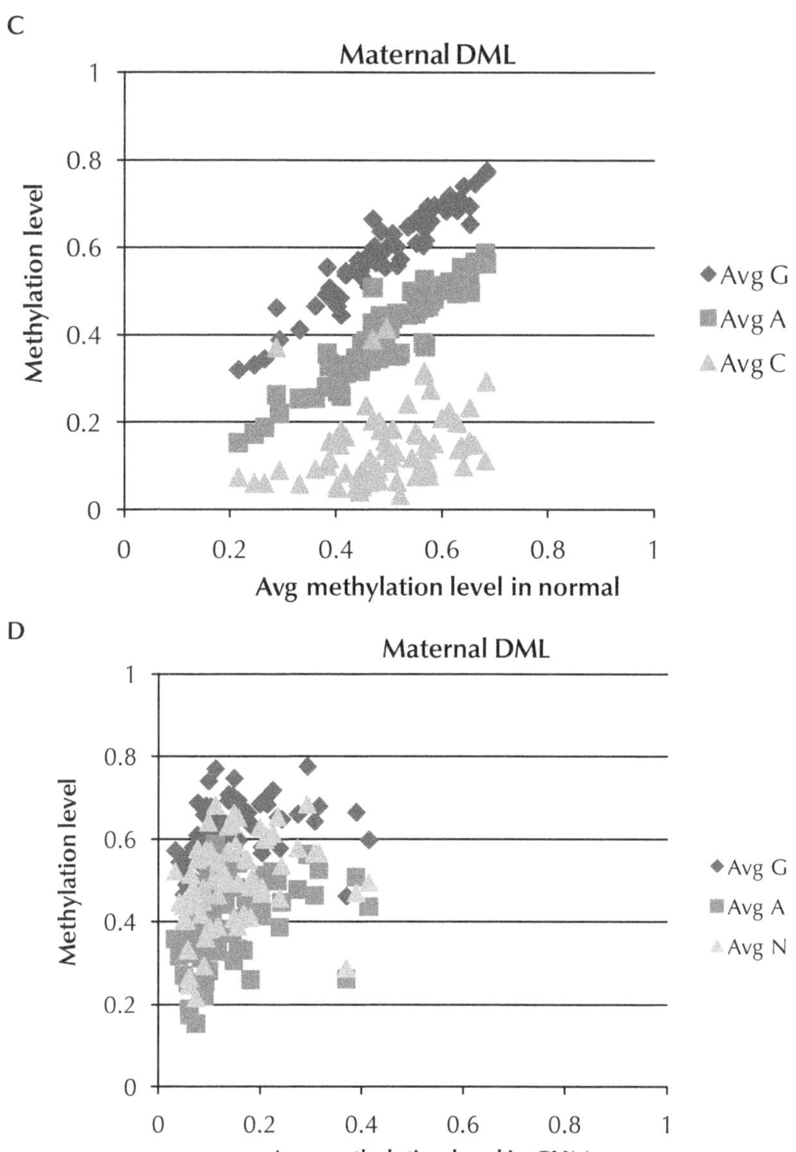

FIGURE 2 Scatterplots of average methylation of paternal and maternal DML. (A) and (C) Average methylation values in normal placentas (*x*-axis) plotted against digynic triploids (Avg G), diandric triploids (Avg A) and CHMs (Avg C) show high correlation. (B) and (D) Average methylation values in CHMs (*x*-axis) plotted against digynic triploids (Avg G), diandric triploids (Avg A) and normal placentas (Avg N).

Fourteen probes failed to follow the expected relative methylation patterns between the between groups (normal placentas with an average methylation level between that for diandric and digynic placentas and CMHs with an average methylation level closer to that in diandric placenta), and these loci were eliminated as candidates for further analysis. This yielded a final list of 108 identified putative DML that are associated with 63 different DMRs from 62 genes (one gene with both paternal and maternal DML). Of the 63 DMRs, 37 are maternally methylated and 26 are paternally methylated (Figure 3). These imprinted DMRs are distributed across the whole genome, with chromosome 7 containing the highest number (nine DMRs), while chromosomes 13, 21 and Y are the only chromosomes for which no DMRs were identified (Figure 3).

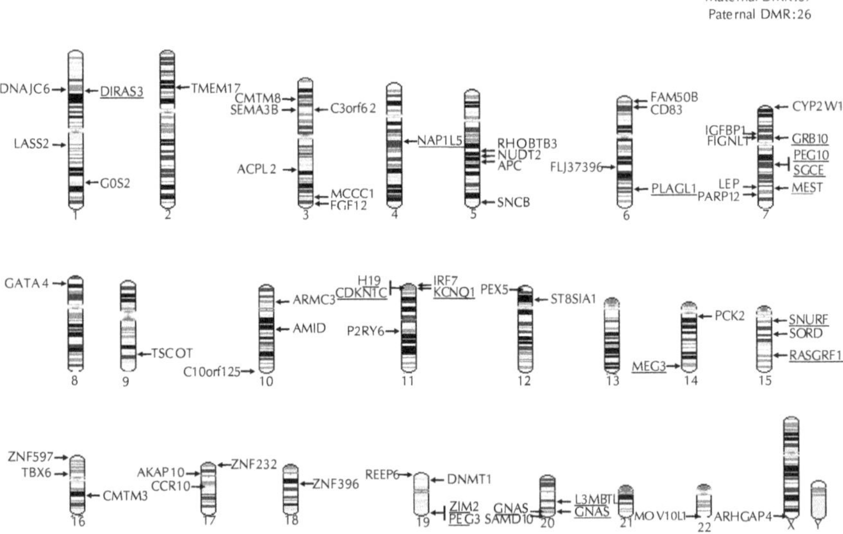

FIGURE 3 Location of the 63 identified DMRs in the genome. Relative location of the identified 37 maternal DMRs and 26 paternal DMRs are shown in the human genome according to the genomic sequence released in 2006 in the UCSC Genome Browser database (hg18). Paternal DMRs are highlighted in blue and maternal DMRs are highlighted in red. Known imprinted genes are boldfaced and underlined. Chromosome 7 contains the highest number of DMRs (nine DMRs), while there are no DMRs identified on chromosomes 13, 21 and Y.

As copy number variation (CNV) can be a potential bias for methylation [49], it was referred to the UCSC Genome Brower database (hg18) (http://www.genome.ucsc.edu webcite) and found that the locations of 37 of the 108 probes overlap with known CNVs. However, any effect of the CNVs on methylation of the candidate sites identified by this criteria was minimal, since the methylation of maternal and paternal DML were clearly separated from each other without much overlap. Similarly, differences between the two groups are unlikely to be caused by differences in genetic sequence polymorphisms that influence methylation, as this would require all ten diandric placentas, by chance, to be of a different genotype from all ten dygynic placentas.

4.4.3 VALIDATION OF DNA METHYLATION PATTERNS OF IDENTIFIED PUTATIVE IMPRINTED DMRS

The microarray included 374 CpG sites in the promoter regions of 59 genes that have previously been reported to be imprinted in humans based on the literature [12] and information in Internet databases (http://igc.otago.ac.nz/ webcite and http://www.geneimprint.com/ webcite). For nine of these genes (*PRIM2A*, *IGF2R*, *TFP12*, *COPG2*, *KLF14*, *ABCA1*, *INPP5F*, *IGF2AS* and *BLCAP*), the included CpG sites were unmethylated (or very lowly methylated) in the normal human placentas, as well as in the triploid placentas. The majority (39 of 50) of the remaining genes showed differences between diandric and digynic triploids at one or more of the associated CpG sites (t-test, $P < 0.05$), though typically not all CpG sites were differentially methylated.

Among the 62 genes identified with parent-of-origin-dependent DMRs (using the stricter criteria of <0.1% FDR and absolute average mean difference >15%), 18 are known imprinted genes associated with 15 distinct DMRs. Two of the identified DMRs, associated with the imprinted genes *CDKN1C* and *RASGRF1*, have been reported only in mice and not in humans [3, 12] (Table 1). While the strict selection criteria yielded only 18 of the 39 known imprinted genes that were statistically significantly different between diandric and digynic triploidies using an uncorrected $P < 0.05$, the missed cases were largely due to the mean difference being less than the 15% average methylation difference cutoff. Eleven of the fifteen imprinted DMRs are known to be ICRs with a parental origin of methylation concordant with what was observed based on the comparison of triploidies (Table 1).

TABLE 1 Identified DMRs with known imprinted DMRs[a].

Location	Gene	Expressed allele	ICR	Known DMR	Identified DMR
1p31	*DIRAS3*	P	-	M	M
4q22.1	*NAP1L5*	P	M	M	M
6q24	*PLAGL1*	P	M	M	M
7p12	*GRB10*	M/P[b]	M	M	M
7q21.3	*PEG10/ SGCE*	P	M	M	M
7q32.2	*MEST*	P	M	M	M
11p15	*CDKN1C*	M	-	P[c]	P
11p15	*H19*	M	P	P	P
11p15	*KCNQ1*[d]	M	M	M	M
14q32	*MEG3*	M	P	P	P
15q11-q12	*SNURF*	P	M	M	M
15q24	*RASGRF1*	P	-	P[c]	M
16p13	*ZNF597*	M	-	-	P
19q13.43	*PEG3/ZIM2*	P	M	M	M
20q13	*GNAS* (NESP)	M	-	P	P
20q13	*GNAS* (XL)	P	M	M	M
20q13	*L3MBTL*	P	-	M	M

[a]DMR, differentially methylated region; ICR, imprinting control region; [b]tissue-specific parental origins of allelic expression; [c]parental origins based on mouse studies; [d]region known as KvDMR1.

To confirm the methylation differences using an independent approach, bisulphite pyrosequencing for a subset of the novel imprinted DMRs was performed. For this purpose, ten DMRs were selected on the basis of their low FDR (*FAM50B*, *MCCC1*, *DNAJC6*, *SORD* and *RHOBTB3*) or their biological significance to the placenta (*APC*, *DNMT1*, *IGFBP1*, *LEP* and *RASGRF1*). A high correlation between the values obtained by microarray and pyrosequencing was observed (r = 0.85 to 0.98; P <0.0001). Specifically, the DNA methylation patterns observed by pyrosequencing were concordant with those found by microarray for both (1) CpG sites analyzed by microarray and their the proximal CpG sites within the pyrosequencing assays and (2) the average methylation levels of all CpG sites covered by pyrosequencing. DNA methylation levels of the selected loci were also assessed in sperm DNA and all were unmethylated (data not shown), suggesting they may be either secondary DMRs or maternal imprinted DMRs.

Further DNA methylation for two genes was evaluated, *FAM50B* and *MCCC1*, which contain SNPs with high average heterozygosity (about 0.4) in the proximal promoter regions that can be used to distinguish alleles (Figures 4A and 4F). Most of the other identified genes do not contain common SNPs in the nearby analyzed regions that could be used for this purpose. Bisulphite cloning and sequencing confirmed monoallelic methylation patterns for both DMRs (Figures 4C and 4H) and maternal origin of allelic methylation that was concordant with that predicted by the triploidy comparison (Figures 4B and 4G). Furthermore, allelic expression analysis showed preferential expression of the unmethylated paternal allele at the proximal promoter regions (Figures 4E and 4I), which is consistent with an inverse correlation between methylation and expression. As allelic methylation can occur in an SNP-dependent manner [50], a methylation-specific pyrosequencing assay was developed for *FAM50B* to evaluate allelic methylation in additional samples. This same approach could not be applied to *MCCC1*, because its interrogated SNP is located at a CpG site. The results of the *FAM50B* assay were concordant with cloning and sequencing results for the same placental sample. As methylation was found in association with either allele (A or G at rs2239713) among 12 heterozygous normal term placental samples and ten heterozygous maternal blood samples, the allelic methylation is not linked to the SNP genotypes, at least for this DMR.

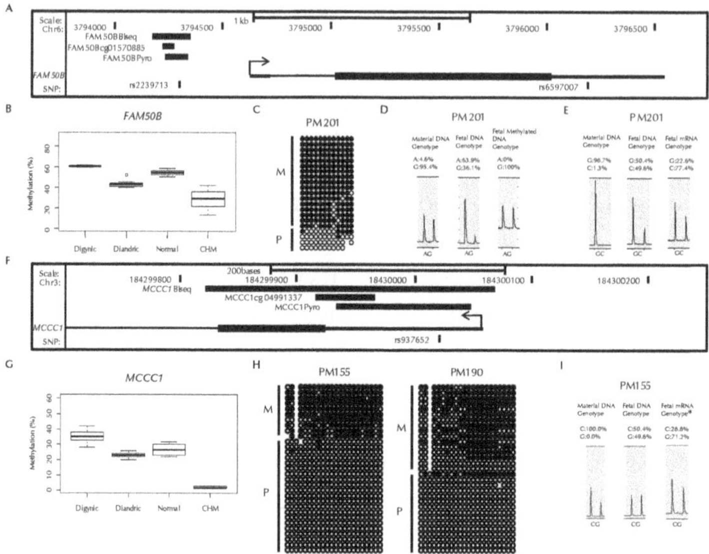

FIGURE 4 Identification of imprinted DMRs at the proximal promoter regions of *FAM50B* and *MCCC1*. (A) and (F) Schematics showing the positions of methylation assays (Biseq: bisulphite cloning and sequencing assay; cg code: probe number of Illumina assay; Pyro: bisulphite pyrosequencing assay) and SNP locations relative to the genes. Arrow directions represent the transcriptional directions for the genes. Genomic coordinates were retrieved from the UCSC Genome Brower database (hg18). (B) and (G) Box plots showing the methylation levels of samples from each placental group for the DMRs analyzed by bisulphite pyrosequencing. Both DMRs in *FAM50B* and *MCCC1* have higher methylation in digynic than diandric triploid placentas, while they have intermediate methylation in normal placentas and particularly low methylation in CHMs. (C) and (H) Bisulphite cloning and sequencing showing parental origins of methylated and unmethylated alleles (M: maternal alleles; P: paternal alleles). Parental origin was determined by genotyping heterozygous informative SNPs for each sample. The DMRs in both *FAM50B* and *MCCC1* are maternally methylated. Each black circle represents a methylated CpG dinucleotide, and each white circle represents an unmethylated CpG dinucleotide. (D) Quantitative genotyping of methylated alleles by pyrosequencing. SNP rs2239713 is homozygous (GG) in maternal DNA and heterozygous (AG) in fetal (placental) DNA (dispensation order: AAG). Genotyping of the placental sample using a methylation-specific pyrosequencing primer shows a homozygous (GG) pattern, indicating that the DMR associated with the maternally inherited 'G' allele is methylated while the one associated with the paternal 'A' allele is not. (E) and (I) Quantitative genotyping of expressed alleles by pyrosequencing. Both SNPs (E) rs6597007 (dispensation order: GGC) and (I) rs937652 (dispensation order for DNA genotyping: CG; dispensation order for RNA genotyping: CCG) are homozygous in maternal DNA and heterozygous in fetal DNA. Genotyping of cDNA shows a bias towards preferential expression of the paternal alleles. *The pyrosequencing primers used for cDNA genotyping (intron-spanning) in *MCCC1* were different from those used for DNA genotyping, so the peak ratio shown in genotyping the pyrogram of cDNA does not correspond to that for DNA.

Since diandric triploid placentas tend to be associated with trophoblast hyperplasia [17], the possibility was considered that some of the identified imprinted DMRs reflect differences in methylation between the trophoblast and mesenchyme, the two components of the chorionic villi [33]. To address this hypothesis, a nonimprinted, trophoblast-specific unmethylated region, *EDNRB,* was used to compare the methylation levels between diandric and digynic triploid placentas. However, a difference in methylation levels between them at this site cannot be found. Likewise, differences were not found in allelic methylation between trophoblast and mesenchyme for the novel identified imprinted gene *MCCC1*. However, *DNAJC6* and *RASGRF1* showed differential methylation between trophoblast and mesenchyme, which may represent cell-type-specific imprinting.

4.4.4 CONFIRMATION OF PARENT-OF-ORIGIN ALLELIC EXPRESSION FOR THE IDENTIFIED PUTATIVE IMPRINTED GENES

While the existence of an imprinted DMR is thought to be predictive of imprinting at the gene expression level, proving this is complicated by the fact that imprinted DMRs may exist in association with imprinted genes even in tissues where the gene is not expressed or is expressed in a biallelic manner. Furthermore, to be informative for demonstrating monoallelic expression, the placental sample must be heterozygous for an expressed SNP. To be informative for parental origin of the expressed allele, the mother must additionally be homozygous for the same SNP.

As many previously reported imprinted genes are expressed in an imprinted manner in placenta, and since it was needed to screen many placenta-mother pairs to find informative cases, it was proceeded to investigate the parental origin of allelic expression for the novel putative imprinted genes using a high-throughput genotyping approach, specifically the iPLEX Gold assay on the MassARRAY platform (Sequenom, Inc., San Diego, CA, USA). Thirty-eight of 45 genes associated with novel imprinted DMRs (the 45 putative imprinted genes including *RASGRF1*, for which imprinted expression has not been reported in humans) were selected on the basis of the availability of an exonic SNP with high average heterozygosity (>0.1) and the presence of expression in the placenta according to the GNF atlas database (http://biogps.gnf.org/ webcite). In addition, two exonic SNPs from *IGF2* were included as positive controls, since *IGF2* is well known to exhibit imprinted expression in human placentas. Thus, a total of 40 SNPs were genotyped in 27 maternal-fetal pairs, including DNA from maternal blood and the corresponding fetal normal term placenta, as well as cDNA from the same placenta.

Of these 40 SNPs, 7 did not pass the quality control criteria (<70% calls or presence of severe allelic bias) and 3 had no informative (heterozygous) genotypes in fetal DNA, leaving a total of 30 SNPs for analysis. The two SNPs from *IGF2* showed the expected paternal allelic expression in all informative cases. Of the 28 novel putative imprinted genes, 11 showed monoallelic expression in at least a portion of informative samples (Table 2). Among these 11 genes, eight had cases informative (homozygous) in maternal blood for parental origin assessment. Since most CpG sites in the microarray are located at the proximal promoter regions of the genes, it was assumed that the DNA methylation would most likely correlate with silencing for all these genes. Six genes (*FAM50B*, *DNMT1*, *RHOBTB3*, *ARMC3*, *AIFM2* and *LEP*) showed parent-of-origin-dependent expression that matched that predicted by the parental origin of the DMRs, while two others (*MOV10L1* and *ST8SIA1*) showed parental expression opposite that predicted in one or more informative cases (Table 2). For *FAM50B* and *RHOBTB3*, monoallelic expression for both reciprocal forms of the SNP was also observed. Some genes with imprinted DMRs may not show allele-specific expression biases because of the presence of tissue-specific or gestational age-specific imprinting that is further regulated by DNA methylation at other nearby sites.

TABLE 2 Eleven genes associated with candidate imprinted DMRs with confirmed monoallelic expression[a].

Gene	DMR	SNP	Monoallelic expression, observed/total (%)	Monoallelic expression observed for reciprocal SNP[b]	Matched expected parental origin, observed/total (%)[c]
FAM50B	M	rs6597007	9/9 (100)	Y	5/5 (100)
DNMT1	M	rs16999593	1/1 (100)	-	1/1 (100)
MOV10L1	P	rs9617066	8/9 (89)	N	1/3 (33)
RHOBTB3	M	rs34896	3/4 (75)	Y	2/2 (100)
SNCB	M	rs2075667	3/4 (75)	N	NI
ARMC3	M	rs12259839	2/3 (67)	N	2/2 (100)
ST8SIA1	M	rs4762737	2/3 (67)	Y	0/1 (0)
ARHGAP4	P	rs2070097	1/2 (50)	-	NI
AIFM2	M	rs7908957	2/8 (25)	N	1/1 (100)
MCCC1	M	rs937652	2/8 (25)	Y	NI
LEP	P	rs2167270	1/15 (7)	-	1/1 (100)

[a]DMR, differentially methylated region; SNP, single-nucleotide polymorphism; NI, not informative. [b]Where both alleles of SNP were observed to be expressed among cases with monoallelic expression; this is impossible to determine if only one case showed monoallelic expression. [c]Number of cases matching the expected parental origin of those cases informative with regard to determining parent of origin.

A number of genes did not consistently show monoallelic expression using the iPLEX Gold assay. For example, for *LEP*, only 1 of 15 samples was scored as monoallelic using this approach. To evaluate the sensitivity of the iPLEX Gold genotyping assay for detecting allelic biases in expression, an RNA-specific genotyping pyrosequencing assay were developed for *LEP*. Although the two methods were correlated ($r = 0.64$; $P < 0.02$), it was found that pyrosequencing was more likely to detect preferential allelic expression, with 5 of 12 informative cases exhibiting a <0.3 allelic ratio by pyrosequencing. Furthermore, in case PM155 for *MCCC1*, preferential paternal allelic expression was found by pyrosequencing, but not by iPLEX Gold genotyping (Table 2). Thus, the iPLEX Gold assay may not be sufficiently sensitive to detect more subtle allelic expression bias, that is, in circumstances where there is a mix of cells with biallelic and monoallelic expression.

4.4.5 TISSUE-SPECIFIC AND GESTATIONAL AGE-SPECIFIC METHYLATION OF IMPRINTED DMRS

To study tissue-specific effects and the effect of gestational age on methylation of the putative imprinted DMRs, further methylation was compared at these sites among three types of fetal somatic tissues (eight brain samples, twelve kidney samples and eleven muscle samples) and two sets of placentas with different gestational ages (ten midgestation and ten term placentas) that had been run in the same Infinium HumanMethylation27 BeadChip array.

For tissue-specific methylation analysis, the DNA methylation levels of the 108 DML (probes) associated with the 63 imprinted DMRs in five tissues (brain, kidney, muscle, midgestation placenta and blood) was compared. Multiclass comparison from SAM was performed with 1,000 permutations. Using a <0.1% FDR cutoff, 53 probes of 46 imprinted DMRs showed differential DNA methylation between tissues. Placenta-specific methylation was observed for 31 of these probes (26 imprinted DMRs), with the average methylation being more than 15% higher in placenta than in any other tissues. A change in methylation of placenta by gestational age was found for 12 probes from ten DMRs using the same statistical

TABLE 3 DNA methylation of identified DMRs in different tissues and gestational ages[a].

Index	Gene	Chromosome	Tissue-specific[b]	Change in gestation[c]	Stable non-tissue-specific[d]	Known imprinted genes[e]
1	*DNAJC6*	1	Y[f]	Y	N	N
2	*LASS2*	1	Y[f]	Y	N	N
3	*PEX5*	12	Y[f]	Y	N	N
4	*RASGRF1*	15	Y[f]	N	N	N
5	*AKAP10*	17	Y[f]	N	N	N
6	*AIFM2*	10	Y[f]	N	N	N
7	*APC*	5	Y[f]	N	N	N
8	*ARHGAP4*	X	Y[f]	N	N	N
9	*ARMC3*	10	Y[f]	N	N	N
10	*C3orf62*	3	Y[f]	N	N	N
11	*CD83*	6	Y[f]	N	N	N
12	*CMTM3*	16	Y[f]	N	N	N
13	*DNMT1*	19	Y[f]	N	N	N
14	*G0S2*	1	Y[f]	N	N	N
15	*GATA4*	8	Y[f]	N	N	N
16	*LEP*	7	Y[f]	N	N	N
17	*MCCC1*	3	Y[f]	N	N	N
18	*NUDT12*	5	Y[f]	N	N	N
19	*PCK2*	14	Y[f]	N	N	N
20	*RHOBTB3*	5	Y[f]	N	N	N
21	*SLC46A2*	9	Y[f]	N	N	N
22	*SNCB*	5	Y[f]	N	N	N
23	*SORD*	15	Y[f]	N	N	N
24	*ST8SIA1*	12	Y[f]	N	N	N
25	*TBX6*	16	Y[f]	N	N	N
26	*TMEM17*	2	Y[f]	N	N	N
27	*ZNF232*	17	Y[f]	N	N	N
28	*ZNF396*	18	Y[f]	N	N	N
29	*AK094715*	6	Y	Y	N	N
30	*DIRAS3*	1	Y	Y	N	Y
31	*CMTM8*	3	Y	Y	N	N
32	*SEMA3B*	3	Y	Y	N	N
33	*CDKN1C*	11	Y	N	N	Y
34	*H19*	11	Y	N	N	Y
35	*KCNQ1*	11	Y	N	N	Y
36	*MEG3*	14	Y	N	N	Y
37	*PEG10*	7	Y	N	N	Y
38	*C10orf125*	10	Y	N	N	N
39	*CCR10*	17	Y	N	N	N
40	*CYP2W1*	7	Y	N	N	N
41	*FIGNL1*	7	Y	N	N	N
42	*IGFBP1*	7	Y	N	N	N
43	*MOV10L1*	22	Y	N	N	N
44	*P2RY6*	11	Y	N	N	N
45	*PARP12*	7	Y	N	N	N
46	*SAMD10*	20	Y	N	N	N
47	*L3MBTL*	20	N	Y	N	Y
48	*ACPL2*	3	N	Y	N	N
49	*REEP6*	19	N	Y	N	N
50	*GNAS(M)*	20	N	N	Y	Y
51	*GNAS(P)*	20	N	N	Y	Y

TABLE 3 *(Continued)*

52	*GRB10*	7	N	N	Y	Y
53	*MEST*	7	N	N	Y	Y
54	*NAP1L5*	4	N	N	Y	Y
55	*PEG3*	19	N	N	Y	Y
56	*PLAGL1*	6	N	N	Y	Y
57	*SGCE*	7	N	N	Y	Y
58	*SNURF*	15	N	N	Y	Y
59	*ZIM2*	19	N	N	Y	Y
60	*ZNF597*	16	N	N	Y	Y
61	*FAM50B*	6	N	N	Y	N
62	*FGF12*	3	N	N	Y	N
63	*IRF7*	11	N	N	Y	N

[a]DMR, differentially methylated region; [b]multiclass comparison of methylation level in brain, kidney, muscle, midgestation placenta and blood with FDR <0.1%; [c]Multiclass comparison of methylation level in early-gestation, midgestation and term placenta, FDR <0.1%; [d]DMRs with no statistically significant changes in methylation level in different tissues and gestational ages; [e]Based on the public databases (http://igc.otago.ac.nz/ and http://www.geneimprint.com/); [f]placenta-specific methylation.

criterion (<0.1% FDR). Thus, imprinted DMRs can show both tissue-specific and gestational age-specific DNA methylation. Nonetheless, 14 of the imprinted DMRs showed constant methylation between different tissues and gestational ages, 11 of which are in ICRs from known imprinted genes. Three identified imprinted DMRs associated with *FAM50B*, *FGF12* and *IRF7* also remained constant across samples and are thus potential ICRs or primary DMRs.

The complexity of DNA methylation associated with imprinted genes can be illustrated by the data for three known imprinted genes, *GNAS*, *CDKN1C* and *MEST*, for which multiple probes were present on the Infinium HumanMethylation27 BeadChip array. For *GNAS*, the array contains probes for 30 CpG sites mapping across three promoter regions of three alternative transcripts (*NESP55*, *GNASXL* and exon 1A of *GNAS*) (Figure 5A). As previously reported, the paternal DMR is located at the promoter of *NESP55* transcript (Figure 5B), while the maternal DMR is located at the promoter of *GNASXL* [27]. While most of the CpG sites have more or less equal average methylation across the locus, cg15160445 to cg1683351 and cg01565918 show clear tissue-specific methylation across different tissues (Figures 5B to 5D). For *CDKN1C*, there are eight probes present in the array (Figure 5E). A previously unidentified paternal DMR was identified at the promoter region of this gene through the comparison of triploids (Figure 5F). Interestingly, not only is the imprinted DMR itself tissue-specific that is, it is a secondary DMR) (Table 3) but there is also a probe (cg20919799) that shows differential methylation across different gestational ages (Figure 5G) and tissues (Figure 5H). (Likewise, for *MEST*, for which ten probes span two regions of the gene, an imprinted DMR can be found in one region, while tissue-specific and gestational age-specific methylation is observed in another region of the *MEST* promoter.

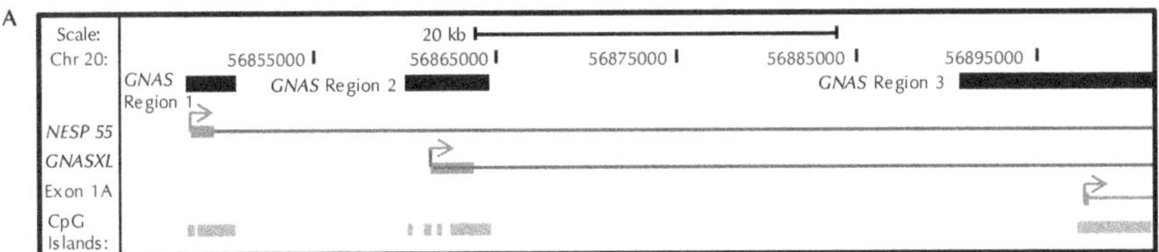

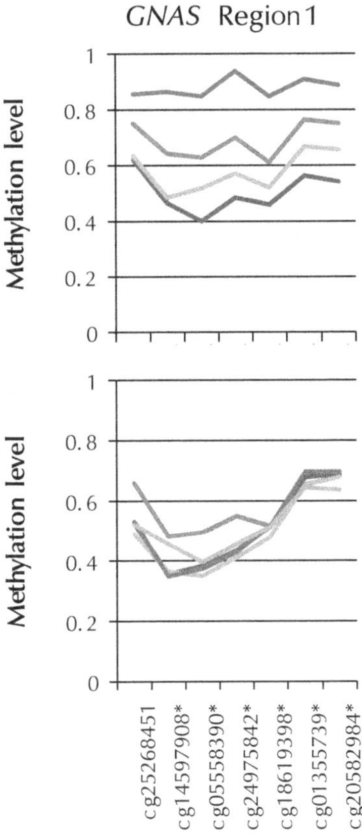

FIGURE 5 *(Continued)*

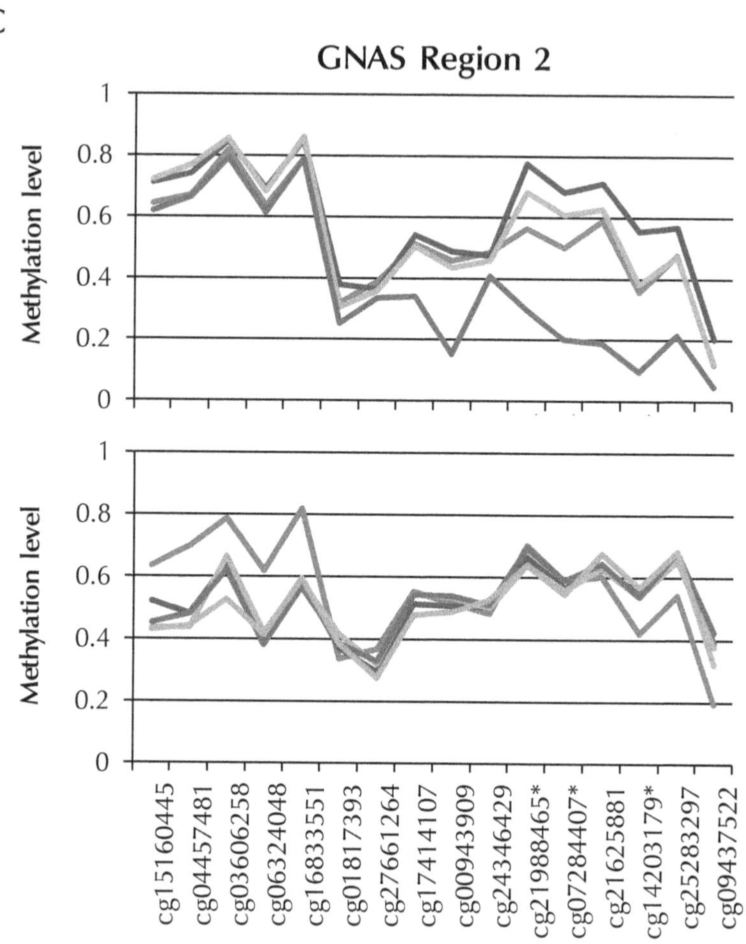

FIGURE 5 *(Continued)*

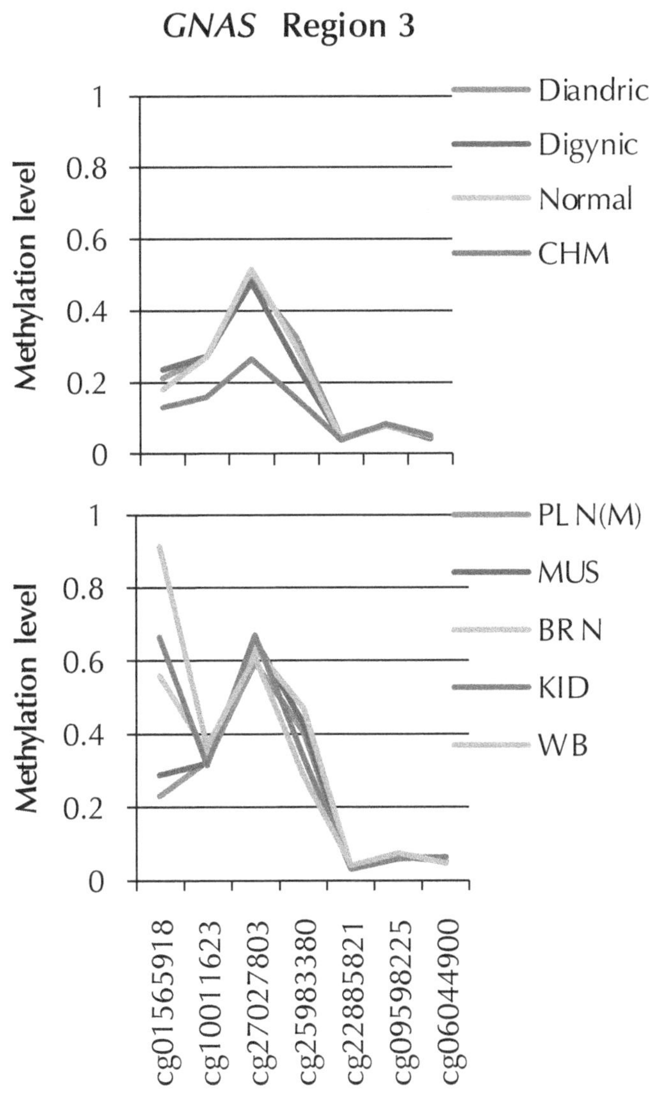

FIGURE 5 *(Continued)*

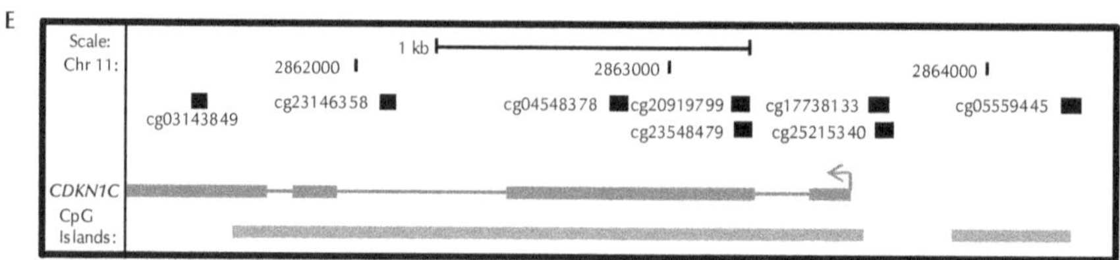

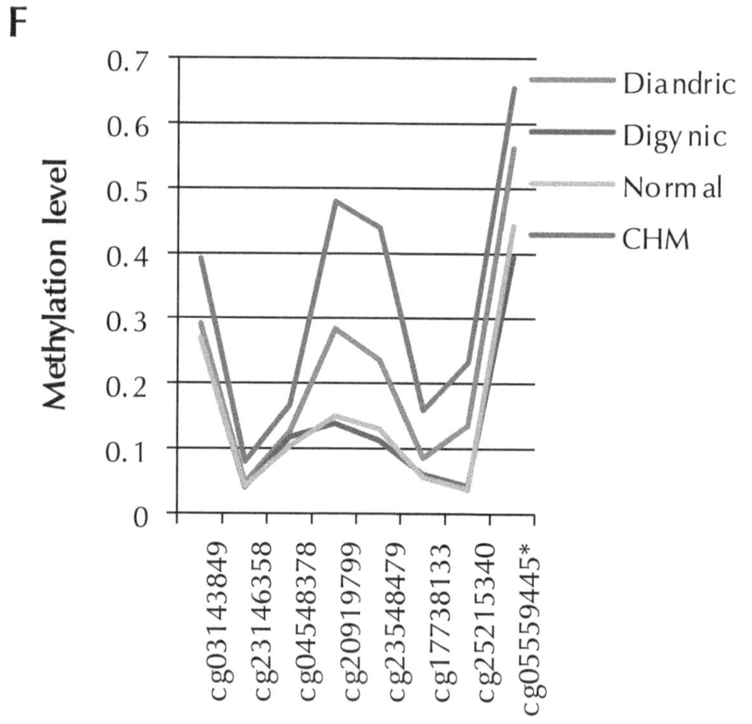

FIGURE 5 *(Continued)*

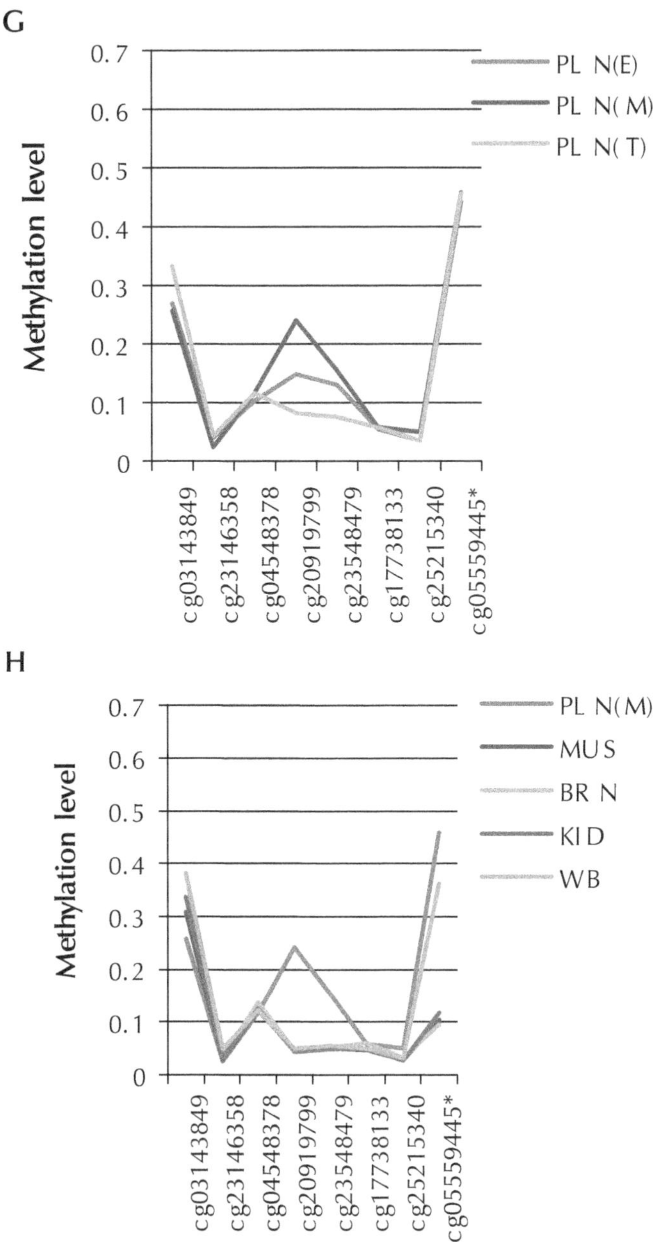

FIGURE 5 Illustration of tissue-specific and gestational age-specific methylation at the proximal promoter regions of *GNAS* and *CDKN1C*. (A) and (E) Schematics showing the positions of the Illumina Infinium probes relative to the genes and transcripts. Arrow

FIGURE 5 *(Caption Continued)*
directions represent the transcriptional directions for the genes or transcripts. Genomic coordinates were retrieved from the UCSC Genome Brower database (hg18). (B) through (D) Average methylation levels of the Illumina Infinium probes in different placental groups (top) and in different tissues (bottom). Probe numbers are shown on the *x*-axes in the bottom panels divided into (B) *GNAS* region 1, (C) *GNAS* region 2 and (D) *GNAS* region 3 according to their proximity to the known transcripts. Tissue-specific methylation can be found from cg15160445 to cg16833551 in *GNAS* region 2 and at cg01565918 in *GNAS* region 3. (F) through (H) Average methylation levels of the Illumina Infinium probes of *CDKN1C* in (F) different placental groups, (G) different gestational ages of placenta and (H) different tissues. Probe numbers are shown on the *x*-axes. Both tissue-specific and gestational age-specific methylation can be found at cg20919799. PLN(E): early gestation placenta; PLN(M): midgestation placenta; PLN(T): term placenta; MUS: muscle; BRN: brain; KID: kidney; WB: whole blood.

4.5 CONCLUSION

In conclusion, it is demonstrated that comparison of diandric and digynic triploids is an effective method for mapping imprinted DMRs in the human genome. This approach can be extended to different tissues, gestational ages or species, thereby generating a comprehensive view of imprinting regulation and evolution. The ability to map novel imprinted DMRs in the human genome should improve our understanding of the causes of placental dysfunction and birth defects. With the rapid advancement of molecular genetics technologies, a complete map of imprinted DMRs may ultimately be generated by the use of whole-genome sequencing. However, the present approach is a convenient, currently available and cost-effective method of imprinted gene mapping.

KEYWORDS

- **Androgenetic Embryos**
- **Imprinted Genes**
- **Placentas**
- **Teratomas**
- **Trophoblast Cells**

ACKNOWLEDGMENT

We thank David Chai and Danny Leung for their technical assistance and advice, Alicia Murdoch and Jennifer Sloan for placenta donor recruitment, Dr Angela Devlin for use of the Biotage PyroMark™ MD system, Dr Michael Kobor for use of the Illumina array and Dr Matthew Lorincz for the use of reagents for cloning. We also thank Dr Louis Lefebvre for critical review of the manuscript. This work was funded by a grant from the Canadian Institutes for Health Research (to WPR) and by a graduate student scholarship from the Child and Family Research Institute (to RKCY).

COMPETING INTERESTS

The authors declare that they have no competing interests.

AUTHORS' CONTRIBUTIONS

RKCY and WPR conceived the study. RKCY designed and performed the experiments. RJ prepared and karyotyped the samples. MSP performed the microarray experiment. RKCY analyzed the data. DEM contributed the tissue samples. RKCY and WPR wrote the paper. All authors read and approved the final manuscript.

REFERENCES

1. Reik, W. and Walter, J. Genomic imprinting: parental influence on the genome. *Nat Rev Genet* **2,** 21–32 (2001).
2. Frost, J. M. and Moore, G. E. The importance of imprinting in the human placenta. *PLoS Genet* **6,** e1001015 (2010).
3. Morison, I. M., Ramsay, J. P., and Spencer, H. G. A census of mammalian imprinting. *Trends Genet* **21,** 457–465 (2005).
4. Albrecht, U., Sutcliffe J. S., Cattanach B. M., Beechey C. V., Armstrong D., Eichele G., and Beaudet A. L. Imprinted expression of the murine Angelman syndrome gene, Ube3a, in hippocampal and Purkinje neurons. *Nat Genet* **17,** 75–78 (1997).

5. Monk, D., Arnaud, P., Apostolidou, S., Hills, F. A., Kelsey, G., Stanier, P., Feil, R., and Moore G. E. Limited evolutionary conservation of imprinting in the human placenta. *Proc Natl Acad Sci USA* **103,** 6623–6628 (2006).
6. Surani, M. A., Barton S. C., and Norris M. L. Development of reconstituted mouse eggs suggests imprinting of the genome during gametogenesis. *Nature* **308,** 548–550 (1984).
7. McGrath J. and Solter D. Completion of mouse embryogenesis requires both the maternal and paternal genomes. *Cell* **37,** 179–183 (1984).
8. Moore, T. and Haig, D. Genomic imprinting in mammalian development: a parental tug-of-war. *Trends Genet* **7,** 45–49 (1991).
9. Delaval, K. and Feil, R. Epigenetic regulation of mammalian genomic imprinting. *Curr Opin Genet Dev* **14,** 188–195 (2004).
10. Mann, J. R. Imprinting in the germ line. *Stem Cells* **19,** 287–294 (2001).
11. Henckel, A. and Arnaud, P. Genome-wide identification of new imprinted genes. *Brief Funct Genomics* **9,** 304–314 (2010).
12. Cooper, W. N. and Constancia, M. How genome-wide approaches can be used to unravel the remaining secrets of the imprintome. *Brief Funct Genomics* **9,** 315–328 (2010).
13. Strichman-Almashanu, L. Z., Lee, R. S. Onyango, P. O., Perlman, E., Flam, F., Frieman, M. B., and Feinberg, A. P. A genome-wide screen for normally methylated human CpG islands that can identify novel imprinted genes. *Genome Res* **12,** 543–554 (2002).
14. Choufani, S., Shapiro, J. S., Susiarjo, M., Butcher, D. T., Grafodatskaya, D., Lou, Y., Ferreira, J. C., Pinto, D., Scherer, S. W., Shaffer, L. G., Coullin, P., Caniggia, I., Beyene, J., Slim, R., Bartolomei, M. S., and Weksberg, R. A novel approach identifies new differentially methylated regions (DMRs) associated with imprinted genes. *Genome Res* **21,** 465–476 (2011).
15. Yuen, R. K., Neumann, S. M., Fok, A. K., Peñaherrera, M. S., McFadden, D. E., Robinson, W. P., and Kobor, M. S. Extensive epigenetic reprogramming in human somatic tissues between fetus and adult. *Epigenetics Chromatin* **4,** 7 (2011).
16. Xue, W. C., Chan, K. Y., Feng, H. C., Chiu, P. M., Ngan, H. Y., Tsao, S. W., and Cheung, A. N. Promoter hypermethylation of multiple genes in hydatidiform mole and choriocarcinoma. *J Mol Diagn* **6,** 326–334 (2004).
17. McFadden, D. E. and Kalousek, D. K. Two different phenotypes of fetuses with chromosomal triploidy: correlation with parental origin of the extra haploid set. *Am J Med Genet* **38,** 535–538 (1991).
18. McFadden, D. E. and Robinson, W. P. Phenotype of triploid embryos. *J Med Genet* **43,** 609–612 (2006).
19. Bourque, D. K., Peñaherrera, M. S., Yuen, R. K., Van Allen, M. I., McFadden, D. E., and Robinson, W. P. The utility of quantitative methylation assays at imprinted genes for the diagnosis of fetal and placental disorders. *Clin Genet* **79,** 169–175 (2011).
20. Bock, C., Tomazou, E. M., Brinkman, A. B., Müller, F., Simmer, F., Gu, H., Jäger, N., Gnirke, A., Stunnenberg, H. G., and Meissner, A. Quantitative comparison of genome-wide DNA methylation mapping technologies. *Nat Biotechnol* **28,** 1106–1114 (2010).
21. Bock, C., Reither, S., Mikeska, T., Paulsen, M., Walter, J., and Lengauer, T. BiQ Analyzer: visualization and quality control for DNA methylation data from bisulfite sequencing. *Bioinformatics* **21,** 4067–4068 (2005).

22. Huang, D. W., Sherman, B. T., and Lempicki, R. A. Systematic and integrative analysis of large gene lists using DAVID bioinformatics resources. *Nat Protoc* **4,** 44–57 (2009).
23. Dennis, G. Jr., Sherman, B. T., Hosack, D. A., Yang, J., Gao, W., Lane, H. C., and Lempicki, R. A. DAVID: Database for Annotation, Visualization, and Integrated Discovery. *Genome Biol* **4,** P3 (2003).
24. Schulz, R., Menheniott, T. R., Woodfine, K., Wood, A. J., Choi, J. D., and Oakey, R. J. Chromosome-wide identification of novel imprinted genes using microarrays and uniparental disomies. *Nucleic Acids Res* **34,** e88 (2006).
25. Sharp, A. J., Migliavacca, E., Dupre, Y., Stathaki, E., Sailani, M. R., Baumer, A., Schinzel, A., Mackay, D. J., Robinson, D. O., Cobellis, G., Cobellis, L., Brunner, H. G., Steiner, B., and Antonarakis, S. E. Methylation profiling in individuals with uniparental disomy identifies novel differentially methylated regions on chromosome 15. *Genome Res* **20,** 1271–1278 (2010).
26. Daelemans, C., Ritchie, M. E., Smits, G., Abu-Amero, S., Sudbery, I. M., Forrest, M. S., Campino, S., Clark, T. G., Stanier, P., Kwiatkowski, D., Deloukas, P., Dermitzakis, E. T., Tavaré, S., Moore, G. E., and Dunham, I. High-throughput analysis of candidate imprinted genes and allele-specific gene expression in the human term placenta. *BMC Genet* **11,** 25 (2010).
27. Kelsey, G. Imprinting on chromosome 20: tissue-specific imprinting and imprinting mutations in the GNAS locus. *Am J Med Genet C Semin Med Genet* 154C, 377–386 (2010).
28. Luedi, P. P., Dietrich, F. S., Weidman, J. R., Bosko, J. M., Jirtle, R. L., and Hartemink, A. J. Computational and experimental identification of novel human imprinted genes. *Genome Res* **17,** 1723–1730 (2007).
29. Kanber, D., Berulava, T., Ammerpohl, O., Mitter, D., Richter, J., Siebert, R., Horsthemke, B., Lohmann, D., and Buiting, K. The human retinoblastoma gene is imprinted. *PLoS Genet* **5,** e1000790 (2009).
30. Cotton, A. M., Avila, L., Peñaherrera, M. S., Affleck, J. G., Robinson, W. P., and Brown, C. J. Inactive X chromosome-specific reduction in placental DNA methylation. *Hum Mol Genet* **18**(19), 3544–3552 (2009).
31. Rossant, J., Guillemot, F., Tanaka, M., Latham, K., Gertenstein, M., and Nagy, A. Mash2 is expressed in oogenesis and preimplantation development but is not required for blastocyst formation. *Mech Dev* **73,** 183–191 (1998).
32. Van Dijk, M., Drewlo, S., and Oudejans, C. B. Differential methylation of STOX1 in human placenta. *Epigenetics* **5,** 736–742 (2010).
33. Avila, L., Yuen, R. K., Diego-Alvarez, D., Peñaherrera, M. S., Jiang, R., and Robinson, W.P. Evaluating DNA methylation and gene expression variability in the human term placenta. *Placenta* **31,** 1070–1077 (2010).
34. Moore, G. and Oakey, R. The role of imprinted genes in humans. *Genome Biol* **12,** 106 (2011).
35. Zhang, A., Skaar, D. A., Li, Y., Huang, D., Price, T. M., Murphy, S. K., and Jirtle, R. L. Novel retrotransposed imprinted locus identified at human 6p25. *Nucleic Acids Res* (in press).
36. Nakabayashi, K., Trujillo, A. M., Tayama, C., Camprubi, C., Yoshida, W., Lapunzina, P., Sanchez, A., Soejima, H., Aburatani, H., Nagae, G., Ogata, T., Hata, K., and Monk,

D. Methylation screening of reciprocal genome-wide UPDs identifies novel human-specific imprinted genes. *Hum Mol Genet* (in press).

37. Wong, N. C., Novakovic, B., Weinrich, B., Dewi, C., Andronikos, R., Sibson, M., Macrae, F., Morley, R., Pertile, M. D., Craig, J. M., and Saffery, R. Methylation of the adenomatous polyposis coli (APC) gene in human placenta and hypermethylation in choriocarcinoma cells. *Cancer Lett* **268,** 56–62 (2008).
38. Novakovic, B., Wong, N. C., Sibson, M., Ng, H. K., Morley, R., Manuelpillai, U., Down, T., Rakyan, V. K., Beck, S., Hiendleder, S., Roberts, C. T., Craig, J. M., and Saffery, R. DNA methylation-mediated down-regulation of DNA methyltransferase-1 (DNMT1) is coincident with, but not essential for, global hypomethylation in human placenta. *J Biol Chem* **285,** 9583–9593 (2010).
39. Guilleret, I., Osterheld, M. C., Braunschweig, R., Gastineau, V., Taillens, S., and Benhattar, J. Imprinting of tumor-suppressor genes in human placenta. *Epigenetics* **4,** 62–68 (2009).
40. Weaver, J. R., Sarkisian, G., Krapp, C., Mager, J., Mann, M. R., and Bartolomei, M. S. Domain-specific response of imprinted genes to reduced DNMT1. *Mol Cell Biol* **30,** 3916–3928 (2010).
41. Sakaue, M., Ohta, H., Kumaki, Y., Oda, M., Sakaide, Y., Matsuoka, C., Yamagiwa, A., Niwa, H., Wakayama, T., and Okano, M. DNA methylation is dispensable for the growth and survival of the extraembryonic lineages. *Curr Biol* **20,** 1452–1457 (2010).
42. Ng, H. K., Novakovic, B., Hiendleder, S., Craig, J. M., Roberts, C. T., and Saffery, R. Distinct patterns of gene-specific methylation in mammalian placentas: implications for placental evolution and function. *Placenta* **31,** 259–268 (2010).
43. Schulz, R., Proudhon, C., Bestor, T. H., Woodfine, K., Lin, C. S., Lin, S. P., Prissette, M., Oakey, R. J., and Bourc'his, D. The parental non-equivalence of imprinting control regions during mammalian development and evolution. *PLoS Genet* **6,** e1001214 (2010).
44. Yuen, R. K., Avila, L., Peñaherrera, M. S., von Dadelszen, P., Lefebvre, L., Kobor, M. S., and Robinson, W. P. Human placental-specific epipolymorphism and its association with adverse pregnancy outcomes. *PLoS One* **4,** e7389 (2009).
45. Papageorgiou, E. A., Fiegler, H., Rakyan, V., Beck, S., Hulten, M., Lamnissou, K., Carter, N. P., and Patsalis, P. C. Sites of differential DNA methylation between placenta and peripheral blood: molecular markers for noninvasive prenatal diagnosis of aneuploidies. *Am J Pathol* **174,** 1609–1618 (2009).
46. Frost, J. M., Monk, D., Stojilkovic-Mikic, T., Woodfine, K., Chitty, L. S., Murrell, A., Stanier, P., and Moore, G. E. Evaluation of allelic expression of imprinted genes in adult human blood. *PLoS One* **5,** e13556 (2010).
47. Yuen, R. K., Peñaherrera, M. S., von Dadelszen, P., McFadden, D. E., and Robinson, W. P. DNA methylation profiling of human placentas reveals promoter hypomethylation of multiple genes in early-onset preeclampsia. *Eur J Hum Genet* **18,** 1006–1012 (2010).
48. Tusher, V. G., Tibshirani, R., and Chu, G. Significance analysis of microarrays applied to the ionizing radiation response. *Proc Natl Acad Sci USA* **98,** 5116–5121 (2001).
49. Robinson, M. D., Stirzaker, C., Statham, A. L., Coolen, M. W., Song, J. Z., Nair, S. S., Strbenac, D., Speed, T. P., and Clark, S. J. Evaluation of affinity-based genome-wide

DNA methylation data: effects of CpG density, amplification bias, and copy number variation. *Genome Res* **20,** 1719–1729 (2010).

50. Kerkel, K., Spadola, A., Yuan, E., Kosek, J., Jiang, L., Hod, E., Li, K., Murty, V. V., Schupf, N., Vilain, E., Morris, M., Haghighi, F., and Tycko, B. Genomic surveys by methylation-sensitive SNP analysis identify sequence-dependent allele-specific DNA methylation. *Nat Genet* **40,** 904–908 (2008).

PART III

HISTONE MODIFICATIONS

CHAPTER 5

GENE SILENCING

IRINA A. MAKSAKOVA, PREETI GOYAL, JLIRN BULLWINKEL, JEREMY P. BROWN, MISHA BILENKY, DIXIE L. MAGER, PRIM B. SINGH, and MATTHEW C. LORINCZ

CONTENTS

5.1 INTRODUCTION

Endogenous retroviral sequences (ERVs) are relics of ancient retroviral integration into the germline. These parasitic elements are abundant in mammals, occupying approximately 8% of the mouse genome and 10% of the human genome [1, 2]. ERVs are subdivided into three diverse classes based on the similarity of their reverse transcriptase genes or their relationship to different genera of exogenous retroviruses. In the mouse, class I ERVs, similar to gammaretroviruses, include active families such as murine leukaemia viruses (MLVs) and murine retroviruses that use $tRNA^{Gln}$ (GLN). Class II ERVs are similar to alpha- and betaretroviruses and include *Mus musculus* ERV using $tRNA^{Lys}$ type 10C (MMERVK10C), the highly retrotranspositionally active intracisternal A-type particles (IAPEz) and early transposon/*Mus musculus* type D retrovirus (ETn/MusD) families. Class III ERVs, the oldest and most abundant ERVs, are most similar to spumaviruses and are represented by mouse endogenous retrovirus type L (MERV-L) and mouse apparent LTR retrotransposons (MaLR) [3, 4]. Numerous regulatory motifs in the ERV long terminal repeats (LTRs) can initiate high levels of transcription in tissues and cell lines [5], and there is extensive evidence of aberrant ERV-driven gene expression in cancers [6–11] and tissues of aging mice [12, 13]. In an effort to counteract the potentially detrimental effects of ERVs, eukaryotic genomes have evolved multiple lines of defence against active exogenous and endogenous retroviruses [14], including DNA methylation and repressive histone modifications.

The DNA methylation was the first epigenetic mark recognized to contribute to ERV silencing, with dramatic upregulation of ERVs observed in DNA methylation-deficient somatic cells [15, 16]. However, genome-wide chromatin immunoprecipitation (ChIP) followed by ChIP sequencing (ChIP-seq) [17-19] or ChIP followed by quantitative PCR (qPCR) [20] revealed that in mouse embryonic stem cells (mESCs), class I and class II ERVs are enriched for the repressive histone H3 lysine 9 trimethylation (H3K9me3) deposited by lysine methyltransferase (KMTase) SETDB1/ESET/KMT1E [20]. SETDB1 is in turn thought to be recruited to ERVs via the obligatory corepressor KRAB-associated protein 1 (KAP-1) [21], presumably through sequence-specific KAP-1-binding zinc finger proteins such as ZFP809 in the case of MLVs [22]. Moreover, it has been recently shown that in mESCs, H3K9me3 and SETDB1 play a greater role than DNA methylation in the silencing of class I and class II ERVs [20, 23]. IAP and ETn/MusD retrotransposons, the two most active class II mouse ERV families and the source of numerous recent germline mutations [24], are among the families with the highest H3K9me3 enrichment levels. Intriguingly, these families are dramatically upregulated in SETDB1 knockout (SETDB1 KO) mESCs [19, 20], confirming that they have a high potential for activation in the absence of H3K9me3. In contrast, the class III MERV-L and MaLR families, which are devoid of the H3K9me3 mark in mESCs, are repressed by the histone lysine-specific demethylase 1 (LSD1/KDM1A) [25], revealing that different ERV classes are regulated by distinct epigenetic modifications in these pluripotent cells.

Acetylation of lysine residues on the N-terminal tails of histones, including H3K9, directly influences the state of chromatin compaction by reducing the affinity of histones for DNA [26, 27]. In contrast, methylation *per se* of such lysine residues is less likely to directly affect chromatin structure, as this modification does not alter their charge. Rather, the prevailing view is that specific proteins, the so-called "readers," bind to methylated lysines and coordinate the biological outcome associated with such covalent histone marks. H3K9me3, for example, which is essential for the establishment and maintenance of the silent chromatin state [28-31], is bound by three isoforms of heterochromatin protein 1 (HP1) in the mouse genome: HP1α (encoded by *Cbx5*), HP1β (encoded by *Cbx1*) and HP1γ (encoded by *Cbx3*) [32]. HP1 is a highly conserved family; its members are frequently present in several copies in eukaryotic genomes and play both structural and gene regulatory roles [33-35]. The chromodomain of HP1 is responsible for binding H3K9me2/3 [36, 37], and a chromoshadow domain is required for HP1 homo- and heterodimerization and the recruitment of other proteins [38, 39].

Although their exact function in transcriptional regulation and crosstalk with histone and DNA methylation varies between species, the ability of HP1s to modulate gene expression via H3K9me2/3 binding has been reported in multiple systems [33, 40-42]. In fission yeast, for example, two HP1 homologues, Swi6 and Chp2, are both required for assembly of repressive chromatin [43]. In mammalian cells, targeting of HP1α, HP1β and HP1γ to heterologous loci is sufficient to induce recruitment of SET-DB1 and deposition of H3K9me3 [44], and HP1 has been implicated in SUV39H1-mediated silencing of euchromatic genes [45].

A role for HP1 proteins in silencing of repetitive and/or transposable elements has been well documented in several model organisms. In *Drosophila,* two families of transposons are derepressed in larvae with mutant HP1a and, to a lesser extent, mutant HP1c [46]. HP1d/Rhino is required for transposon silencing in the female germline of *Drosophila,* but this silencing seems to stem from Rhino's role in Piwi-interacting RNA (piRNA) production rather than establishment of repressive chromatin [47]. At transposable elements in *Neurospora,* DNA methylation is dependent on methylated H3K9 bound by HP1 [48, 49]. In *Arabidopsis,* however, H3K9me3-directed DNA methylation applies only to CpNpG methylation, not to CpG methylation, of transposons [50, 51]. HP1γ is a negative regulator of HIV in human cell lines [52] and of non-LTR LINE1 retrotransposons in male mouse germ cells [53]. On the contrary, HP1γ has also been implicated in activating gene expression through its association with elongating RNA polymerase II [54, 55]. The latter example notwithstanding, HP1 proteins are excellent candidates for the role of downstream effectors of H3K9me3-dependent silencing affecting ERVs in mESCs. Indeed, an intact HP1-binding domain of KAP-1 is essential for complete restriction of MLV in mouse embryonic carcinoma cells [56]. Furthermore, direct interaction of HP1 and KAP-1, as well as binding of HP1 to H3K9me3, is necessary for the full extent of silencing mediated by these factors [57-61]. Moreover, it has been recently demonstrated by ChIP-qPCR that HP1α, HP1β and HP1γ are enriched on IAPEz, MusD and MLV ERV sequences in mESCs, albeit at modest levels, and that this binding is partially dependent on SETDB1-deposited H3K9me3 [20]. On the basis of these observations, it is hypothesized that HP1s might play a role in H3K9me3-mediated ERV silencing in mESCs and possibly in early embryos.

In addition to their reported roles in transcriptional silencing, HP1 proteins are required for heterochromatin spreading in specific genomic contexts in *Drosophila* [62, 63], yeast [64] and mammals [42, 57]. The presence of both chromodomains and chromoshadow domains suggests that HP1 proteins may bind H3K9me3 and recruit additional proteins, such as SUV39H1/2 or SETDB1-bound KAP-1 [61, 65, 66], to facilitate the spreading of the repressive H3K9me3 mark [67, 68]. Intriguingly, repetitive elements may act as foci of *de novo* heterochromatin formation and spreading, as H3K9me3 is enriched at sequences flanking ERVs [18, 19]. Conversely, in *Neurospora*, HP1 is a component of a histone demethylase-containing complex that prevents spreading of heterochromatin [69].

In addition to HP1s, many other mouse chromodomain proteins [70] are reported to bind H3K9me3 *in vitro*, including CDYL, CDYL2, CBX2, CBX4, CBX7 and M-phase phosphoprotein 8 (MPP8) [71-78]. Furthermore, nonchromodomain proteins with affinity for H3K9me3 have also been identified [79]. Although MPP8 and CBX7 have been shown to negatively influence transcription of specific genes [71, 80], the functional and biological significance of the interaction of most of these H3K9me3 readers with H3K9me3 remains poorly understood.

To determine what role, if any, H3K9me3 readers play in silencing of ERVs and spreading of repressive chromatin from these repetitive elements, first *Cbx1* (HP1β) knock-out (KO) and *Cbx5* (HP1α) KO mESCs was generated [40, 81]. Surprisingly, no upregulation of ERVs were observed in *Cbx5*$^{-/-}$ mESCs and only modest upregulation of several ERV families in *Cbx1*$^{-/-}$ mESCs compared to that seen in *Setdb1* KO mESCs. It is found that both HP1α and HP1β are dispensable for DNA methylation of the ETnII/MusD family of ERVs, although HP1α has a modest influence on DNA methylation of IAP elements. Furthermore, it is demonstrated that while deposition of H4K20me3 at major satellite repeats is dependent in part on HP1α, as reported previously [82], HP1α and HP1β are dispensable for deposition of H4K20me3 at ERVs and play only a modest role in spreading of H4K20me3 into sequences flanking these elements. Finally, employing RNAi and newly derived mESC lines harbouring silenced IAP, MusD and exogenous MLV-based reporters, it is shown that depletion of all of the HP1 proteins, alone or in combination, or each of the remaining known H3K9me3-binding proteins, has only a modest effect on ERV derepression, indicating that at classes I and II ERVs, H3K9me3 inhibits transcription independently of HP1 and other known H3K9me3 readers.

5.2 MATERIALS AND METHODS

5.2.1 CELL CULTURE, CONSTRUCTS AND RECOMBINASE-MEDIATED CASSETTE EXCHANGE

To produce the *Cbx5*$^{-/-}$ and *Cbx1* mESC lines, each *Cbx* allele was targeted sequentially using two different targeting vectors. mESCs were cultured in DMEM supplemented with 15% fetal bovine serum (HyClone Laboratories, Logan, UT, USA), 20 mM 4-(2-hydroxyethyl)-1-piperazineethanesulfonic acid, 0.1 mM nonessential amino acids, 0.1 mM 2-mercaptoethanol, 100 U/mL penicillin, 0.05 mM streptomycin, leukaemia-inhibitory factor and 2 mM L-glutamine on gelatinized plates. For RMCE into HA36 cells, CMV was cut out of the L1-CMV-GFP-1L vector [127] by restriction with *Cla*I and *Nhe*I restriction enzymes. IAP and MusD LTR, together with the downstream sequence, were cloned into the resulting *Cla*I-*Nhe*I site upstream of the enhanced green fluorescent protein (*EGFP*) gene. MusD from the C57BL/6 genomic DNA on chr8:131270355-131277831 (*mm9*) was cloned, an element similar in sequence to those commonly expressed in wt cells [20]. The *Nhe*I site at nt 444 prevented us from including a longer fragment. However, this sequence still included the 319 bp 5'-LTR and 125 bp immediately downstream of it. A fragment containing a LTR and a downstream sequence, approximately 800 bp in total, was cloned for an IAP reporter. The element chosen was the one at the site of a novel insertion into the A/WySn mouse strain [128, 129] and was cloned from the DNA of the respective strain. All inserts were confirmed by sequencing.

5.2.2 RECOMBINASE-MEDIATED CASSETTE EXCHANGE, TRANSFECTION, AND TRANSGENE SELECTION

For targeting of the ERV reporter constructs into the genome, Cre RMCE was used [87, 130]. The HA36 mESC line contains a cassette with the *HyTK* fusion gene at the random integration site, which allows CMV-GFP expression for multiple passages (cell line a gift from F Lienert and D Schübeler). This selectable marker allows for positive selection through resistance to hygromycin B and for negative selection through sensitivity to ganciclovir. HA36 mESCs were cultured in 25 μg/mL hygromycin B for 14 days before transfection to select for cells expressing the fusion gene. Cells were transfected with Lipofectamine 2000 (Invitrogen, Carlsbad, CA, USA) in a 24-well plate according to the manufacturer's recommendations. Briefly, 1.5 μg of a cassette with a MusD, IAP or MFG insert was co-transfected with 0.5 μg of CMV-Cre plasmid using 2 μL of Lipofectamine 2000 per well. After 3 days, cells were transferred to medium containing 3 μM ganciclovir to select against cells still expressing the *HyTK* fusion gene. Cells were grown in ganciclovir-containing medium for 5 or more days, with subculturing performed when necessary.

5.2.3 siRNA-MEDIATED KNOCKDOWN

For reporter assays, 10,000 mESCs per well of a 96-well plate were seeded into antibiotic-free mESC medium the day before transfection. Transfection was performed according to the manufacturer's protocol using 100 nM concentrations of each siRNA (siGENOME SMARTpool reagent Dharmacon, Lafayette, CO, USA) and 0.4 μL of DharmaFECT 1 siRNA transfection reagent (Dharmacon) per well. On the first day after transfection, approximately one-fifth of the cells were transferred into another 96-well plate containing antibiotic-free mESC medium, and the KD was repeated on the third day. The next day, approximately 1/2 of the cells were transferred into a 24-well or 12-well plate, and flow cytometry was performed on day 4 or 5 after the second KD. For RNA or protein collection, the first KD was performed in a 12-well plate and the cells were transferred to two 6-cm dishes the next day. The day after the second KD in 6-cm dishes, three-fourths of the cells were collected for RNA for confirmation of KD efficiency, and the rest were plated onto two 10-cm dishes for expansion and collection for RNA or protein on day 4 after the second KD.

5.2.4 PREPARATION OF GENOMIC DNA, BISULPHITE TREATMENT, T/A CLONING, AND SEQUENCING

Genomic DNA was extracted using DNAzol reagent (Invitrogen), and bisulphite conversion of DNA was performed using the EZ DNA Methylation

Kit (Zymo Research, Orange, CA, USA) according to the manufacturer's protocol. The approximately 370 bp of IAP and approximately 590 bp of ETnII/MusD element sequence containing the LTR and the downstream region were amplified from converted DNA by PCR using Platinum *Taq* (Invitrogen). PCR products from three separate PCRs for each sample were cloned using the pGEM-T Easy Vector System kit (Promega, Madison, WI, USA). All sequences had a conversion rate of > 98%. QUMA http://quma.cdb.riken.jp/top/index.html webcite, with some follow-up processing, was used for analysis of bisulphite data [131].

5.2.5 NATIVE CHROMATIN IMMUNOPRECIPITATION ASSAY AND QUANTITATIVE PCR

Briefly, 1×10^7 mESCs for each cell line were resuspended in douncing buffer and homogenized through a 25-gauge 5/8-inchneedle syringe for 20 repetitions. A quantity of 1.875 µL of 20 U/µL micrococcal nuclease (MNase; Worthington Biochemical Corp., Lakewood, NJ, USA) was added and incubated at 37°C for 7 min. The reaction was quenched with 0.5 M ethylenediaminetetraacetic acid and incubated on ice for 5 min; then 1 mL of hypotonic buffer was added and incubated on ice for 1 hr. Cellular debris was pelleted, and the supernatant was recovered. Protein A/G Sepharose beads were blocked with single-stranded salmon sperm DNA and BSA, washed and resuspended in immunoprecipitation buffer. Blocked protein A/G Sepharose beads were added to the digested chromatin fractions and rotated at 4°C for 2 hr to preclear chromatin. A quantity of 100 µL of the precleared chromatin was purified by phenol-chloroform extraction, and DNA fragment sizes were analyzed and confirmed to correspond to one to three nucleosome fragments. Chromatin was subdivided into aliquots for each immunoprecipitated sample. Antibodies specific for unmodified H3 (H9289; Sigma-Aldrich, St Louis, MO, USA), H3K9me3 (Active Motif 39161, Carlsbad, CA, USA), H4K20me3 (Active Motif 39180) and control immunoglobulin G (I8140; Sigma-Aldrich, St Louis, MO, USA) were added to each tube and rotated at 4°C for 1 hr. The antibody-protein-DNA complex was precipitated by adding 20 µL of the blocked protein A/G Sepharose beads and rotated at 4°C overnight. The complex was washed and eluted, and immunoprecipitated material was purified using the QIA-quick PCR Purification Kit (Qiagen, Germantown, MD, USA). The purified DNA was analyzed by qPCR with respect to input using EvaGreen dye (Biotium, Hayward, CA, USA) and Maxima Hot Start *Taq* DNA Polymerase (Fermentas, Vilnius, Lithuania).

5.2.6 RNA ISOLATION, REVERSE TRANSCRIPTION AND QUANTITATIVE RT-PCR

The RNA was isolated using GenElute™ Mammalian Total RNA Miniprep Kit (Sigma-Aldrich) and reverse-transcribed using SuperScript III Reverse Transcriptase (Invitrogen) as per the manufacturers' instructions. Quantitative RT-PCR was carried out using SsoFAST™ EvaGreen Supermix (Bio-Rad Laboratories, Hercules, CA, USA) on StepOne™ version 2.1 software (Applied Biosystems, Foster City, CA, USA) in a total volume of 20 μL. Data are presented as means ± standard deviations of three technical replicates. Primer efficiencies were around 100%. Dissociation curve analysis was performed after the end of the PCR to confirm the presence of a single and specific product.

5.2.7 WHOLE-CELL PROTEIN EXTRACTS AND WESTERN BLOT ANALYSIS

Briefly, cells were resuspended in 2 ± Laemmli buffer and incubated at 100°C for 10 min. Cells were then homogenized through a 25-gauge needle syringe 10 to 15 times. Extracts were run on SDS-PAGE gels and transferred onto a membrane. Primary antibodies used were α-HP1α (05-684, 1:200 dilution; Upstate Biotechnology, Lake Placid, NY, USA), α-HP1β (MCA 1946, 1:100 dilution; AbD Serotec, Burlington, ON, Canada) and α-H3 (Active Motif 39163, 1:200 dilution, Carlsbad, CA, USA). Secondary antibodies used at 1:10,000 dilutions were IRdye 800CW (926-32210) and IRdye 680 (926-32221), both from LI-COR Biosciences (Lincoln, NE, USA). Membranes were analyzed using the Odyssey Infrared Imaging System (LI-COR Biosciences).

5.2.8 NORTHERN BLOT ANALYSIS

For each lane, 6 mg of RNA were denatured, electrophoresed in 1% agarose/3.7% formaldehyde gel in 1 ± 3-(N-morpholino)propanesulfonic acid buffer, transferred overnight onto a Zeta-Probe nylon membrane (Bio-Rad Laboratories, Hercules, CA, USA) and baked at 80°C. ETnII/MusD-, IAP- and *Gapdh*-specific probes were synthesized by PCR. An amplified DNA fragment was α-^{32}P-labeled using the Random Primers DNA Labeling System (Invitrogen). Membranes were prehybridized in ExpressHyb hybridization solution (Clontech, Mountain View, CA, USA) for 2–4 hr at 68°C, hybridized overnight at the same temperature in fresh ExpressHyb solution, washed according to the manufacturer's instructions and exposed to film.

5.2.9 FLUORESCENCE-ACTIVATED CELL SORTING AND ANALYSIS OF CASSETTE INTEGRATION

FACS analysis was performed using BD FACSAria III cell sorter with BD FACSDiva software (BD Biosciences), and flow analysis was performed using a BD LSR II flow cytometer. Viable cells were gated on the basis of propidium iodide exclusion. At least 10,000 propidium iodide-negative events were analyzed. Untransfected cells were used as a control for baseline EGFP fluorescence.

5.2.10 H3K9ME3 PROFILING OF ENDOGENOUS RETROVIRUSES

To determine the H3K9me3 status of ERVs in TT2 wt versus *Setdb1* KO mESC lines, average H3K9me3 profiles were generated for representative ERVs upregulated in the latter [19], including MusD, MMERVK10C, IAPEz, MLV and GLN elements. For each ERV family, all sequenced 50-bp reads from our previously published TT2 and *Setdb1* KO H3K9me3 native ChIP-seq data sets [19] were aligned to the corresponding consensus sequences (including internal regions and corresponding

LTRs) from Repbase http://www.girinst.org/repbase/ website [132] for all ERVs except MusD. For MusD, a representative active element was used (139824 to 132348 nt of AC084696, reverse strand). The Burrows-Wheeler Aligner http://bio-bwa.sourceforge.net/ website [133] was employed with default parameters (allowing up to two mismatches in the 32-bp seed and one gap). Reads were directionally extended by 150 bp, and extended reads were profiled along the element. All mapped reads were taken into account, and the profiles for each library were normalized by the total number of reads uniquely mapped to the *mm9* genome. For reads that were aligned into multiple locations (LTRs), only one randomly selected alignment location was considered. The irregular nature of the profile is most likely attributable to SNPs and insertions and/or deletions in the consensus vs. genomic reads.

5.2.11 H3K9ME3 PROFILING IN THE SEQUENCES FLANKING ENDOGENOUS RETROVIRUSES

To compare the average density of H3K9me3 in the genomic regions flanking ERVs, H3K9me3 N-ChIP-seq data sets for TT2 wt and *Setdb1* KO mESCs [19] were used. Intact elements were selected for three ERV families: MusD, IAPEz, and MLV. For MusD, IAPEz and MLV, 195, 599 and 51 elements, respectively, satisfied the length and sequence similarity criteria that are applied [19]. All H3K9me3 reads that passed the quality score threshold above 7 were aligned to the mouse genome (*mm9*) using the Burrows-Wheeler Aligner [133] and directionally extended by 150 bp [19]. Only reads uniquely aligned to the regions within 7 kb on either side of intact elements were considered. If multiple reads were mapped to the same location, only one copy of the read was counted. To generate the profiles shown for the TT2 wt and *Setdb1* KO cell lines, extended reads were first agglomerated for 5'- and 3'-flanks. Subsequently, the data were normalized to the total number of included elements and weighted by the total number of aligned reads to the genome for each sample.

5.3 DISCUSSION

5.3.1 ROLE OF H3K9ME3 AND HP1 IN SILENCING OF ERVS

It has been recently shown that the H3K9me3 KMTase SETDB1 is critical for silencing of class I and class II ERVs in mESCs [20, 21]. However, the mechanism by which H3K9me3 modification leads to their transcriptional repression is currently unclear. In the present study, it is shown that HP1α plays a modest role in maintaining DNA methylation of IAPEz ERVs, while HP1β plays a modest role in promoting the spreading of H4K20me3 into the regions flanking these elements. HP1β also contributes to silencing of select IAPEz and MMERVK10C elements, but has no effect on DNA methylation of the ERVs analyzed. However, individual depletion of HP1α, HP1β and all other candidate H3K9me3 readers does not result in upregulation of ERVs or ERV reporters to a level observed in *Setdb1* KO or *Setdb1* KD mESCs, indicating that these factors either play only a modest role in silencing or act redundantly in this process. Strikingly, robust proviral derepression was not observed, even after simultaneous depletion of all three HP1 proteins, ruling out the latter, at least for these readers. Nevertheless, the possibility cannot be excluded that an as yet unidentified H3K9me3-binding protein and/or functional redundancy between H3K9me readers other than HP1 proteins may be required for H3K9me3-mediated ERV repression.

5.3.2 H3K9ME3-DEPENDENT, H3K9ME3 READER-INDEPENDENT PROVIRAL SILENCING?

Consistent with our observation that HP1s do not play a major role in transcriptional silencing of ERVs in mESCs, tethering of HP1 proteins in *Drosophila* inactivates only a limited number of reporter lines [102, 103]. Indeed, H3K9me3-dependent silencing may occur through mechanisms independent of H3K9me3 readers, such as by preventing the binding of transcription factors essential for transcription and/or the recruitment of the RNA polymerase II complex itself.

Specifically, H3K9me3 may directly or indirectly inhibit the deposition of active covalent histone marks. Acetylation of H3 at lysine 9 (H3K9Ac), for example, which is incompatible with methylation at this residue, promotes recruitment of chromatin remodelers and binding of RNA polymerase II in promoter regions [104-108], and the histone acetyltransferases GCN5 and PCAF, which acetylate H3K9 [109,110], are required for expression of specific genes [110] and retroviral elements [111]. Furthermore, hyperacetylation of H3 and H4 occurs concomitantly with IAP upregulation in MEFs and early embryos deficient in lymphoid-specific helicase (LSH) [112], implicating histone acetylation in ERV transcriptional activity. Intriguingly, in *Xenopus* oocytes expressing human H3K9 KMTases and HP1, H3K9me3 mediates transcriptional repression independently of HP1 recruitment through a mechanism that involves histone deacetylation [59].

Alternatively (or in addition), H3K9me3 may block transcription by indirectly inhibiting phosphorylation at serine 10 (H3S10ph) in the proviral promoter region. Intriguingly, transcriptional activation of the mouse mammary tumor retrovirus is dependent on H3S10ph and hyperacetylation of H3, mediated by binding of the nuclear factor 1 (NF-1) transcription factor to the proviral LTR [113-115]. Predicted NF-1-binding sites are also found in the LTRs of other ERVs, including ETn elements [116], implicating a broad role for H3S10ph in transcription of these elements. While experiments directly addressing whether H3K9me3 blocks phosphorylation of H3S10ph have not been conducted in mammalian cells, H3K9me3 severely inhibits H3S10ph mediated by the Ipl1/aurora kinase in yeast [117].

Finally, H3K4 di- and trimethylation, marks also associated with the promoter regions of transcriptionally active genes, may also be inhibited by the presence of H3K9me3. Indeed, the H3K4 methyltransferases ASH1L [118] and SET7 [119] are less efficient in depositing H3K4me on histones marked with H3K9me in human cell lines, and H3K9me3 and H3K4me3 are mutually exclusive marks in mESCs [18]. While H3K4me2 is detected in the promoter region of IAP elements in *Lsh*$^{-/-}$ MEFs concomitant with their upregulation [120], the appearance of such active marks may be a consequence of, rather than a prerequisite for, transcriptional activation.

To directly address the role of H3K4 methylation in retroviral expression, it is sought to determine whether KD of *Wdr5*, a subunit of MLL/SET1 H3K4 methyltransferase complexes, inhibits *Setdb1* KD-induced activation of the ERV reporters. It is found that simultaneous KD of *Setdb1* and *Wdr5*, reduced the level of reactivation of all ERV reporters, especially MusD and IAP, indicating that H3K4me3, the catalytic product of WDR5-containing complexes [121], is indeed required for optimal transcription of ERVs. Thus, the presence of H3K9me3 may effectively block transcription by inhibiting deposition of H3K4me3 and/or the other active marks mentioned above.

5.3.3 HETEROCHROMATIN SPREADING INTO SEQUENCES FLANKING ERV

Heterochromatin spreading is thought to involve a reiterative process of HP1 proteins binding to H3K9me2/3 [36, 37] followed by the recruitment of protein complexes with H3K9me2/3 catalytic activity, such as SUV39H1/2 [65] and SETDB1 [61]. Consistent with this model, HP1 proteins have been implicated in heterochromatin spreading in *Drosophila* [62, 63], yeast [64] and mammals [42, 57]. Moreover, H3K9me3, a hallmark of silent chromatin, is abundant in the vicinity of ERVs [18, 19]. However, our results indicate that HP1α and HP1β play only a modest role, if any, in the spreading of H3K9me3 into the sequences flanking ERVs. In contrast, HP1β is required for efficient spreading of H4K20me3 at the IAP ERVs analyzed. Although the biological role of H4K20me3 spreading is still unclear, recent studies have indicated that this covalent mark is involved in the maintenance of genomic stability [122-124]. Intriguingly, a role for HP1 in the DNA damage response independent of H3K9me3 has also been reported [125, 126]. The availability of HP1β KO embryos will allow for studies aimed at addressing whether the distribution of H4K20me3 is dependent upon this protein and whether DNA damage repair pathways are perturbed *in vivo*.

5.4 RESULTS

5.4.1 CATALYTIC ACTIVITY OF SETDB1 IS LARGELY REQUIRED FOR ERV SILENCING

It is recently showed by ChIP-qPCR [20] and ChIP-seq [19] analyzes that numerous class I and class II ERV families are marked by H3K9me3. Furthermore, it is demonstrated that the critical role of SETDB1, the KMTase that deposits this mark, in transcriptional repression of these ERVs. Mapping all H3K9me3 ChIP-seq reads along the span of the consensus sequences of class I and class II ERVs, including IAPEz, MusD, MMERVK10C, MLV and GLN, confirms a high but nonuniform level of H3K9me3 along these elements in wt mESCs and a significantly lower level of H3K9me3 in *Setdb1* KO mESCs (Figure 1). Consistent with these data and those published in a previous studies [18], analysis of the uniquely mapped ChIP-seq reads reveals a high level of H3K9me3 in the regions flanking IAPEz, MusD and MLV ERVs.

To confirm that the KMTase activity of SETDB1 is critical for ERV silencing in mESCs [20], the *Setdb1* conditional KO mESC line was analyzed, either unmodified (SETDB1 KO) or stably expressing wild-type (wt) (SETDB1 KO TG+) or KMTase-defective (SETDB1 KO C1243A) SETDB1 transgenes, the latter harbouring a single amino acid change in the catalytic domain [20]. As expected, robust derepression of ERVs is observed in the SETDB1 KO line. Despite the fact that the SETDB1 C1243A line expresses an approximately threefold higher level of *Setdb1* than wt cells, derepression of several of these ERVs is also observed in this transgenic line, confirming that SETDB1 KMTase activity is essential for ERV silencing. Interestingly, the extent of derepression was dependent on the ERV family. The level of upregulation of MusD and IAPEz elements was equivalent in the SETDB1 KO and catalytic mutant lines, suggesting that silencing of these elements depends on the KMTase activity of SETDB1. MMERVK10C and GLN show a lower level of derepression in the SETDB1 C1243A line than the SETDB1 KO line, and MLV remains completely restricted in the SETDB1 C1243A line. Similar results were noted previously in Northern blot analyzes [20]. Taken together, these results indicate that different ERV families are subject to SETDB1-mediated silencing generally dependent on SETDB1 catalytic activity.

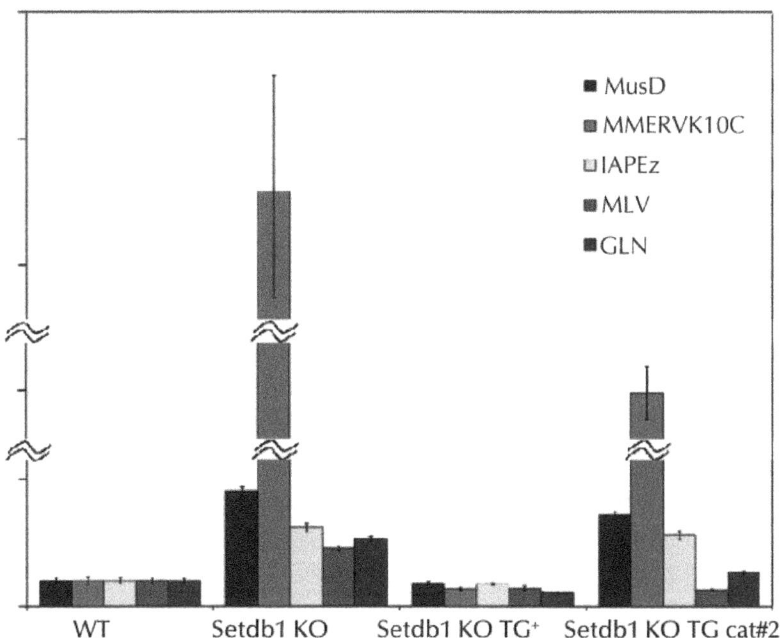

FIGURE 1 Catalytically active SETDB1 is required for endogenous retrovirus silencing. **(A)** Profiling of trimethylated lysine 9 of histone 3 (H3K9me3) along the length of IAPEz endogenous retroviruses (ERVs) in the TT2 wild type (TT2 wt) and *Setdb1* knockout (*Setdb1* KO) mouse embryonic stem cells (mESCs). The profile was generated by aligning chromatin immunoprecipitation assay sequencing (ChIP-seq) reads from TT2 wt and *Setdb1* KO mESCs [19] to the consensus sequence of IAPEz. H3K9me3 enrichment levels are presented as reads per kilobase per million mapped reads values (RPKM). **(B)** Profiling of H3K9me3 and H4K20me3 in the genomic regions flanking 599 IAPEz elements in TT2 wt and *Setdb1* KO mESCs. H3K9me3 ChIP-seq reads from TT2 wt (C57BL/6 ± CBA) and *Setdb1* KO mESCs [19] were used, along with H4K20me3 ChIP-seq from the wt V6.5 mESCs (129SvJae ± C57BL/6) [18]. Reads were aligned to the mouse genome (*mm9*), and the density of reads mapping to the 7-kb regions flanking intact IAPEz ERV families was plotted for H3K9me3 in TT2 wt and *Setdb1* KO mESCs and for H4K20me3 in V6.5 wt mESCs. Vertical lines indicate the 5' and 3' boundaries of the ERV. The average mappability for 50-bp reads was confirmed to be, on average, uniform in the assayed 7 kb region (data not shown), ruling out the possibility of mapping bias. **(C)** *Setdb1* deletion was induced with 4-hydroxytamoxifen (4-OHT) in mESCs containing no transgene (KO), a wt transgene (KO TG+) or a transgene with a mutation rendering SETDB1 catalytically inactive (KO C1243A) [20]. Expression is normalized to *β-actin* relative to wt. Data are presented as means ± standard deviations (SD) for three technical replicates. **(D)** To establish the expression levels of *Setdb1* in the KO and transgenic lines, quantitative RT-PCR (qRT-PCR) was performed with *Setdb1*-specific primers, and expression was normalized to *β-actin* relative to wt. Data are presented as means ± SD for three technical replicates.

5.4.2 DEPLETION OF HP1B BUT NOT HP1A LEADS TO MODEST UPREGULATION OF SELECT ERV FAMILIES

Having confirmed that the KMTase activity of SETDB1 is required for efficient silencing of MMERVK10C, MusD and IAPEz, next it was sought to determine whether the archetypal heterochromatic H3K9me2/3 readers HP1α and HP1β [40], both of which are enriched on IAPEz, MusD and MLV ERVs [20], are the effectors of transcriptional suppression of these elements. *Cbx5* (HP1α) KO mESCs and *Cbx1* (HP1β) KO mESCs were generated and confirmed downregulation of the corresponding genes at the mRNA level by qRT-PCR and at the protein level by Western blot analysis. Equivalent levels of expression of the pluripotency factor *Nanog* were detected in these lines, indicating that deletion of HP1 proteins does not stimulate differentiation (Figure 2A). Interestingly, while compensatory upregulation of the *Cbx1* and *Cbx3* genes was observed at the mRNA level in the $Cbx5^{-/-}$ line, upregulation of these genes was not observed at the protein level (Figure 2B).

Surprisingly, unlike deletion of *Setdb1*, deletion of *Cbx5* does not lead to upregulation of any members of the ERV families analyzed, as determined by qRT-PCR or Northern blot analysis. Similarly, deletion of *Cbx1* has no effect on MusD, MLV or GLN elements. Although *Cbx1* deletion does result in modest derepression of MMERVK10C (approximately 3-fold) and IAPEz (approximately 1.5-fold) relative to the parental HM1 line, these ERVs show approximately 47-fold and approximately 3-fold upregulation respectively, in the *Setdb1* KO line, relative to the parental TT2 line, (see Figure 1). Taken together, these results indicate that in contrast to SETDB1, HP1α and HP1β play no role or a relatively minor role, respectively, in class II ERV silencing in mESCs.

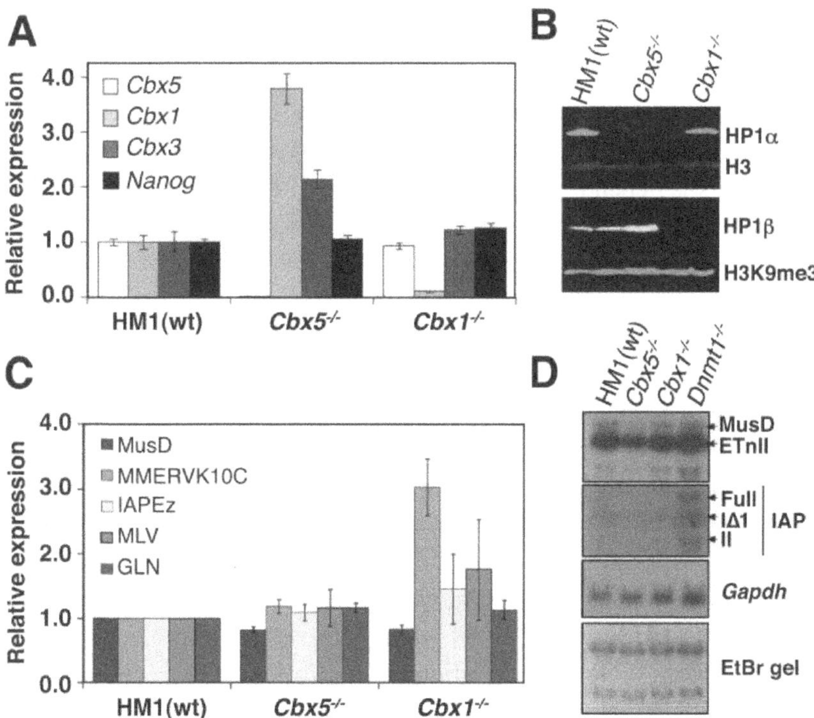

FIGURE 2 Expression of heterochromatin protein 1 genes and ERVs in the *Cbx5*$^{-/-}$ **(A)** qRT-PCR with primers specific for *Cbx5* (encoding HP1α) *Cbx1* (encoding HP1β), *Cbx3* (encoding HP1γ) and the pluripotency factor *Nanog* in the *Cbx5*$^{-/-}$ and *Cbx1*$^{-/-}$ mESC lines confirms the KOs and reveals compensatory upregulation of the genes encoding the remaining HP1 proteins in the *Cbx5*$^{-/-}$ line. Expression levels were normalized to *β-actin* relative to wt, and the data are presented as means ± SD for three technical replicates. **(B)** Western blot analysis of whole-cell lysates confirms the lack of expression of HP1α and HP1β in the *Cbx5*$^{-/-}$ and *Cbx1*$^{-/-}$ mESC lines, respectively. **(C)** Expression of representative ERV families in the HM1 (wt), *Cbx5*$^{-/-}$ and *Cbx1*$^{-/-}$ mESCs was determined by qRT-PCR. Expression levels were normalized to *β-actin* relative to wt. Data are presented as means ± SD of four independent experiments, each of which was performed in triplicate. **(D)** Northern blot analysis of RNA isolated from the parental HM1,*Cbx5*$^{-/-}$ and *Cbx1*$^{-/-}$ mESC lines using probes specific for ETnII, MusD and IAP ERVs are shown. RNA from *Dnmt1*$^{-/-}$ mESCs, in which IAP elements are upregulated approximately fourfold [20, 86] and MusD elements are upregulated approximately 1.5-fold [86], was used as a control.

5.4.3 DEPLETION OF HP1A RESULTS IN A MODEST REDUCTION OF DNA METHYLATION AT IAPEZ ERVS

It was recently demonstrated that while G9a is dispensable for silencing of ERVs, this H3K9 KMTase is required for efficient DNA methylation of these elements in mESCs [83]. Similarly, DNA methylation of major satellite repeats is dependent upon the H3K9 KMTase SUV39H1/2 in mESCs [84]. Intriguingly, HP1 proteins are required for DNA methylation of repetitive elements in *Neurospora* [48, 85], but the role of HP1 proteins in DNA methylation of ERVs in mESCs has not been explored. To address this question, ETnII/MusD and IAPEz families, shown previously to be densely DNA methylated in mESCs [20, 83, 86], were analyzed by bisulphite sequencing using genomic DNA isolated from wt, *Cbx1*$^{-/-}$ and *Cbx5*$^{-/-}$ mESCs. In wt cells, several copies of ETnII and MusD were either completely unmethylated or hypomethylated specifically at the 5' end of the LTR (Figures 3A and 3C) as observed previously [86]. The number of methylated CpG sites per element of this family remained similar in either of the *Cbx* KO lines. In contrast, while the level of DNA methylation was very high at IAPEz elements in wt cells, several IAP molecules showed reduced levels of DNA methylation in the *Cbx5*$^{-/-}$ cell line (Figures 3B and 3C), indicating that HP1α plays a role in DNA methylation of a subset of IAP elements, presumably dependent upon their genomic location. Nevertheless, as discussed above, this modest decrease in DNA methylation did not result in derepression of IAP elements in these cells.

5.4.4 NEITHER HP1A NOR HP1B ARE ESSENTIAL FOR H4K20ME3 DEPOSITION AT ERVS

Although H4K20me3 is dispensable for proviral silencing in mESCs, its deposition by SUV4-20H at ERVs requires SETDB1-deposited H3K9me3 [20]. On the basis of the fact that in mouse embryonic fibroblasts (MEFs), H4K20me3 at satellite repeats is dependent on SUV39H1/2-deposited H3K9me3 and subsequent binding of HP1 to this mark [82], it was investigated whether H4K20me3 at ERVs is also dependent upon HP1 proteins in mESCs. Native ChIP (N-ChIP) followed by qPCR revealed that H4K-20me3 enrichment was reduced by more than 50% at major satellite repeats in the *Cbx5*$^{-/-}$ line, demonstrating that as in MEFs [82], HP1 proteins are required for efficient H4K20me3 deposition at pericentric heterochromatin in mESCs. However, this mark is not entirely lost in either of the KO lines, presumably due to partial redundancy of HP1 proteins at major satellites. In line with these findings, it was recently shown that HP1β is dispensable for H4K20me3 and H3K9me3 deposition and localization in heterochromatin of mouse neurons [81]. Similarly, H4K20me3 levels at IAPEz, ETnII/MusD and MLV ERVs in the *Cbx1*$^{-/-}$ and *Cbx5*$^{-/-}$ lines remained at levels similar to the wt parent line, demonstrating either that H4K20me3 is deposited independently of HP1 binding or that these proteins act redundantly to promote deposition of H4K20me3 at these elements. As expected, H3K9me3 also remained unaltered in the absence of HP1α or HP1β (Figure 3D).

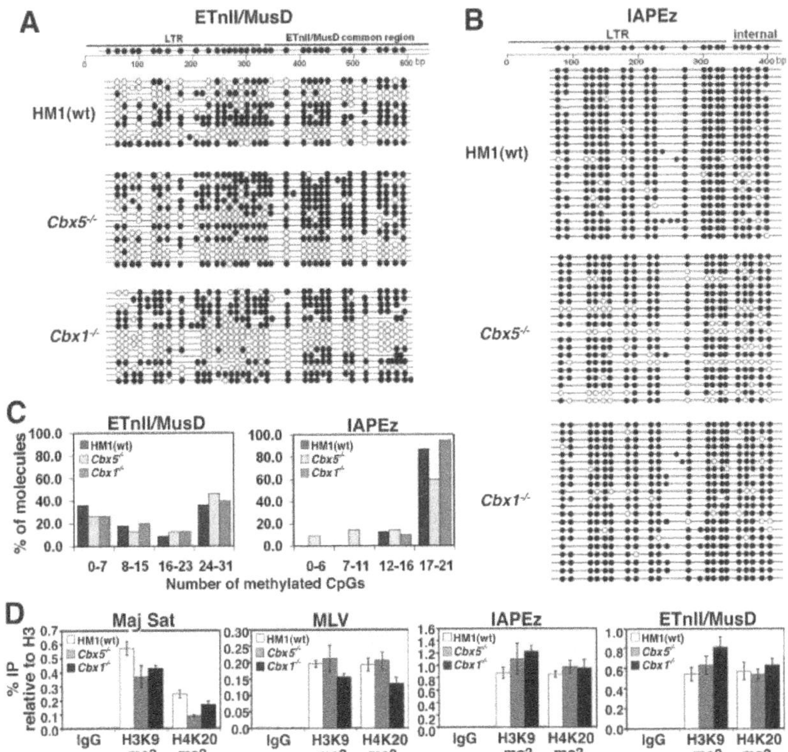

FIGURE 3 DNA methylation and chromatin marks at ERVs in *Cbx5*$^{-/-}$. **(A)** An approximately 600-bp fragment of the LTR and downstream region of ETnII/MusD ERVs was analyzed by bisulphite sequencing using primers that detect 105 ETnII/MusD

FIGURE 3 *(Caption Continued)*
elements, according to *in silico* PCR analysis (UCSC Genome Browser, http://genome.ucsc.edu/ website). **(B)** An approximately 400-bp fragment of the LTR and downstream region of IAP ERVs was analyzed by bisulphite sequencing using primers that detect 1,461 IAP elements. **(C)** Bisulphite-sequenced molecules were binned into four categories, depending on the number of methylated CpG sites detected, and the data are presented as the percentage of all clones for each cell line in each bin. While the ETnII/MusD family shows no difference in DNA methylation, several IAP molecules in the *Cbx5*$^{-/-}$ cell line exhibit partial demethylation. **(D)** Native ChIP (N-ChIP) with antibodies against H3K9me3, H4K20me3 and pan-H3 was followed by qPCR using primers specific for major satellite repeats and IAP, MLV and MusD ERVs, and the data are presented as means ± SD for three technical replicates. The level of H4K20me3 was reduced by more than 50% at major satellite repeats in the *Cbx5*$^{-/-}$ mESCs but remained at the same level at ERVs.

5.4.5 HP1B PLAYS A ROLE IN THE SPREADING OF H4K20ME3 BUT NOT H3K9ME3 FROM ERVS INTO FLANKING GENOMIC REGIONS

Intriguingly, while HP1 homologs play a positive role in heterochromatin spreading in *Drosophila* [62, 63] and mammals [42, 57], HP1 plays a critical role in inhibiting aberrant spreading of heterochromatin in *Neurospora* [69]. Genome-wide analysis of H3K9me3 in wt mESCs reveals high levels of H3K9me3 in the immediate flanks of ERVs, including IAPEz, MusD and MLV elements, with progressively lower levels of this mark at distances farther from the ERV integration site [18, 19]. As expected, deposition of H3K9me3 in these regions is SETDB1-dependent [19]. Notably, the profile of H4K20me3 in the genomic regions flanking IAP, MusD and MLV elements is similar to that of H3K9me3 and the relative levels of both marks are consistent with their abundance in each ERV family (that is, IAP > MusD > MLV [20]). To determine whether spreading of H3K9me3 and/or H4K20me3 is affected in *Cbx1*$^{-/-}$ and/or *Cbx5*$^{-/-}$ mESCs, these marks were examined at the flanks of two randomly chosen full-length IAPEz elements and three genomic locations distal to the integration sites of these ERVs by ChIP-qPCR (Figure 4). IAPEz elements were chosen because, among the ERVs analyzed, on average, this family showed the highest mean H3K9me3 density in flanking genomic regions. As expected, in wt cells, the levels of both H3K9me3 and

H4K20me3 generally declined as the distance from the IAP increased, dropping substantially at approximately 3.5 kb. Depletion of either HP1 protein did not show a consistent effect on the spreading of H3K9me3 into the flanks of the selected IAP elements, since at the majority of regions surveyed enrichment was not statistically significantly different in each of the KO lines from the wt control. HP1β may be involved in propagation of H3K9me3 beyond 2 kb from the IAP assayed on chromosome 2, however, suggesting that at least at some loci, HP1β may facilitate the spreading of H3K9me3.

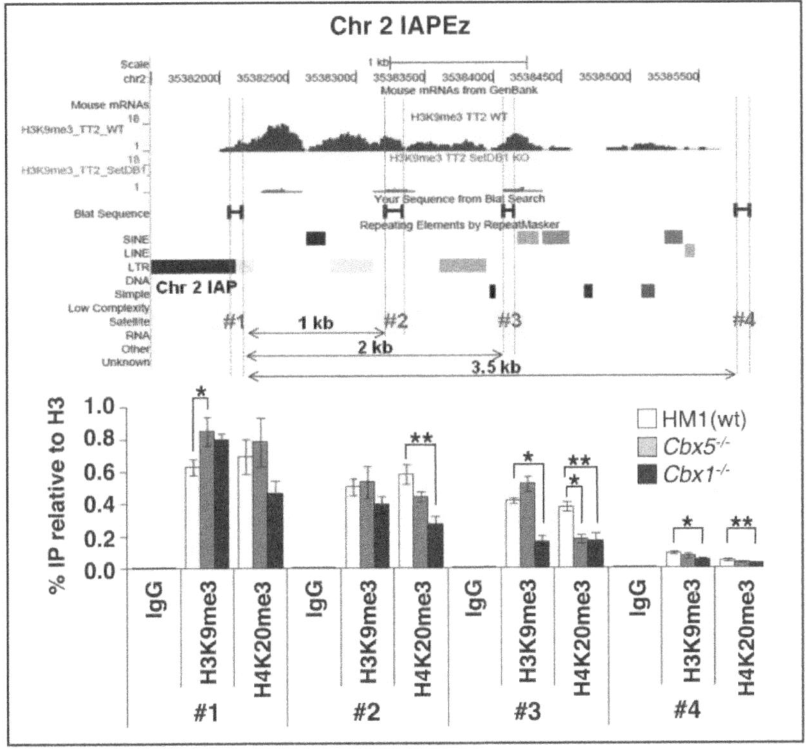

FIGURE 4 *(Continued)*

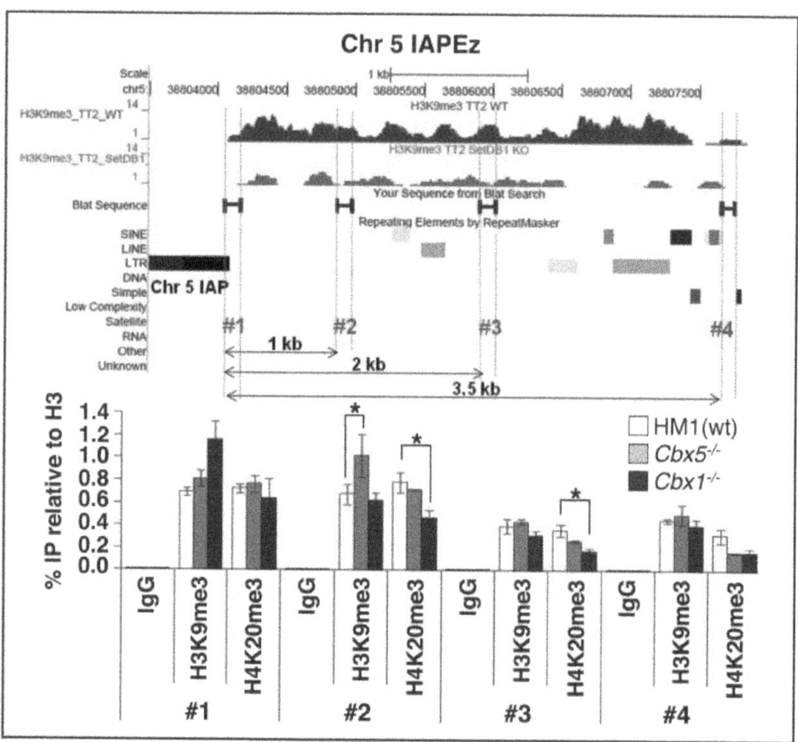

FIGURE 4 HP1β plays a role in H4K20me3 but not H3K9me3 spreading from ERVs into flanking genomic DNA. N-ChIP was performed with H3K9me3-, H4K20me3- and pan-H3-specific antibodies using chromatin isolated from HM1, *Cbx5*$^{-/-}$ and *Cbx1*$^{-/-}$ mESCs. The level of enrichment of these modifications at the flanks of two full-length IAP elements on chromosomes 2 and 5 as well as at positions approximately 1 kb, 2 kb and 3.5 kb distal to these flanking regions, was determined by qPCR. Data are presented as means ± SD of three technical replicates, and pairs of control and experimental samples with $*P < 0.05$ and $**P < 0.01$ (two-tailed Student's *t*-test) are shown. H3K9me3 enrichment levels across these genomic regions as determined using our previously published ChIP-seq data sets [19] are also shown for wt and *Setdb1* KO mESCs.

Analysis of H4K20me3 in the same regions revealed no decrease in this mark in the *Cbx5*$^{-/-}$ line. In contrast, relative to the HM1 parent line, the *Cbx1*$^{-/-}$ line showed a consistent, approximately 1.5- to 2-fold decrease ($P < 0.05$, two-tailed Student's *t*-test) in H4K20me3 at both loci in distal regions 2 and 3 (Figure 4). Thus, while neither HP1 protein is required for deposition of H4K20me3 at the ERVs themselves, HP1β may generally be involved in the spreading of this covalent mark into the genomic regions flanking these repetitive elements.

5.4.6 APPLICATION OF NOVEL ERV REPORTER LINES IN A SIRNA-BASED SCREEN OF H3K9ME3-BINDING PROTEINS

In addition to HP1 proteins, a number of other chromodomain proteins, including CDYL2, CBX2, CBX4, CBX7 and MPP8, as well as the Tudor domain-containing protein TDRD7, were recently shown to bind H3K-9me3 *in vitro* [71-73, 75, 76]. To address whether any of the H3K9me3 readers expressed in mESCs (all of those mentioned above with the exception of *Cbx4* and *Tdrd7*) play a role in SETDB1-dependent silencing, recombinase-mediated cassette exchange (RMCE) [87, 88] (Figure 5A) was used to derive novel mESC lines with single-copy proviral reporters integrated at a specific genomic site. Specifically, constructs harbouring the green fluorescent protein (*GFP*) gene downstream of the MusD or IAP LTR promoters were generated and introduced into the same genomic site in the mESC line HA36 (a gift from F Lienert and D Schübeler) via RMCE. In parallel, the MFG-*GFP* construct [89] derived from the Moloney murine leukaemia virus (MMLV) and efficiently silenced in mESCs and embryonic carcinoma cells [20, 90-93], and a cytomegalovirus (CMV)-*GFP* cassette were introduced into the same site. Following Cre-mediated recombination and a five-day negative selection with ganciclovir to exclude cells harbouring the original hygromycin B-herpes simplex virus thymidine kinase fusion (HyTK) cassette, each of the LTR reporters became silenced, while the CMV promoter maintained expression (data not shown). To select cells that contain the ERV-driven *GFP* gene silenced via the SETDB1 pathway, it was transiently transfected the GFP-negative ganciclovir-resistant pools with siRNA specific for *Setdb1*. Depending on the cassette, GFP expression was induced in 20% to 65% of viable cells and these GFP-positive cells were isolated by fluorescence-activated cell-sorting (FACS). The LTR reporter cassettes were progressively resilenced over approximately 3 weeks in culture (Figure 5B), and the negative populations were sorted at day 12 to be used as reporters. Importantly, GFP was efficiently reactivated in each population upon subsequent treatment of these pools with *Setdb1* siRNA (Figure 5C), confirming that silencing of these LTR reporters is SETDB1-dependent at this integration site.

A

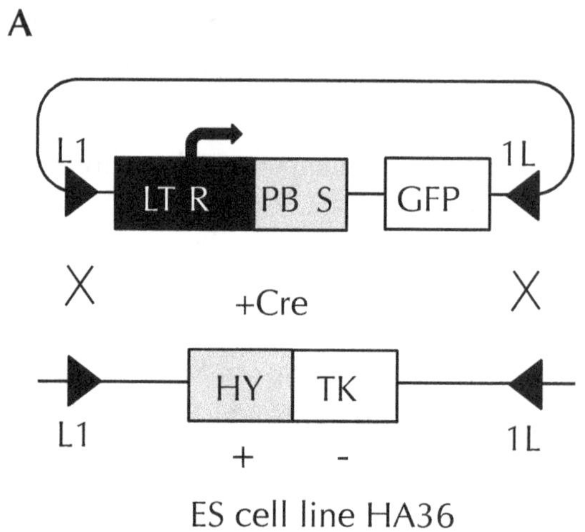

B

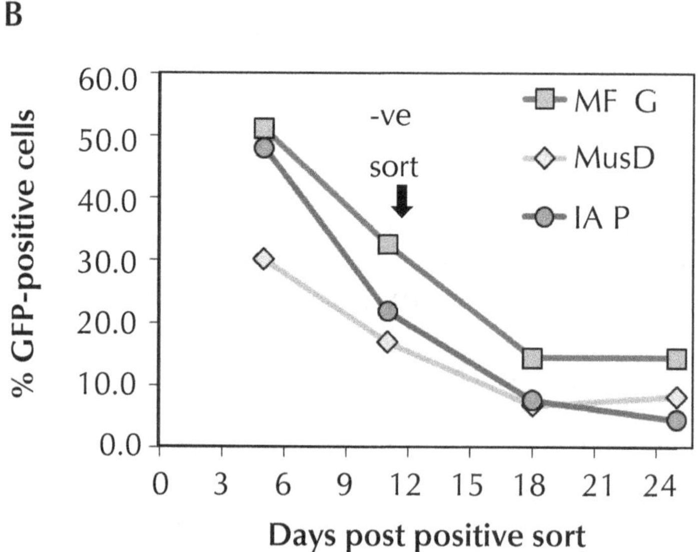

FIGURE 5 *(Continued)*

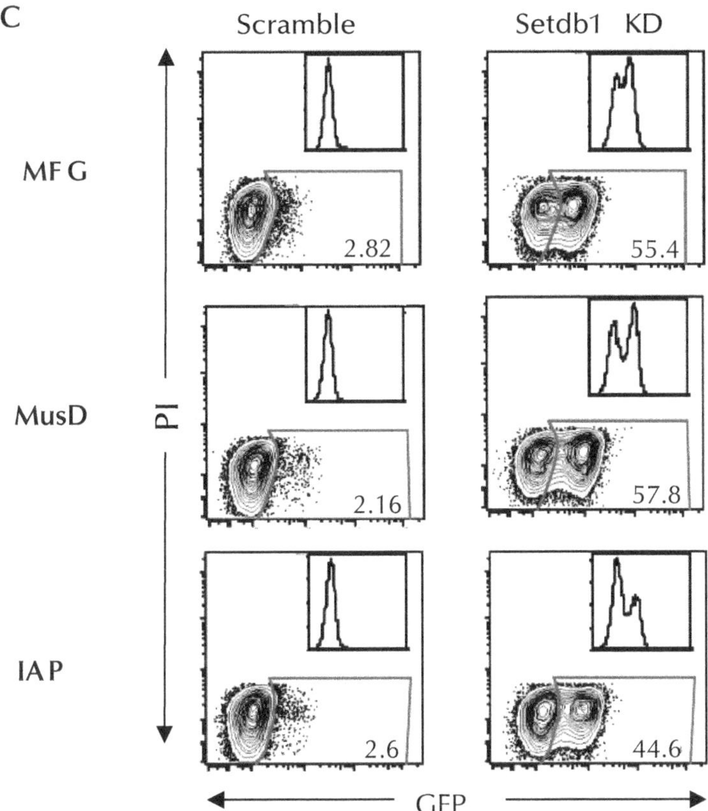

FIGURE 5 Silencing kinetics and reactivation of ERV reporters integrated in a specific genomic site. **(A)** Scheme for targeting of ERV reporters into a specific genomic site in mESCs via recombinase-mediated cassette exchange (RMCE). The mESC line HA36 contains a hygromycin B and herpes simplex virus thymidine kinase (HyTK) cassette between inverted Lox sites (L1 and 1L). MFG-green fluorescent protein (*GFP*), MusD-*GFP* and IAP-*GFP* proviral reporter cassettes, which contain the Moloney murine leukaemia virus, MusD (approximately +130 bp of downstream sequence) and IAP (approximately +450 bp of downstream sequence) LTRs, respectively, flanked by L1 and 1L sites, were cotransfected into the HA36 line with a Cre recombinase expression vector. Negative selection with ganciclovir eliminated cells with the original HyTK cassette, yielding pools of cells harbouring the proviral reporter cassettes predominantly integrated at the same site. **(B)** The kinetics of silencing of the MFG, MusD and IAP cassettes after reactivation of the RMCE pool with siRNA against *Setdb1* are shown. **(C)** GFP-negative cells were sorted at day 12 postinduction with *Setdb1* siRNA. Robust reactivation of GFP } expression from each of these pools of cells was observed upon secondary *Setdb1* knockdown (KD). Flow cytometry data are presented as contour plots and histograms of 10,000 viable (propidium iodide (PI)-negative) cells.

To determine whether any of the remaining chromodomain-containing H3K9me3 readers expressed in mESCs are required for SETDB1-mediated silencing, *Cbx3* (HP1γ) as well as *Cdyl2*, *Cbx1*, *Cbx2*, *Cbx5*, *Cbx7* and *Mpp8* were knocked down in the above-described reporter lines and a previously described pool of mESCs harbouring the silent murine stem cell virus (MSCV) provirus [20]. As expected, treatment of the MFG, MSCV, IAP and MusD reporter lines with *Setdb1* and *Kap1* specific siRNAs induced GFP expression in approximately 45% and approximately 25% of cells, respectively (Figure 6). In contrast, knockdown (KD) of each of the H3K9me3-binding proteins failed to induce GFP expression to the levels seen upon KD of *Setdb1* or *Kap1*, despite efficient depletion of the target mRNAs. KD of genes encoding other chromodomain-encoding proteins with H3K9me-binding properties, such as *Cdyl* [72] and *Chd4* [94, 95], also did not result in reporter reactivation.

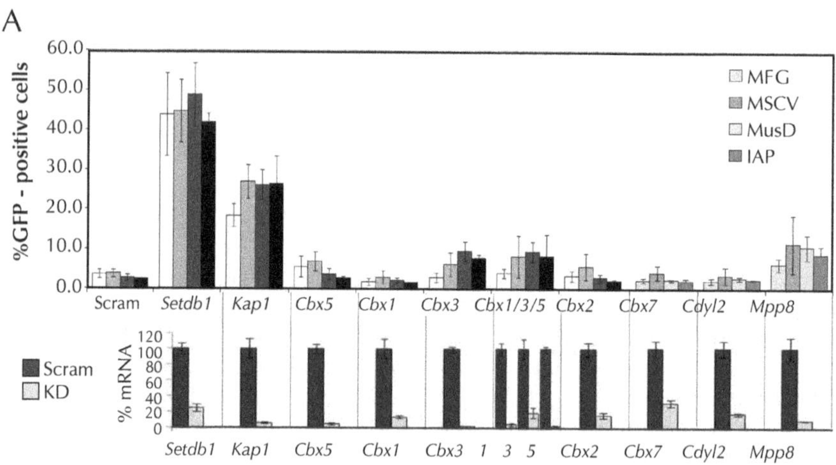

FIGURE 6 *(Continued)*

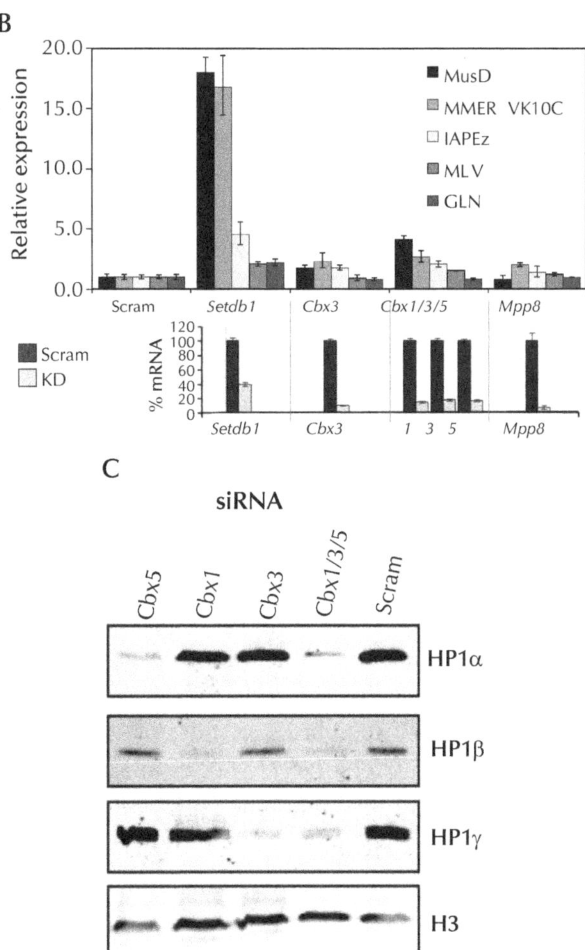

FIGURE 6 Reactivation of ERV reporters and ERVs upon siRNA-mediated KD of H3K9me3-binding proteins. **(A)** The percentage of enhanced green fluorescent protein-positive mESCs with reactivated ERV reporters was determined by flow cytometry (upper panel) on day 5 after the second transfection with siRNA against specified H3K9me3 readers. At least 10,000 cells were collected for each sample. Data are presented as means ± SD of three biological replicates. KD efficiency was determined by qRT-PCR at 30 hr after the second siRNA transfection (lower panel). Data are presented as means ± SD of three technical replicates. **(B)** Relative expression of ERVs at day 5 after the second KD with the indicated siRNA pool (upper panel), along with the efficiency of each KD, as determined by qRT-PCR at 30 hr after the second siRNA transfection (lower panel) is presented as means ± SD of three technical replicates. For each amplicon, expression was normalized to *β-actin* relative to scramble siRNA KD. **(C)** Western blot analysis of HP1 proteins in single- and triple-KD cells at day 5 after the second siRNA transfection is shown. H3 was used as a loading control.

The KD of *Cbx3* and *Mpp8* did induce GFP expression in about 10% of treated cells, raising the possibility that these H3K9me3 readers act in a redundant manner to maintain these ERV reporters in a silent state. However, simultaneous KD of *Cbx3* in combination with *Mpp8* or of *Cbx3* in combination with *Cbx1* and *Cbx5* (*Cbx1/3/5*) did not significantly increase the percentage of GFP-positive cells over that observed with individual KD, despite efficient depletion of each mRNA. KD of *Cbx1* and *Cbx3* in the *Cbx5*$^{-/-}$ mESCs and KD of *Cbx3* and *Cbx5* in the *Cbx1*$^{-/-}$ mESCs showed similar results (data not shown). Thus, none of the assayed chromodomain-encoding proteins with H3K9-binding activity are essential for proviral silencing.

The H3K4me2/3 demethylase JARID1C (SMCX, which does not harbor a chromodomain, is capable of binding H3K9me3 via its plant homeodomain (PHD) [96], and its yeast homologue, Lid2, interacts directly with the H3K9 HMTase Clr4 [97]. These interactions suggest that JARID1C may direct H3K4 demethylation to loci marked by H3K9me3, promoting silencing. However, KD of the *Jarid1* genes expressed in mESCs, including *Jarid1a*, *Jarid1b* and *Jarid1c*, either alone or in combination, leads to only modest reactivation of the proviral reporters, indicating that H3K-9me3-recognizing H3K4 demethylases are not critical for maintenance of ERV silencing. Similarly, KD of *Uhrf1* (NP95 in mouse and ICBP90 in human), which was recently shown to bind H3K9me3 via its PHD or SRA (SET- and RING-associated) domain [79, 98-100], and/or KD of a related gene, *Uhrf2* yields minimal upregulation of the four ERV reporters. Based on its pericentric localization [98], the main function of ICBP90 may lie in replication of heterochromatin and transcriptional regulation of major satellites [101], which show SUV39H1/2-dependent H3K9me3. Taken together, these data reveal that none of the known H3K9me3 readers are essential for silencing of ERVs that are repressed by the SETDB1 pathway.

To determine whether *Cbx3* and *Mpp8*, the H3K9me3 readers, which showed the highest reactivation of the LTR reporters, are required for silencing of ERVs, we performed qRT-PCR on cDNA isolated from wt TT2 and siRNA-treated mESCs. In *Setdb1* KD cells, the MMERVK10C and MusD families showed the highest level of derepression, as expected. The same families, however, are only modestly upregulated upon KD of *Cbx3* or *Mpp8*, despite reduction of the target mRNA to 9% to 30% of wt levels and dramatic downregulation of *Cbx3* at the protein level, as determined by Western blot analysis.

Finally, to determine whether HP1 proteins act redundantly to silence ERVs, simultaneous KD of *Cbx1*, *Cbx3 and Cbx5* were performed. Strikingly it was observed only modest reactivation of each of the ERVs analyzed. Similar levels of upregulation of MMERVK10C and IAPEz ERVs in the *Cbx1*$^{-/-}$ mESCs (3.0- and 1.5-fold, respectively), the *Cbx3* KD (2.3- and 1.8-fold, respectively) and the triple *Cbx1/3/5* KD (2.6- and 2.1-fold, respectively) suggest that *Cbx1* and *Cbx3* account for most of the HP1-mediated silencing of these ERV families. However, MusD elements, which are upregulated approximately fourfold in the triple KD, were not upregulated in any of the KOs, suggesting that all three HP1 proteins must be depleted to generate the relatively modest level of derepression observed for this family. Although it cannot be ruled out the possibility that an insufficient level or duration of HP1 depletion upon KD is responsible for these negative results, Western blot analysis revealed almost complete loss of all HP1 proteins in cells simultaneously depleted of *Cbx1*, *Cbx3* and *Cbx5*. On the basis of the lack of ERV upregulation upon simultaneous KD of all three HP1 isoforms, it is postulated that redundancy in HP1 function might not be the major factor preventing broad ERV reactivation. Similarly, as the maximum level of ERV reactivation upon KD of the remaining H3K-9me3 readers is considerably lower than that observed in *Setdb1* KO or *Setdb1* KD cells, it is concluded that none of these H3K9me3 readers play a major role in SETDB1-mediated ERV silencing in mESCs.

5.5 CONCLUSION

In this chapter, the surprising finding was demonstrated that despite the accepted function of HP1 proteins in H3K9me-dependent gene silencing and the critical role of H3K9me3 in transcriptional repression of class I and class II ERVs, HP1α and HP1β are not required for silencing of these repetitive elements. Furthermore, while neither HP1α nor HP1β is essential for DNA methylation or the deposition of H3K9me3 or H4K20me3 within ERVs, HP1β plays a role in the spreading of the latter into sequences flanking these elements. Using a RNAi-based screen with newly derived mESCs harbouring novel ERV reporters, it has been shown that the remaining proteins reported to bind H3K9me3 *in vitro*, including HP1γ, CDYL, CDYL2, CBX2, CBX7, MPP8, UHRF1 and JARID1A-C, are also dispensable for ERV silencing. The lack of proviral derepression in these experiments may be explained by functional redundancy of these or as yet unidentified H3K9me3 readers. Alternatively, H3K9me3 may repress ERV transcription via inhibiting deposition of covalent histone modifications required for transcription. The ERV reporter cell lines generated here should be useful in future screens of factors predicted to play a role in proviral expression. Regardless, additional studies aimed at delineating the functional significance of H3K9 readers, including nuclear processes not directly related to transcription, are clearly warranted.

KEYWORDS

- **Acetylation**
- **Chromodomain**
- **Retroviruses**
- **RNA Isolation**
- **Neurospora**

ACKNOWLEDGMENTS

We thank Danny Leung, Sandra Lee, Lucia Lam and the University of British Columbia flow cytometry facility for technical support. We also thank Florian Lienert and Dirk Schübeler for providing the HA36 ES cell line, En Li for providing the *DNMT1*-KO line, Yoichi Shinkai for providing the *Setdb1* KO line and Mark Bedford for helpful suggestions. This work was supported by CIHR grant 77805 (to ML) and CIHR grant 92090 (to ML and DM). This work was also supported by Biotechnology and Biological Sciences Research Council core strategic grants and Deutsche Forschungsgemeinschaft grant SI 1209/2-1 (to PS). ML is a Scholar of the MSFHR and a CIHR New Investigator.

AUTHORS' CONTRIBUTIONS

IM carried out most of the research. PG performed ChIP-qPCR. JB and JPB derived the KO mESCs. MB performed bioinformatics analysis. DM and PS contributed reagents. IM and ML designed the study, analyzed the data and wrote the manuscript. All authors read and approved the final manuscript.

REFERENCES

1. International Mouse Genome Sequencing Consortium: Initial sequencing and comparative analysis of the mouse genome. *Nature* **420,** 520–562 (2002).
2. International Human Genome Sequencing Consortium: Initial sequencing and analysis of the human genome. *Nature* **409,** 860–921 (2001).
3. Stocking, C. and Kozak, C. Endogenous retroviruses. *Cell Mol Life Sci* **65,** 3383–3398 (2008).
4. Gifford, R., Kabat, P., Martin, J., Lynch, C., and Tristem, M. Evolution and distribution of class II-related endogenous retroviruses. *J Virol* **79,** 6478–6486 (2005).
5. Gimenez, J., Montgiraud, C., Pichon, J. P., Bonnaud, B., Arsac, M., Ruel, K., Bouton, O., and Mallet, F. Custom human endogenous retroviruses dedicated microarray identifies self-induced HERV-W family elements reactivated in testicular cancer upon methylation control. *Nucleic Acids Res* **38,** 2229–2246 (2010).
6. Lamprecht, B., Walter, K., Kreher, S., Kumar, R., Hummel, M., Lenze, D., Köchert, K., Bouhlel, M. A., Richter, J., Soler, E., Stadhouders, R., Jöhrens, K., Wurster, K. D., Callen, D. F., Harte, M. F., Giefing, M., Barlow, R., Stein, H., Anagnostopoulos, I., Janz, M., Cockerill, P. N., Siebert, R., Dörken, B., Bonifer, C., and Mathas, S. Derepression of an endogenous long terminal repeat activates the CSF1R proto-oncogene in human lymphoma. *Nat Med* **16,** 571–579 (2010).
7. Moyes, D., Griffiths, D. J., and Venables, P. J. Insertional polymorphisms: a new lease of life for endogenous retroviruses in human disease. *Trends Genet* **23,** 326–333 (2007).
8. McLaughlin-Drubin, M. E. and Munger, K. Viruses associated with human cancer. *Biochim Biophys Acta* **1782,** 127–150 (2008).
9. Howard, G., Eiges, R., Gaudet, F., Jaenisch, R., and Eden, A. Activation and transposition of endogenous retroviral elements in hypomethylation induced tumors in mice. *Oncogene* **27,** 404–408 (2008).
10. Lee, J. S., Haruna, T., Ishimoto, A., Honjo, T., and Yanagawa, S. Intracisternal type A particle-mediated activation of the Notch4/int3 gene in a mouse mammary tumor: generation of truncated Notch4/int3 mRNAs by retroviral splicing events. *J Virol* **73,** 5166–5171 (1999).

11. Romanish, M. T., Cohen, C. J., and Mager, D. L. Potential mechanisms of endogenous retroviral-mediated genomic instability in human cancer. *Semin Cancer Biol* **20,** 246–253 (2010).
12. Puech, A., Dupressoir, A., Loireau, M. P., Mattei, M. G., and Heidmann, T. Characterization of two age-induced intracisternal A-particle-related transcripts in the mouse liver: transcriptional read-through into an open reading frame with similarities to the yeast ccr4 transcription factor. *J Biol Chem* **272,** 5995–6003 (1997).
13. Barbot, W., Dupressoir, A., Lazar, V., and Heidmann, T. Epigenetic regulation of an IAP retrotransposon in the aging mouse: progressive demethylation and de-silencing of the element by its repetitive induction. *Nucleic Acids Res* **30,** 2365–2373 (2002).
14. Goff, S. P. Retrovirus restriction factors. *Mol Cell* **16,** 849–859 (2004).
15. Yoder, J. A., Walsh, C. P., and Bestor, T. H. Cytosine methylation and the ecology of intragenomic parasites. *Trends Genet* **13,** 335–340 (1997).
16. Walsh, C. P., Chaillet, J. R., and Bestor, T. H. Transcription of IAP endogenous retroviruses is constrained by cytosine methylation. *Nat Genet* **20,** 116–117 (1998).
17. Martens, J. H., O'Sullivan, R. J., Braunschweig, U., Opravil, S., Radolf, M., Steinlein, P., and Jenuwein, T. The profile of repeat-associated histone lysine methylation states in the mouse epigenome. *EMBO J* **24,** 800–812 (2005).
18. Mikkelsen, T. S., Ku, M., Jaffe, D. B., Issac, B., Lieberman, E., Giannoukos, G., Alvarez, P., Brockman, W., Kim, T. K., Koche, R. P., Lee, W., Mendenhall, E., O'Donovan, A., Presser, A., Russ, C., Xie, X., Meissner, A., Wernig, M., Jaenisch, R., Nusbaum, C., Lander, E. S., and Bernstein, B.E. Genome-wide maps of chromatin state in pluripotent and lineage-committed cells. *Nature* **448,** 553–560 (2007).
19. Karimi, M. M., Goyal, P., Maksakova, I. A., Bilenky, M., Leung, D., Tang, J. X., Shinkai, Y., Mager, D. L., Jones, S., Hirst, M., and Lorincz, M. C. DNA methylation and SETDB1/H3K9me3 regulate predominantly distinct sets of genes, retroelements, and chimeric transcripts in mESCs. *Cell Stem Cell* **8,** 676–687 (2011).
20. Matsui, T., Leung, D., Miyashita, H., Maksakova, I. A., Miyachi, H., Kimura, H., Tachibana, M., Lorincz, M. C., and Shinkai, Y. Proviral silencing in embryonic stem cells requires the histone methyltransferase ESET. *Nature* **464,** 927–931 (2010).
21. Rowe, H. M., Jakobsson, J., Mesnard, D., Rougemont, J., Reynard, S., Aktas, T., Maillard, P. V., Layard-Liesching, H., Verp, S., Marquis, J., Spitz, F., Constam, D. B., and Trono, D. KAP1 controls endogenous retroviruses in embryonic stem cells. *Nature* **463,** 237–240 (2010).
22. Wolf, D. and Goff, S. P. Embryonic stem cells use ZFP809 to silence retroviral DNAs. *Nature* **458,** 1201–1204 (2009).
23. Hutnick, L. K., Huang, X., Loo, T. C., Ma, Z., and Fan, G. Repression of retrotransposal elements in mouse embryonic stem cells is primarily mediated by a DNA methylation-independent mechanism. *J Biol Chem* **285,** 21082–21091 (2010).
24. Maksakova, I. A., Romanish, M. T., Gagnier, L., Dunn, C. A., van de Lagemaat, L. N., and Mager, D. L. Retroviral elements and their hosts: insertional mutagenesis in the mouse germ line. *PLoS Genet* **2,** e2 (2006).
25. Macfarlan, T., Gifford, W. D., Agarwal, S., Driscoll, S., Lettieri, K., Wang, J., Andrews, S. E., Franco, L., Rosenfeld, M. G., Ren, B., and Pfaff, S. L. Endogenous retroviruses and neighboring genes are coordinately repressed by LSD1/KDM1A. *Genes Dev* **25,** 594–607 (2011).

26. Tse, C., Sera, T., Wolffe, A. P., and Hansen, J. C. Disruption of higher-order folding by core histone acetylation dramatically enhances transcription of nucleosomal arrays by RNA polymerase III. *Mol Cell Biol* **18,** 4629–4638 (1998).
27. Krajewski, W. A. Histone hyperacetylation facilitates chromatin remodelling in a Drosophila embryo cell-free system. *Mol Gen Genet* **263,** 38–47 (2000).
28. Strahl, B. D. and Allis, C. D. The language of covalent histone modifications. *Nature* **403,** 41–45 (2000).
29. Jenuwein, T. and Allis, C. D Translating the histone code. *Science* **293,** 1074–1080 (2001).
30. Wysocka, J. Identifying novel proteins recognizing histone modifications using peptide pull-down assay. *Methods* **40,** 339–343 (2006).
31. Daniel, J. A., Pray-Grant, M. G., and Grant, P. A. Effector proteins for methylated histones: an expanding family. *Cell Cycle* **4,** 919–926 (2005).
32. Zeng, W., Ball, A. R. Jr., and Yokomori, K. HP1: heterochromatin binding proteins working the genome. *Epigenetics* **5,** 287–292 (2010).
33. Vermaak, D. and Malik, H. S. Multiple roles for heterochromatin protein 1 genes in Drosophila. *Annu Rev Genet* **43,** 467–492 (2009).
34. Kwon, S. H. and Workman, J. L. The changing faces of HP1: from heterochromatin formation and gene silencing to euchromatic gene expression. HP1 acts as a positive regulator of transcription. *Bioessays* **33,** 280–289 (2011).
35. Singh, P. B. and Georgatos, S. D. HP1: facts, open questions, and speculation. *J Struct Biol* **140,** 10–16 (2002).
36. Bannister, A. J., Zegerman, P., Partridge, J. F., Miska, E. A., Thomas, J. O., Allshire, R. C., and Kouzarides, T. Selective recognition of methylated lysine 9 on histone H3 by the HP1 chromo domain. *Nature* **410,** 120–124 (2001).
37. Lachner, M., O'Carroll, D., Rea, S., Mechtler, K., and Jenuwein, T. Methylation of histone H3 lysine 9 creates a binding site for HP1 proteins. *Nature* **410,** 116–120 (2001).
38. Nielsen, A. L., Oulad-Abdelghani, M., Ortiz, J. A., Remboutsika, E., Chambon, P., and Losson, R. Heterochromatin formation in mammalian cells: interaction between histones and HP1 proteins. *Mol Cell* **7,** 729–739 (2001).
39. Thiru, A., Nietlispach, D., Mott, H. R., Okuwaki, M., Lyon, D., Nielsen, P. R., Hirshberg, M., Verreault, A., Murzina, N. V., and Laue, E. D. Structural basis of HP1/PXVXL motif peptide interactions and HP1 localisation to heterochromatin. *EMBO J* **23,** 489–499 (2004).
40. Singh, P. B. HP1 proteins: what is the essential interaction? *Genetika* **46,** 1424–1429 (2010).
41. Rountree, M. R. and Selker, E. U. DNA methylation and the formation of heterochromatin in Neurospora crassa. *Heredity* **105,** 38–44 (2010).
42. Groner, A. C., Meylan, S., Ciuffi, A., Zangger, N., Ambrosini, G., Dénervaud, N., Bucher, P., and Trono, D. KRAB-zinc finger proteins and KAP1 can mediate long-range transcriptional repression through heterochromatin spreading. *PLoS Genet* **6,** e1000869 (2010).
43. Sadaie, M., Kawaguchi, R., Ohtani, Y., Arisaka, F., Tanaka, K., Shirahige, K., and Nakayama, J. Balance between distinct HP1 family proteins controls heterochromatin assembly in fission yeast. *Mol Cell Biol* **28,** 6973–6988 (2008).

44. Kourmouli, N., Sun, Y. M., van der Sar, S., Singh, P. B., and Brown, J. P. Epigenetic regulation of mammalian pericentric heterochromatin in vivo by HP1. *Biochem Biophys Res Commun* **337,** 901–907 (2005).
45. Nielsen, S. J., Schneider, R., Bauer, U. M., Bannister, A. J., Morrison, A., O'Carroll, D., Firestein, R., Cleary, M., Jenuwein, T., Herrera, R. E. and Kouzarides, T. Rb targets histone H3 methylation and HP1 to promoters. *Nature* **412,** 561–565 (2001).
46. Kwon, S. H., Florens, L., Swanson, S. K., Washburn, M. P., Abmayr, S. M., and Workman, J. L. Heterochromatin protein 1 (HP1) connects the FACT histone chaperone complex to the phosphorylated CTD of RNA polymerase II. *Genes Dev* **24,** 2133–2145 (2010).
47. Klattenhoff, C., Xi, H., Li, C., Lee, S., Xu, J., Khurana, J. S., Zhang, F., Schultz, N., Koppetsch, B. S., Nowosielska, A., Seitz, H., Zamore, P. D., Weng, Z., and Theurkauf, W. E. The Drosophila HP1 homologue Rhino is required for transposon silencing and piRNA production by dual-strand clusters. *Cell* **138,** 1137–1149 (2009).
48. Freitag, M., Hickey, P. C., Khlafallah, T. K., Read, N. D., and Selker, E. U. HP1 is essential for DNA methylation in Neurospora. *Mol Cell* **13,** 427–434 (2004).
49. Tamaru, H. and Selker, E. U. A histone H3 methyltransferase controls DNA methylation in Neurospora crassa. *Nature* **414,** 277–283 (2001).
50. Johnson, L. M., Cao, X., and Jacobsen, S. E. Interplay between two epigenetic marks: DNA methylation and histone H3 lysine 9 methylation. *Curr Biol* **12,** 1360–1367 (2002).
51. Jackson, J. P., Lindroth, A. M., Cao, X., and Jacobsen, S. E. Control of CpNpG DNA methylation by the KRYPTONITE histone H3 methyltransferase. *Nature* **416,** 556–560 (2002).
52. Du Chéné, I., Basyuk, E., Lin, Y. L., Triboulet, R., Knezevich, A., Chable-Bessia, C., Mettling, C., Baillat, V., Reynes, J., Corbeau, P., Bertrand, E., Marcello, A., Emiliani, S., Kiernan, R., and Benkirane, M. Suv39H1 and HP1γ are responsible for chromatin-mediated HIV-1 transcriptional silencing and post-integration latency. *EMBO J* **26,** 424–435 (2007).
53. Brown, J. P., Bullwinkel, J., Baron-Lühr, B., Billur, M., Schneider, P., Winking, H., and Singh, P. B. HP1γ function is required for male germ cell survival and spermatogenesis. *Epigenetics Chromatin* **3,** 9 (2010).
54. Vakoc, C. R., Mandat, S. A., Olenchock, B. A., and Blobel, G. A. Histone H3 lysine 9 methylation and HP1γ are associated with transcription elongation through mammalian chromatin. *Mol Cell* **19,** 381–391 (2005).
55. Mateescu, B., Bourachot, B., Rachez, C., Ogryzko, V., and Muchardt, C. Regulation of an inducible promoter by an HP1β-HP1γ switch. *EMBO Rep* **9,** 267–272 (2008).
56. Wolf, D., Cammas, F., Losson, R., and Goff, S. P. Primer binding site-dependent restriction of murine leukemia virus requires HP1 binding by TRIM28. *J Virol* **82,** 4675–4679 (2008).
57. Sripathy, S. P., Stevens, J., and Schultz, D. C. The KAP1 corepressor functions to coordinate the assembly of de novo HP1-demarcated microenvironments of heterochromatin required for KRAB zinc finger protein-mediated transcriptional repression. *Mol Cell Biol* **26,** 8623–8638 (2006).
58. Schultz, D. C., Ayyanathan, K., Negorev, D., Maul, G. G., and Rauscher, F. J. SETDB1: a novel KAP-1-associated histone H3, lysine 9-specific methyltransferase that

contributes to HP1-mediated silencing of euchromatic genes by KRAB zinc-finger proteins. *Genes Dev* **16,** 919–932 (2002).
59. Stewart, M. D., Li, J., and Wong, J. Relationship between histone H3 lysine 9 methylation, transcription repression, and heterochromatin protein 1 recruitment. *Mol Cell Biol* **25,** 2525–2538 (2005).
60. Nielsen, A. L., Ortiz, J. A. You, J, Oulad-Abdelghani, M., Khechumian, R., Gansmuller, A., Chambon, P., and Losson, R. Interaction with members of the heterochromatin protein 1 (HP1) family and histone deacetylation are differentially involved in transcriptional silencing by members of the TIF1 family. *EMBO J* **18,** 6385–6395 (1999).
61. Ryan, R. F., Schultz, D. C., Ayyanathan, K., Singh, P. B., Friedman, J. R., Fredericks, W. J., and Rauscher, F. J. KAP-1 corepressor protein interacts and colocalizes with heterochromatic and euchromatic HP1 proteins: a potential role for Krüppel-associated box-zinc finger proteins in heterochromatin-mediated gene silencing. *Mol Cell Biol* **19,** 4366–4378 (1999).
62. Danzer, J. R. and Wallrath, L. L. Mechanisms of HP1-mediated gene silencing in Drosophila. *Development* **131,** 3571–3580 (2004).
63. Hines, K. A., Cryderman, D. E., Flannery, K. M., Yang, H., Vitalini, M. W., Hazelrigg, T., Mizzen, C. A., and Wallrath, L. L. Domains of heterochromatin protein 1 required for Drosophila melanogaster heterochromatin spreading. *Genetics* **182,** 967–977 (2009).
64. Partridge, J. F., Borgstrom, B., and Allshire, R. C. Distinct protein interaction domains and protein spreading in a complex centromere. *Genes Dev* **14,** 783–791 (2000).
65. Aagaard, L., Laible, G., Selenko, P., Schmid, M., Dorn, R., Schotta, G., Kuhfittig, S., Wolf, A., Lebersorger, A., Singh, P. B., Reuter, G., and Jenuwein, T. Functional mammalian homologues of the Drosophila PEV–modifier Su(var)3–9 encode centromere-associated proteins which complex with the heterochromatin component M31. *EMBO J* **18,** 1923–1938 (1999).
66. Li, Y., Kirschmann, D. A., and Wallrath, L. L. Does heterochromatin protein 1 always follow code? *Proc Natl Acad Sci USA* **99**(Suppl 4), 16462–16469 (2002).
67. Locke, S. M. and Martienssen, R. A. Slicing and spreading of heterochromatic silencing by RNA interference. *Cold Spring Harb Symp Quant Biol* **71,** 497–503 (2006).
68. Talbert, P. B. and Henikoff, S. Spreading of silent chromatin: inaction at a distance. *Nat Rev Genet* **7,** 793–803 (2006).
69. Honda, S., Lewis, Z. A., Huarte, M., Cho, L. Y., David, L. L., Shi, Y., and Selker, E. U. The DMM complex prevents spreading of DNA methylation from transposons to nearby genes in Neurospora crassa. *Genes Dev* **24,** 443–454 (2010).
70. Tajul-Arifin, K., Teasdale, R., Ravasi, T., Hume, D. A., and Mattick, J. S. Identification and analysis of chromodomain-containing proteins encoded in the mouse transcriptome. *Genome Res* **13,** 1416–1429 (2003).
71. Kokura, K., Sun, L., Bedford, M. T., and Fang, J. Methyl-H3K9-binding protein MPP8 mediates E-cadherin gene silencing and promotes tumor cell motility and invasion. *EMBO J* **29,** 3673–3687 (2010).
72. Bua, D. J., Kuo, A. J., Cheung, P., Liu, C. L., Migliori, V., Espejo, A., Casadio, F., Bassi, C., Amati, B., Bedford, M. T., Guccione, E., and Gozani, O. Epigenome microarray platform for proteome-wide dissection of chromatin-signaling networks. *PLoS One* **4,** e6789 (2009).

73. Liu, H., Galka, M., Iberg, A., Wang, Z., Li, L., Voss, C., Jiang, X., Lajoie, G., Huang, Z., Bedford, M. T., and Li, S. S. Systematic identification of methyllysine-driven interactions for histone and nonhistone targets. *J Proteome Res* **9,** 5827–5836 (2010).
74. Fischle, W., Franz, H., Jacobs, S. A., Allis, C. D., and Khorasanizadeh, S. Specificity of the chromodomain Y chromosome family of chromodomains for lysine-methylated ARK(S/T) motifs. *J Biol Chem* **283,** 19626–19635 (2008).
75. Kim, J., Daniel, J., Espejo, A., Lake, A., Krishna, M., Xia, L., Zhang, Y., and Bedford, M. T. Tudor, MBT and chromo domains gauge the degree of lysine methylation. *EMBO Rep* **7,** 397–403 (2006).
76. Mulligan, P., Westbrook, T. F., Ottinger, M., Pavlova, N., Chang, B., Macia, E., Shi, Y. J., Barretina, J., Liu, J., Howley, P. M., Elledge, S. J., and Shi, Y. CDYL bridges REST and histone methyltransferases for gene repression and suppression of cellular transformation. *Mol Cell* **32,** 718–726 (2008).
77. Bernstein, E., Duncan, E. M., Masui, O., Gil, J., Heard, E., Allis, C. D. Mouse polycomb proteins bind differentially to methylated histone H3 and RNA and are enriched in facultative heterochromatin. *Mol Cell Biol* **26,** 2560–2569 (2006).
78. Quinn, A. M., Bedford, M. T., Espejo, A., Spannhoff, A., Austin, C. P., Oppermann, U., and Simeonov, A. A homogeneous method for investigation of methylation-dependent protein-protein interactions in epigenetics. *Nucleic Acids Res* **38,** e11 (2009).
79. Rottach, A., Frauer, C., Pichler, G., Bonapace, I. M., Spada, F., and Leonhardt, H. The multi-domain protein Np95 connects DNA methylation and histone modification. *Nucleic Acids Res* **38,** 1796–1804 (2010).
80. Bernard, D., Martinez-Leal, J. F., Rizzo, S., Martinez, D., Hudson, D., Visakorpi, T., Peters, G., Carnero, A., Beach, D., and Gil, J. CBX7 controls the growth of normal and tumor-derived prostate cells by repressing the Ink4a/Arf locus. *Oncogene* **24,** 5543–5551 (2005).
81. Aucott, R., Bullwinkel, J., Yu, Y., Shi, W., Billur, M., Brown, J. P., Menzel, U., Kioussis, D., Wang, G., Reisert, I., Weimer, J., Pandita, R. K., Sharma, G. G., Pandita, T. K., Fundele, R., and Singh, P. B. HP1-β is required for development of the cerebral neocortex and neuromuscular junctions. *J Cell Biol* **183,** 597–606 (2008).
82. Schotta, G., Lachner, M., Sarma, K., Ebert, A., Sengupta, R., Reuter, G., Reinberg, D., Jenuwein, T. A silencing pathway to induce H3-K9 and H4-K20 trimethylation at constitutive heterochromatin. *Genes Dev* **18,** 1251–1262 (2004).
83. Dong, K. B., Maksakova, I. A., Mohn, F., Leung, D., Appanah, R., Lee, S., Yang, H. W., Lam, L. L., Mager, D. L., Schübeler, D., Tachibana, M., Shinkai, Y., Lorincz, M. C. DNA methylation in ES cells requires the lysine methyltransferase G9a but not its catalytic activity. *EMBO J* **27,** 2691–2701 (2008).
84. Lehnertz, B., Ueda, Y., Derijck, A. A., Braunschweig, U., Perez-Burgos, L., Kubicek, S., Chen, T., Li, E., Jenuwein, T., and Peters, A. H. Suv39h-mediated histone H3 lysine 9 methylation directs DNA methylation to major satellite repeats at pericentric heterochromatin. *Curr Biol* **13,** 1192–1200 (2003).
85. Honda, S. and Selker, E. U. Direct interaction between DNA methyltransferase DIM-2 and HP1 is required for DNA methylation in Neurospora crassa. *Mol Cell Biol* **28,** 6044–6055 (2008).

86. Maksakova, I. A., Zhang, Y., and Mager, D. L. Preferential epigenetic suppression of the autonomous MusD over the nonautonomous ETn mouse retrotransposons. *Mol Cell Biol* **29,** 2456–2468 (2009).
87. Feng, Y. Q., Seibler, J., Alami, R., Eisen, A., Westerman, K. A., Leboulch, P., Fiering, S., and Bouhassira, E. E. Site-specific chromosomal integration in mammalian cells: highly efficient CRE recombinase-mediated cassette exchange. *J Mol Biol* **292,** 779–785 (1999).
88. Schübeler, D., Lorincz, M. C., and Groudine, M. Targeting silence: the use of site-specific recombination to introduce in vitro methylated DNA into the genome. *Sci STKE* **2001**(83), 11 (2001).
89. Lorincz, M. C., Schübeler, D., and Groudine, M. Methylation-mediated proviral silencing is associated with MeCP2 recruitment and localized histone H3 deacetylation. *Mol Cell Biol* **21,** 7913–7922 (2001).
90. Teich, N. M., Weiss, R. A., Martin, G. R., and Lowy, D. R. Virus infection of murine teratocarcinoma stem cell lines. *Cell* **12,** 973–982 (1977).
91. Pannell, D., Osborne, C. S., Yao, S., Sukonnik, T., Pasceri, P., Karaiskakis, A., Okano, M., Li, E., Lipshitz, H. D., and Ellis, J. Retrovirus vector silencing is de novo methylase independent and marked by a repressive histone code. *EMBO J* **19,** 5884–5894 (2000).
92. Niwa, O., Yokota, Y., Ishida, H., and Sugahara, T. Independent mechanisms involved in suppression of the Moloney leukemia virus genome during differentiation of murine teratocarcinoma cells. *Cell* **32,** 1105–1113 (1983).
93. Cheng, L., Du, C., Murray, D., Tong, X., Zhang, Y. A., Chen, B. P., and Hawley, R. G. A GFP reporter system to assess gene transfer and expression in human hematopoietic progenitor cells. *Gene Ther* **4,** 1013–1022 (1997).
94. Musselman, C. A., Mansfield, R. E., Garske, A. L., Davrazou, F., Kwan, A. H., Oliver, S. S., O'Leary, H., Denu, J. M., Mackay, J. P., and Kutateladze, T. G. Binding of the CHD4 PHD2 finger to histone H3 is modulated by covalent modifications. *Biochem J* **423,** 179–187 (2009).
95. Mansfield, R. E., Musselman, C. A., Kwan, A. H., Oliver, S. S., Garske, A. L., Davrazou, F., Denu, J. M., Kutateladze, T. G., and Mackay, J. P. The plant homeodomain (PHD) fingers of CHD4 are histone H3-binding modules with preference for unmodified H3K4 and methylated H3K9. *J Biol Chem* **286,** 11779–11791 (2011).
96. Iwase, S., Lan, F., Bayliss, P., de la Torre-Ubieta, L., Huarte, M., Qi, H. H., Whetstine, J. R., Bonni, A., Roberts, T. M., and Shi, Y. The X-linked mental retardation gene SMCX/JARID1C defines a family of histone H3 lysine 4 demethylases. *Cell* **128,** 1077–1088 (2007).
97. Li, F., Huarte, M., Zaratiegui, M., Vaughn, M. W., Shi, Y., Martienssen, R., Cande, W. Z. Lid2 is required for coordinating H3K4 and H3K9 methylation of heterochromatin and euchromatin. *Cell* **135,** 272–283 (2008).
98. Karagianni, P., Amazit, L., Qin, J., and Wong, J. ICBP90, a novel methyl K9 H3 binding protein linking protein ubiquitination with heterochromatin formation. *Mol Cell Biol* **28,** 705–717 (2008).
99. Bartke, T., Vermeulen, M., Xhemalce, B., Robson, S. C., Mann, M., and Kouzarides, T. Nucleosome-interacting proteins regulated by DNA and histone methylation. *Cell* **143,** 470–484 (2010).

100. Nady, N., Lemak, A., Walker, J. R., Avvakumov, G. V., Kareta, M. S., Achour, M., Xue, S., Duan, S., Allali-Hassani, A., Zuo, X., Wang, Y. X., Bronner, C., Chédin, F., Arrowsmith, C. H., and Dhe-Paganon, S. Recognition of multivalent histone states associated with heterochromatin by UHRF1 protein. *J Biol Chem* **286,** 24300–24311 (2011).
101. Papait, R., Pistore, C., Negri, D., Pecoraro, D., Cantarini, L., and Bonapace, I. M. Np95 is implicated in pericentromeric heterochromatin replication and in major satellite silencing. *Mol Biol Cell* **18,** 1098–1106 (2007).
102. Seum, C., Delattre, M., Spierer, A., and Spierer, P. Ectopic HP1 promotes chromosome loops and variegated silencing in Drosophila. *EMBO J* **20,** 812–818 (2001).
103. Seum, C., Spierer, A., Delattre, M., Pauli, D., Spierer, P. A GAL4-HP1 fusion protein targeted near heterochromatin promotes gene silencing. *Chromosoma* **109,** 453–459 (2000).
104. Agalioti, T., Lomvardas, S., Parekh, B., Yie, J., Maniatis, T., and Thanos, D. Ordered recruitment of chromatin modifying and general transcription factors to the IFN-β promoter. *Cell* **103,** 667–678 (2000).
105. Kasten, M., Szerlong, H., Erdjument-Bromage, H., Tempst, P., Werner, M., and Cairns, B. R. Tandem bromodomains in the chromatin remodeler RSC recognize acetylated histone H3 Lys14. *EMBO J* **23,** 1348–1359 (2004).
106. Vicent, G. P., Zaurin, R., Nacht, A. S., Li, A., Font-Mateu, J., Le Dily, F., Vermeulen, M., Mann, M., and Beato, M. Two chromatin remodeling activities cooperate during activation of hormone responsive promoters. *PLoS Genet* **5,** e1000567 (2009).
107. Hassan, A. H., Neely, K. E. and Workman, J. L. Histone acetyltransferase complexes stabilize SWI/SNF binding to promoter nucleosomes. *Cell* **104,** 817–827 (2001).
108. Wang, Z., Zang, C., Cui, K., Schones, D. E., Barski, A., Peng, W., and Zhao, K. Genome-wide mapping of HATs and HDACs reveals distinct functions in active and inactive genes. *Cell* **138,** 1019–1031 (2009).
109. Jin, Q., Yu, L. R., Wang, L., Zhang, Z., Kasper, L. H., Lee, J. E., Wang, C., Brindle, P. K., Dent, S. Y., and Ge, K. Distinct roles of GCN5/PCAF-mediated H3K9ac and CBP/p300-mediated H3K18/27ac in nuclear receptor transactivation. *EMBO J* **30,** 249–262 (2011).
110. Nagy, Z., Riss, A., Fujiyama, S., Krebs, A., Orpinell, M., Jansen, P., Cohen, A., Stunnenberg, H. G., Kato, S., and Tora, L. The metazoan ATAC and SAGA coactivator HAT complexes regulate different sets of inducible target genes. *Cell Mol Life Sci* **67,** 611–628 (2010).
111. Nagy, Z. and Tora, L. Distinct GCN5/PCAF-containing complexes function as co-activators and are involved in transcription factor and global histone acetylation. *Oncogene* **26,** 5341–5357 (2007).
112. Huang, J., Fan, T., Yan, Q., Zhu, H., Fox, S., Issaq, H. J., Best, L., Gangi, L., Munroe, D., and Muegge, K. Lsh, an epigenetic guardian of repetitive elements. *Nucleic Acids Res* **32,** 5019–5028 (2004).
113. Vicent, G. P., Ballaré, C., Nacht, A. S., Clausell, J., Subtil-Rodríguez, A., Quiles, I., Jordan, A., and Beato, M. Induction of progesterone target genes requires activation of Erk and Msk kinases and phosphorylation of histone H3. *Mol Cell* **24,** 367–381 (2006).

114. Hebbar, P. B. and Archer, T. K. Nuclear factor 1 is required for both hormone-dependent chromatin remodeling and transcriptional activation of the mouse mammary tumor virus promoter. *Mol Cell Biol* **23,** 887–898 (2003).
115. Vicent, G. P., Zaurin, R., Nacht, A. S., Font–Mateu, J., Le Dily, F., and Beato, M. Nuclear factor 1 synergizes with progesterone receptor on the mouse mammary tumor virus promoter wrapped around a histone H3/H4 tetramer by facilitating access to the central hormone-responsive elements. *J Biol Chem* **285,** 2622–2631 (2010).
116. Baust, C., Gagnier, L., Baillie, G. J., Harris, M. J., Juriloff, D. M., and Mager, D. L. Structure and expression of mobile ETnII retroelements and their coding-competent MusD relatives in the mouse. *J Virol* **77,** 11448–11458 (2003).
117. Rea, S., Eisenhaber, F., O'Carroll, D., Strahl, B. D., Sun, Z. W., Schmid, M., Opravil, S., Mechtler, K., Ponting, C. P., Allis, C. D., and Jenuwein, T. Regulation of chromatin structure by site-specific histone H3 methyltransferases. *Nature* **406,** 593–599 (2000).
118. Gregory, G. D., Vakoc, C. R., Rozovskaia, T., Zheng, X., Patel, S., Nakamura, T., Canaani, E., and Blobel, G. A. Mammalian ASH1L is a histone methyltransferase that occupies the transcribed region of active genes. *Mol Cell Biol* **27,** 8466–8479 (2007).
119. Wang, H., Cao, R., Xia, L., Erdjument-Bromage, H, Borchers, C, Tempst, P, and Zhang, Y. Purification and functional characterization of a histone H3-lysine 4-specific methyltransferase. *Mol Cell* **8,** 1207–1217 (2001).
120. Yan, Q., Huang, J., Fan, T., Zhu, H., and Muegge, K. Lsh, a modulator of CpG methylation, is crucial for normal histone methylation. *EMBO J* **22,** 5154–5162 (2003).
121. Wysocka, J., Swigut, T., Milne, T. A., Dou, Y., Zhang, X., Burlingame, A. L., Roeder, R. G., Brivanlou, A. H., and Allis, C. D. WDR5 associates with histone H3 methylated at K4 and is essential for H3 K4 methylation and vertebrate development. *Cell* **121,** 859–872 (2005).
122. Balakrishnan, L. and Milavetz, B. Decoding the histone H4 lysine 20 methylation mark. *Crit Rev Biochem Mol Biol* **45,** 440–452 (2010).
123. Gonzalo, S., García-Cao, M., Fraga, M. F., Schotta, G., Peters, A. H., Cotter, S. E., Eguía, R., Dean, D. C., Esteller, M., Jenuwein, T., and Blasco, M. A. Role of the RB1 family in stabilizing histone methylation at constitutive heterochromatin. *Nat Cell Biol* **7,** 420–428 (2005).
124. Schotta, G., Sengupta, R., Kubicek, S., Malin, S., Kauer, M., Callén, E., Celeste, A., Pagani, M., Opravil, S., De La Rosa-Velazquez, I. A., Espejo, A., Bedford, M. T., Nussenzweig, A., Busslinger, M., and Jenuwein, T. A chromatin-wide transition to H4K20 monomethylation impairs genome integrity and programmed DNA rearrangements in the mouse. *Genes Dev* **22,** 2048–2061 (2008).
125. Ayoub, N., Jeyasekharan, A. D., and Venkitaraman, A. R. Mobilization and recruitment of HP1β: a bimodal response to DNA breakage. *Cell Cycle* **8,** 2945–2950 (2009).
126. Luijsterburg, M. S., Dinant, C., Lans, H., Stap, J., Wiernasz, E., Lagerwerf, S., Warmerdam, D. O., Lindh, M., Brink, M. C., Dobrucki, J. W., Aten, J. A., Fousteri, M. I., Jansen, G., Dantuma, N. P., Vermeulen, W., Mullenders, L. H., Houtsmuller, A. B., Verschure, P. J., and van Driel, R. Heterochromatin protein 1 is recruited to various types of DNA damage. *J Cell Biol* **185,** 577–586 (2009).

127. Feng, Y. Q., Lorincz, M. C., Fiering, S., Greally, J. M., and Bouhassira, E. E. Position effects are influenced by the orientation of a transgene with respect to flanking chromatin. *Mol Cell Biol* **21,** 298–309 (2001).
128. Juriloff, D. M., Harris, M. J., Dewell, S. L., Brown, C. J., Mager, D. L., Gagnier, L., and Mah, D. G. Investigations of the genomic region that contains the clf1 mutation, a causal gene in multifactorial cleft lip and palate in mice. *Birth Defects Res A Clin Mol Teratol* **73,** 103–113 (2005).
129. Plamondon, J. A., Harris, M. J., Mager, D. L., Gagnier, L., and Juriloff, D. M. The clf2 gene has an epigenetic role in the multifactorial etiology of cleft lip and palate in the A/WySn mouse strain. *Birth Defects Res A Clin Mol Teratol* (in press).
130. Bouhassira, E. E., Westerman, K., and Leboulch, P. Transcriptional behavior of LCR enhancer elements integrated at the same chromosomal locus by recombinase-mediated cassette exchange. *Blood* **90,** 3332–3344 (1997).
131. Kumaki, Y., Oda, M., and Okano, M. QUMA: quantification tool for methylation analysis. *Nucleic Acids Res* (36 Web Server), W170–W175 (2008).
132. Jurka, J., Kapitonov, V. V., Pavlicek, A., Klonowski, P., Kohany, O., and Walichiewicz, J. Repbase Update, a database of eukaryotic repetitive elements. *Cytogenet Genome Res* **110,** 462–467 (2005).
133. Li, H. and Durbin, R. Fast and accurate short read alignment with Burrows-Wheeler transform. *Bioinformatics* **25,** 1754–1760 (2009).

CHAPTER 6

HISTONE (H1) PHOSPHORYLATION

ANNA GRÉEN, BETTINA SARG, HENRIK GRÉEN, ANITA LÖNN, HERBERT H. LINDNER, and INGEMAR RUNDQUIST

CONTENTS

6.1 INTRODUCTION

Cell division is a complex process, in which correct passage through the cell cycle is essential for cell survival and correct transmission of genetic information to the daughter cells. During the cell cycle, the cell nucleus undergoes dramatic structural changes. DNA, which is compacted into chromatin by various proteins, is locally decondensed in S phase but condenses in prophase. In metaphase, highly condensed chromosomes are visible which start to segregate during anaphase. Segregation is completed during telophase, and two daughter cells are produced. Before re-entry into G_1, the chromatin again becomes dispersed.

In the nucleosome, the basic unit of chromatin, approximately 146 bp of DNA are wrapped 1.65 turns around an octamer consisting of two copies of each core histone: H2A, H2B, H3, and H4 [1]. A fifth histone, histone H1 (also referred to as linker histone), binds at or near to the entry/exit point of DNA and to linker DNA [2]. Histone H1 has a central globular domain and hydrophilic tails in the N and C terminals. Histone H1 is a protein family with at least eight members in mammals. Some of these are present only in highly specialized cell types. In most somatic cells, histones H1.2, H1.3, H1.4 and H1.5 are present [3]. The function of histone H1 in the cell and the purpose of several H1 subtypes remain to be determined in detail; however, histone H1 is implicated in the compaction of chromatin into higher-order structures [4] and in transcriptional regulation [3, 5-7]. Knockout experiments in mice have identified a remarkable redundancy and overlapping functionalities of the different subtypes, but have also proved that histone H1 is indispensable in mouse development [8]. In chapter, some subtypes seem to have specialized functions [9]; a particular example is H1.2, which is a part of the apoptosis signaling process as a response to DNA double-strand breaks [10].

In chapter to the complexity of multiple subtypes, H1 subtypes are post-translationally modified, primarily by phosphorylation at multiple sites. The significance of this modification is unclear, but is believed to reduce the affinity of histone H1 for chromatin [11, 12]. Histone H1 phosphorylation has been implicated in various physiological processes, for example in gene regulation, chromatin condensation/decondensation, and cell cycle progression [12]. Regulation of gene expression may be executed through chromatin remodeling, regulated by histone H1 phosphorylation [13, 14].

The H1 phosphorylation was initially connected to mitotic condensation of chromatin [15], but other studies have shown that H1 phosphorylation can also be involved in decondensation of chromatin [11]. Increasing evidence suggests that histone H1 phosphorylation is involved in both chromatin condensation and decondensation during the cell cycle. In mid to late G_1 and S phase, increased H1 phosphorylation, Cdk2 activation and local chromatin decondensation occur [16, 17]. This may be performed by disassembly of heterochromatin, as H1 phosphorylation by Cdk2 disrupts the interaction between histone H1 and heterochromatin protein 1α [18]. The phosphorylation of histone H1 and chromatin decondensation in mid to late G_1 and S phases have been suggested to be a prerequisite for DNA replication competence [12, 16, 19, 20].

The phosphorylation of H1 histones in the cell cycle has been described as a sequential event. In Chinese hamster cell and in rat and mice synchronized cell cultures, H1 phosphorylation was shown to start during mid to late G_1, increase during S, and reach its maximum at mitosis [21, 22]. The major phosphorylation sites in human somatic H1 histones have been mapped and are located on serines in SP(K/A)K motifs in H1.2, H1.3, H1.4, and H1.5 during interphase [23]. Mitotic up phosphorylation takes place on threonine residues only [23, 24].

Increased H1 phosphorylation in ras-transformed G_1 mouse fibroblasts, compared with their normal counterparts, has been detected [25]. The increase in detected mouse H1b (homologous to human H1.5) phosphorylation in the transformed cells was concluded to be caused by increased Cdk2 activity in the transformed mouse fibroblasts [26]. Furthermore, in Rb-deficient fibroblasts, increased H1 phosphorylation was detected in G_1 along with less condensed chromatin and increased Cdk2 activity [17].

In the search for cell cycle-specific phosphorylation of histone H1, human cancer cells or cells from species other than human have been used. To knowledge, no normal human cells have been investigated to date. Because many signaling pathways are dysregulated in cancer cells, especially within the cell cycle control system, it is of interest to use normal cells when studying cell cycle events. In chapter, most other studies have used various chemical agents to arrest or synchronize the cycling cells in the different cell cycle stages. Because such methods may affect H1 phosphorylation, we used activated T cells and fluorescence-activated cell sorting for studying the cell cycle-dependent phosphorylation of human H1 histones. To detect differences in the phosphorylation pattern between malignant and normal cells, the cell cycle-dependent H1 phosphorylation of Jurkat T lymphoblastoid cells was also examined. Histone H1 subtypes composition and phosphorylation was analyzed by reversed phase high performance liquid chromatography (RP-HPLC) and capillary electrophoresis. We found a substantial increase in H1.5 content after activation of T cells. Furthermore, the major part of interphase H1 phosphorylation took place in G1 or early S phase, and was preserved during S and G2/M phases. We also found enhanced H1 phosphorylation, in particular for H1.5, in the G_1 phase of T lymphoblastoid cells compared with activated normal T cells.

6.2 METHODS

6.2.1 ISOLATION OF PERIPHERAL BLOOD LYMPHOCYTES

Leukocyte-enriched buffy coats from three healthy blood donors were obtained (Blood Bank, Linköping University Hospital, Sweden). Peripheral blood mononuclear cells (PBMCs) were isolated by density gradient centrifugation (Ficoll-Paque PLUS; GE Heathcare Bio-Sciences, Uppsala, Sweden). Monocytes were removed by plastic adherence during incubation for 1 hr at 37°C and 5% CO_2, and peripheral blood lymphocytes (PBLs) were then collected from the supernatants.

6.2.2 ACTIVATION OF PERIPHERAL BLOOD LYMPHOCYTES, CELL CULTURE, AND STAINING

All media and chemicals were obtained from Gibco (Paisley, Renfrewshire, UK) unless otherwise indicated. After isolation, PBLs were resuspended in RPMI 1640 medium supplemented with 10% v/v fetal bovine serum (FBS), 60 μg/ml penicillin, 100 μg/mL streptomycin, 10 mmol/L HEPES and 2 mmol/L L-glutamine at a concentration of 1 × 10^6 cells/mL. The cells were activated by addition of 150 U/ml interleukin (IL)-2 (Proleukin; Chiron Corporation, Emeryville, CA, USA) and1 μg/mL phytohemagglutinin (PHA-M) (Sigma, St Louis, MO, USA).

The cells were counted by trypan blue exclusion and recultured daily or after 23 days, depending on cell concentration. Cells were recultured to a cell concentration of 0.50.6 × 10^6 cells/mL in culture medium supplemented with150 U/mL IL-2. The PBLs were cultured for 69 days, depending on the number of cells. The day before sorting, the cells were reconstituted to a concentration of 0.50.6 × 10^6 cells/mL. Before sorting, about 200300 × 10^6 PBLs were stained by addition of 10 μg/mL Hoechst 33342 (Molecular Probes, Eugene, OR, USA) into the medium, and incubated in the dark at 37°C and 5% CO_2 for 30 min. The stained cells were subsequently separated by centrifugation at 300 g for 10 min at 4°C, and the cell pellet resuspended in fresh culture medium to obtain approximately 60 × 10^6 cells/mL. The resuspended cells were kept on ice until sorting. This staining procedure was performed in batches 23 times during the cell sorting to minimize the effects of dye exposure, agitation of tubes in the flow cytometer, and prolonged incubation on ice.

6.2.3 JURKAT CELL CULTURE AND STAINING

Jurkat cells (clone E6.1, ECACC, UK) were cultured in RPMI 1640 supplemented with 10% v/v FBS, 60 μg/ml penicillin, 100 μg/mL streptomycin, 2 mmol/l L-glutamine at 37°C and 5% CO_2. The cells were split 3 times per week, and kept at a level of 0.1 to 1 × 10^6 cells per mL. The day before sorting, the cells were seeded to 0.25 × 10^6 cells per mL so that they were in log growth phase with approximately 0.5 × 10^6cells per mL upon sorting. The cells were stained with10 μg/mL Hoechst 33342 for 30 min in the dark at 37°C and 5% CO_2. Stained cells were separated by centrifugation at 300 g for 10 min at 4°C. The cell pellet was resuspended in fresh medium to obtain a cell concentration of approximately 60 × 10^6 cells per mL, and kept on ice until sorted. The staining was performed in batches during sorting until sorting was completed.

6.2.4 T CELL ASSESSMENTS

T cells were assessed for purity, activation, and viability by flow cytometry. T cell purity was determined by measurements of the fraction of CD3+ cells. Cell growth was assessed through cell cycle analysis of propidium iodide (PI)-stained cell nuclei; by cell tracing using carboxyfluorescein diacetate N-succinimidyl ester (CFSE) labeling. The amount of CFSE becomes divided between the daughter cells at cell division, which enables determination of the fraction of cycling cells. Cell viability was measured by Annexin V staining. All measurements were done immediately after isolation, at 13day intervals post-activation until cell sorting, and at cell sorting.

6.2.5 THE CFSE STAINING AND FLOW CYTOMETRY MEASUREMENTS

After isolation of PBLs, 25 × 10^6 cells were separated by centrifugation at 300 g, and washed with phosphate-buffered saline (PBS) supplemented with 1% Bovine serum albumin (BSA). Cells were resuspended in 25 ml of 10 μmol/L CFSE (Fluka; Sigma) in PBS plus 1% BSA, and incubated for 10 min at 37°C in the dark. After labeling, the cells were washed twice in cold culture medium, and once with cold PBS. After centrifugation, the PBLs were resuspended in culture medium, and activated with IL-2 and PHA-M as described. For CFSE measurements, 1 × 10^6 cells were separated by centrifugation at 300 g for 10 min and resuspended in 1 ml PBS, after which 15,000 cells were analyzed using a flow cytometer (excitation 488 nm, emission *via* a BP 530/28 filter) (BD LSR; BD Biosciences). Histograms of CFSE fluorescence, after excluding debris in forward scatter (FSC) and side scatter (SSC), were obtained using CellQuest™ Pro software (BD Biosciences, San Jose, CA, USA). Cells with lower fluorescence than the original fluorescence channel FL1 peak appearing at 1st day were considered as cycling cells.

6.2.6 CD3 STAINING AND FLOW CYTOMETRY MEASUREMENTS

The fraction of CD3+ cells in the cell culture was measured using monoclonal antihuman CD3 phycoerythrin (PE) conjugate (Sigma) according to the manufacturer's recommendations. The cells were analyzed for PE fluorescence intensity using a flow cytometer (excitation at 488 nm, emission *via* a BP 575/26 filter) (BD LSR; BD Biosciences). The FSC and SSC were registered, and 15,000 non-gated events were collected. Histograms of PE fluorescence were acquired using CellQuest™ Pro (BD Biosciences).

6.2.7 CELL CYCLE ANALYSIS USING PI

Cell cycle distribution of activated PBLs was determined at various time points using a method developed by Vindelöv [35]. The PI fluorescence was measured on a flow cytometer (BD LSR; BD Biosciences) using a BP 575/26 filter. The FSC and SSC were also measured after excitation with the argon 488 nm laser; 15,000 non-gated events were collected. Fluorescence histograms of PI were obtained and analyzed with ModFit LT (Verity Software House, Topsham, ME, USA) after gating cell nuclei by FSC and SSC to exclude cell debris of low FSC and SSC.

6.2.8 DETECTION OF APOPTOTIC PERIPHERAL BLOOD LYMPHOCYTES

To determine the fraction of apoptotic PBLs in the cell cultures, apoptosis was assessed with Annexin V staining (Annexin V-PE Apoptosis Detection Kit I; BD Biosciences Pharmingen, San Diego, CA, USA) as described [27].

6.2.9 CELL SORTING

After Hoechst 33342 staining of activated T cells and Jurkat cells, cells were sorted using a cell sorter (FACSAria Special Order System Cell Sorter; BD Biosciences). During sorting, the samples were kept at 4°C and were continuously agitated. Sorted cells were kept at below 4°C. Hoechst 33342 fluorescence was detected using a 450/50 filter after excitation of a 355 nm UV laser (yttrium-aluminum-garnet (YAG) 20 mW from Coherent, BD Biosciences). The FSC and SSC of cells were detected using a 488/10 filter after excitation with a 488 nm laser (Sapphire 100 mW from Coherent, BD BioSciences). Scatter plots of FSC versus SSC and of width (calculated from height of signal) versus area of the Hoechst 33342 signal, and histograms of Hoechst 33342 fluorescence were obtained using FACSDiva software (BD Biosciences). Gating was performed in the FSC/SSC plot to exclude debris and in the Hoechst 33342 area/width plot to exclude cell doublets. The area or height of the Hoechst 33342 fluorescence from the cells present in both these gates was plotted in a DNA histogram, in which sorting gates were created to achieve sorting of the cells into G_1, S, and G_2/M phase cells with the highest recovery and purity possible. To assess the purity of sorted cells, reanalysis of sorted cell populations was performed at various times during sorting. Continuous sorting using a yield mask was performed, resulting in a sort rate of about 20,000–25,000 cells/s and an efficiency of more than 98%. During each experiment, 400600 × 10^6 cells were passed through the high-speed sorter, and about 70150 × 10^6 G_1 phase, 1235 × 10^6 S phase, and 1030 × 10^6 G_2/M phase cells were sorted out.

EXTRACTION OF H1 HISTONES

H1 histones were extracted from whole cells with perchloric acid as described in past [36].

6.2.10 CAPILLARY ELECTROPHORESIS

The high-performance capillary electrophoresis (HPCE) was performed on an electrophoresis system (P/ACE 2100; Beckman Instruments) and System Gold software (Beckman Instruments, Palo Alto, CA, USA). This software was also used for determination of peak heights. An untreated capillary was used in all experiments. Protein samples were injected under pressure, and detection was performed by measuring UV absorption at 200 nm. Separation of H1 histones was performed as described [37, 38]. All runs were performed at constant voltage (12 kV) and at a capillary temperature of 25°C. The peaks in the electropherograms were identified and designated as described [27], using the same types of cells.

6.2.11 REVERSED PHASE HIGH PERFORMANCE LIQUID CHROMATOGRAPHY

Separation of whole linker histones was performed on a column (250 × 3 mm I.D.; 5 μm particle pore size; 30 nm pore size; end-capped) (Nucleosil 3005 C_{18}; Machery-Nagel, Düren, Germany). The lyophilized proteins were dissolved in water containing 20 mmol/L 2-mercaptoethanol, and whole samples were injected onto the column. The histone H1 sample was separated by chromatography within 30 min at a constant flow of 0.35 ml/min with a linear acetonitrile gradient starting (solvent A: solvent B 30:70; solvent A being water containing 0.1% trifluoroacetic acid (TFA), and solvent B being 70% acetonitrile and 0.1% TFA). The concentration of solvent B was increased from 30% to 60% during a period 30 min. The peaks in the chromatograms were identified and designated as described [23].

6.3 DISCUSSION

Cell cycle regulation is important in normal tissue homeostasis and both in the origin and progression of cancer. A vital part of cell cycle regulation and progression is the preparation of chromatin for replication. We and others believe that H1 histones and their phosphorylation are important in these processes. In this study, we found that the interphase phosphorylation pattern of H1 histones was established in G_1 or early S phase in activated human T cells and Jurkat cells. This pattern was largely preserved during S and G_2/M phases. Unfortunately, because of a lack of cells, we were not able to introduce separate sorting windows in early and late S phase, but because H1 phosphorylation has been shown to occur site-specifically in a certain order [23], it is unlikely that rapid dephosphorylation/ rephosphorylation events affecting different phosphorylation sites can be an alternative explanation for the preserved phosphorylation patterns. Activation of T cells altered the H1 subtype composition; in particular, we detected a significant increase in the relative H1.5 content in cycling T cells compared with resting T cells.

The pattern of H1.5 mono- and diphosphorylation and of H1.2 and H1.3 monophosphorylation (and most probably of H1.4 mono- and diphosphorylation) became to a large extent established in G_1 phase or early S phase, and remained virtually preserved in G_2/M in both activated T cells and Jurkat cells. The similarity between S phase and G_2/M phase phosphorylation patterns also indicate that the newly synthesized H1 histones in S phase became phosphorylated to the same extent as the preexisting ones, in line with data. The small differences in G_2/M phosphorylation patterns between T cells and Jurkat cells can be explained by the higher content of contaminating G_1 cells in the T cell G_2/M populations. The G_1 phosphorylation pattern differed between Jurkat and activated T cells, with more extended phosphorylation in G_1 Jurkat cells. We expect that all these phosphorylations occur on serine residues, because it has been shown that only serines in SP(K/A)K motifs were phosphorylated in interphase [23, 24]. The number of S/TPXK sites, and their phosphorylation, in the present H1 subtypes has been thoroughly investigated, and results did not deviate from those results [23]. No influence on other sites was detected.

The observations are partly in contrast with earlier data describing a sequential increase of H1 phosphorylation across the cell cycle [21, 22, 28]. In mouse NIH 3T3 fibroblasts, H1 phosphorylation began during late G_1, increased during the S phase, and in late S phase 03 phosphate groups were detected on various mouse H1 subtypes [22]. In the G_2/M transition, H1 phosphorylation levels increased, and reached their maximum at M phase [22]. Using Chinese hamster cells, with one predominant histone H1 subtype, histone H1 was shown to have no phosphate groups in early G_1 [28]. Phosphorylation began in mid G_1 [21], and one phosphate group was detected in the beginning of S phase [28]. During the S and G_2 phases, up to 3 phosphates were seen, and maximum was reached at M phase, with up to 6 phosphates [21, 28].

In agreement with past data, results indicate that in normal T cells, H1.2, H1.3, H1.4, and H1.5 are mainly unphosphorylated at the beginning of the G_1 phase of the cell cycle. This is probably true also after T cell activation, as H1 histones from slow-growing populations of T cells contained very few phosphorylated variants (data not shown). However, some caution should be taken in data interpretation from such T cell cultures because these cells may be on the way to become apoptotic, even though only viable cells were sorted. We have recently shown that apoptosis may affect the H1 phosphorylation pattern [27].

H1 histones are conserved proteins, and require strongly resolving analytical techniques for their separation [29]. In study, the presence of differentially phosphorylated subtypes further complicates the separation of all variants. However, the combination of RP-HPLC and HPCE allows complete resolution of H1.5 and its phosphorylated forms. From data, we thus propose the following model of cell cycle-dependent serine phosphorylation of histone H1.5 (Figure 1), but it may also be valid for other H1 subtypes.

The main kinase activity takes place during late G_1 and early S phase as indicated by the red segment. The relative phosphorylation level, shown as the fraction of phosphorylated H1.5 serines per DNA unit, is indicated by the width of the gray segment. For clarity, the threonine phosphorylation taking place during mitosis is shown as a widening of the gray segment at the end of the cell cycle. The obtained results on H1.5 phosphorylation fit with, but do not prove, this model.

This model indicates that H1.5 is unphosphorylated during the first part of the G_1 phase, and becomes mono- and diphosphorylated on serine residues later in G_1 and in early S phase. It is possible that some H1.5 monophosphorylated at Ser17 is present already in earlier stages of the G_1 phase, as indicated by recent data [24]. Besides the complementary phosphorylation of the newly synthesized H1 molecules during S phase (Figure 1), some further up phosphorylation takes place during S, G_2 and M phases. In particular, threonine phosphorylation during mitosis results in a slight widening of the relative amount of phosphorylated H1.5 sites at the end of the cell cycle (Figure 1, gray segment) before the expected complete dephosphorylation takes place before the next G_1 phase is entered.

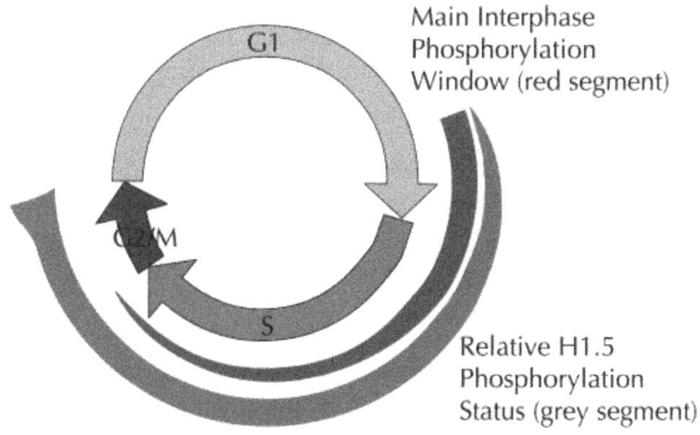

FIGURE 1 Hypothetical model for cell cycle-dependent histone H1.5 serine phosphorylation in T cells and Jurkat cells.

The differences we detected in G_1 phosphorylation between T cells and Jurkat cells may be explained either by the shorter cell cycling time in Jurkat cells, and/or by increased kinase activity. Rapid cell growth usually correlates with a shorter G_1 phase, and thereby more G_1 cells within the phosphorylation window for Jurkat cells compared with T cells, and a higher degree of phosphorylated H1.5. The fraction of cells in S phase is often used as a measurement of cell cycle velocity. In activated T cells from the three donors, the average fraction of S phase cells was 31%, compared with 39% in the three Jurkat cell samples. Therefore, the difference in cell cycling time is probably not sufficiently large to be the sole explanation for the differences in G_1 phosphorylation.

Another explanation is the presence of an overactive H1 kinase in the G_1 phase of Jurkat cells. H1 from ras-transformed mouse fibroblasts exhibited higher phosphorylation than did their untransformed counterparts [25]. This was not a result of cell cycle changes upon transformation, because transformed G_1/S arrested cells had higher levels of phosphorylated H1 than G_1/S arrested untransformed cells [25]. The ras-transformed cells also had less condensed chromatin than in untransformed cells [25]. In further studies, the increased H1b (homologue to human H1.5) phosphorylation after ras-transformation of mouse fibroblasts was found to be derived from overactivity of Cdk2, rather than from reduced activity of H1 phosphatases [26]. In the ras-transformed mouse fibroblasts, ras expression resulted in an initial increase in $p21^{cip1}$ (a Cdk2 inhibitor) levels and inhibition of Cdk2 activity, followed by a decrease in $p21^{cip1}$ and activation of Cdk2, producing increased H1b phosphorylation [26]. Transformation of mouse fibroblasts with other oncogenes affecting the Ras-mitogen-activated protein kinase signal transduction pathway (for example mos, raf, fes and myc) also resulted in increased H1 phosphorylation [25]. Therefore, we suggest that a part of the increased H1.5 phosphorylation in G_1 in Jurkat cells is a result of overactive H1 kinases, either within an unchanged phosphorylation window, or during an extended phosphorylation window occupying a larger part of the G_1 phase. An alternative explanation for the extended G_1 H1.5 phosphorylation would be a defective H1 phosphatase. In agreement with data [26], this is less likely, because the sorted G_1 cells contained substantially reduced levels of phosphorylated H1 compared with G_2/M populations.

In G_1 and S phases, H1 phosphorylation is coupled to less condensed chromatin [12, 17, 25]. Extended H1 phosphorylation may then lead to facilitated S phase entry of malignant cells, as part of a disturbed cell cycle control. Increasing evidence indicates that histone H1 phosphorylation in S phase is important for chromatin decondensation in the replication process [16]. Possibly, the specific serine phosphorylation pattern established in late G_1/early S phase, as described here, takes place to partially displace certain parts of the H1 histones to allow access for, or to recruit, other proteins that are involved in chromatin decondensation and S phase progression, as described for cdc45 [16]. The fine-tuning of replication timing during S phase may then be regulated by small additional local variations in the H1 phosphorylation pattern, in line with recent observations [20].

The precise physiological role of histone H1, its phosphorylation, and the significance of having multiple H1 subtypes remain to be determined. Histone H1 subtypes are evolutionarily conserved, and are therefore predicted to have different roles [30], even though H1 subtypes can compensate for one another [8]. During the time between activation of the T cells and cell sorting, we found that the relative amounts of the individual subtypes altered, and that the relative content of H1.5 was more than doubled compared with G_0 T cells (Figure 3). From the same figures, it is also evident that H1.4 was decreased in activated T cells. However, because of comigration in HPCE, it is more difficult to state anything about the other subtypes. The subtype composition is believed to be tissue-specific, developmental-specific, and differentiation-specific [31, 32]. Alterations in H1 subtype composition have also been connected to the proliferative activity of mouse cells, in which H1a and H1b (corresponding to human H1.5) were synthesized in large amounts in dividing cells only [33]. Studies of mRNA expression indicated that the levels of H1a, H1b and H1d were reduced in terminally differentiated cells and G_0-arrested cells [34]. In line with these observations, results suggest that the H1.5 increase upon T cell activation is coupled to initiation of proliferative capacity, possibly by priming of chromatin for DNA replication. An intriguing possibility is that a major physiological function of the entire histone H1 protein family and their phosphorylation is to participate in the regulation of local chromatin structure during the cell cycle. If this is true, further exploration of the biological mechanisms behind the extended H1 phosphorylation in G_1 of malignant cells may provide new targets for cancer therapy in the future.

6.4 RESULTS

6.4.1 T CELL ACTIVATION RESULTS IN RAPIDLY PROLIFERATING T CELL POPULATIONS

After isolation, the PBLs from all three donors consisted of over 94% viable cells, as measured by Annexin V. They appeared to contain a normal T cell ratio, which was confirmed by measuring the fraction of CD3+ cells. Cell division started after 2 days of activation, and was evident at 3rd day (Figure 2). Upon sorting, more than 97% of the cells were passing through the cell cycle (Figure 2).

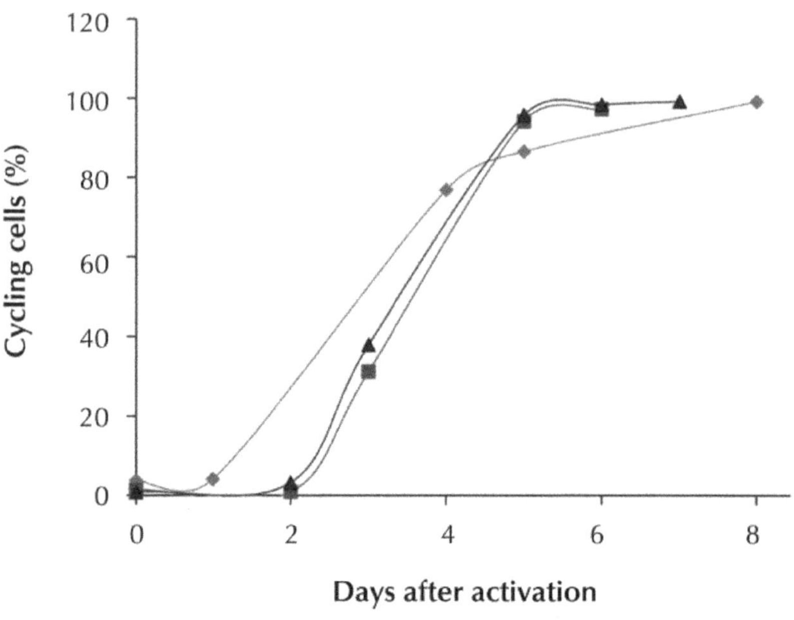

FIGURE 2 The T cell activation assessed by 5(6)-CFSE tracing.

The proportion of cycling cells, with decreased CFSE fluorescence, was measured on days 0, 1, 4, 5, and 8 for sample 1 (green); on days 0, 2, 3, 5, and 6 for sample 2 (blue); on days 0, 2, 3, 5, 6, and 7 for sample 3 (red). Sorting of activated T cells was performed on day 8, 6, and 7, respectively.

At this point most cells had very low levels of 5(6)-CFSE fluorescence, as a result of multiple cell divisions. In this study, cell cycle analysis using PI staining showed appearance of S and G_2 cells at 2nd day and thereafter. During cultivation of PBLs, the fraction of CD3-positive cells increased, and at the time for cell sorting, all cell populations consisted almost solely of T cells (data not shown). When stimulated, CD4+ and CD8+ T cells proliferate, whereas other cell types die through apoptosis or become diluted *via* recultivation of the growing T cells. On examination of the activated PBLs under the microscope, samples from all donors were found to have a similar cell appearance (data not shown). At sorting, the T cell cultures consisted of more than 90% viable cells. The cell cycle distributions of activated T cells are presented in Table 1.

TABLE 1 Cell cycle phase distributions of activated T cell populations, determined by flow cytometry using PI staining, at the time of cell sorting.

Sample	G1, %	S, %	G2/M, %
T cells 1	63.8	29.6	6.6
T cells 2	72.6	23.1	4.3
T cells 3	72.8	23.4	3.8

6.4.2 H1.5 EXPRESSION IS INCREASED IN PROLIFERATING T CELLS COMPARED WITH RESTING LYMPHOCYTES

The H1 subtype composition in non-activated, resting (G_0) PBLs was analyzed by HPCE (Figure 3). The migration order coincided exactly with earlier published data [27], and no other peaks were detected. The relative subtype compositions were then determined by measuring the height of the peaks containing H1.2, H1.3, H1.4, and H1.5 in the electropherograms, and normalizing these to the sum of these peak heights. The relative H1 subtype composition in PBLs from the three donors was (mean ± SD): 18.8 ± 2.1% for H1.2, 25.9 ± 2.7% for H1.3, 39.7 ± 3.9% for H1.4, and, 15.6 ± 0.7% for H1.5. This subtype composition is presumed to be approximately the same as in pure T cell populations, because PBLs from normal donors generally contain more than 80% T cells, with the major part of the contaminating cells known to be B cells. We have investigated the H1 subtype distribution in purified human B cells, and these results showed an almost identical H1 subtype composition to that of the PBLs described.

At the time for cell sorting, a significant relative increase in H1.5 content was seen in activated T cells from all donors, compared with G_0 cells. The areas of the peaks containing H1.5 and the peaks containing the remaining subtypes were determined for both activated T cells and Jurkat cells. The small peak between peaks 1 and 2, most probably containing H1x, was omitted from the calculations. The relative H1.5 content was determined to be 36 ± 2% (n = 3) for activated T cells, and 47 ± 1% (n = 3) for Jurkat cells. The available number of resting T cells from each donor was not sufficiently large for growth stimulation and RP-HPLC fractionation, but because both RP-HPLC and HPCE use UV absorption for protein detection, and we only report the fractions of each subtype or group of subtypes, these results can be compared.

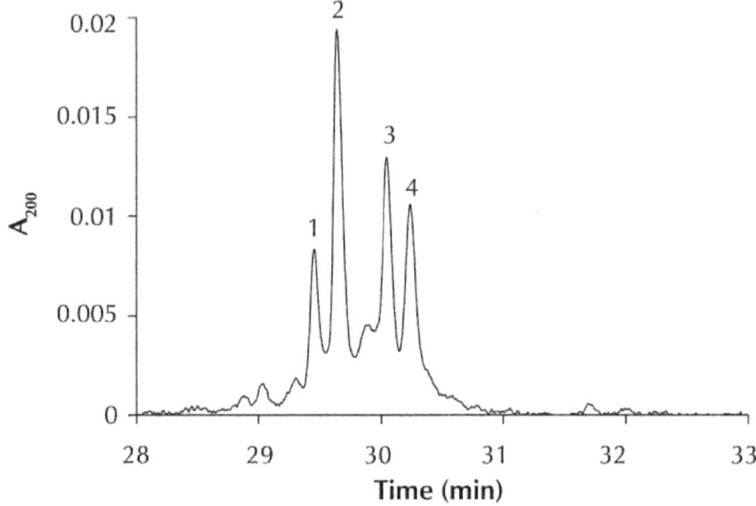

FIGURE 3 The HPCE separation of perchloric acid extracted H1 histones from non-activated PBLs. Only unphosphorylated subtypes were detected. The peak designations are (1) H1.5, (2) H1.4, (3) H1.3, and (4) H1.2.

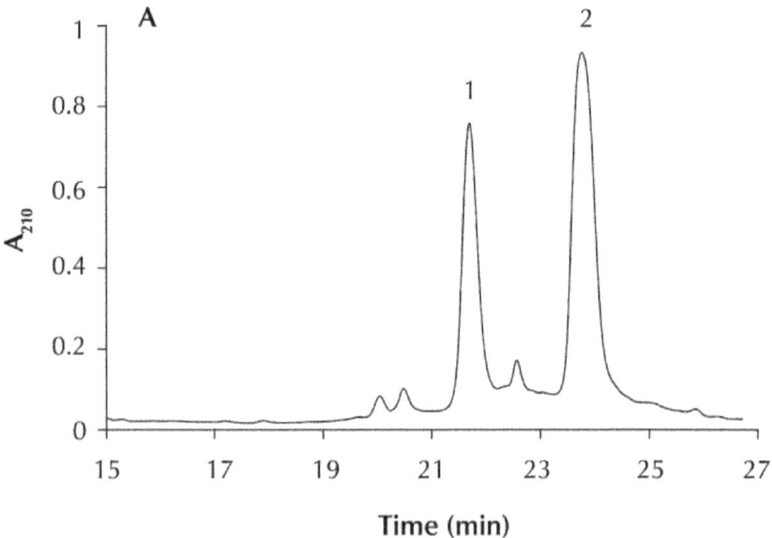

FIGURE 4 *(Continued)*

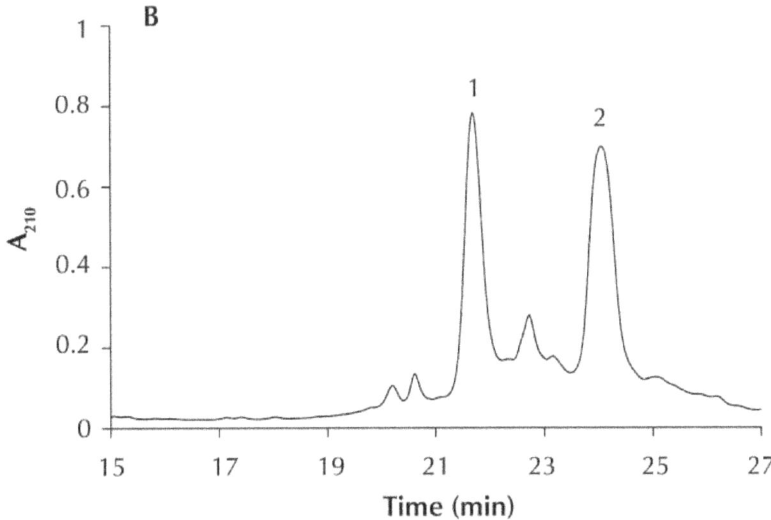

FIGURE 4 The RP-HPLC fractionation of H1 histones from (A) activated T cells and (B) Jurkat cells. Peak 1 contained H1.5 and phosphorylated variants thereof, and peak 2 contained subtypes H1.2, H1.3, H1.4, and their phosphorylated counterparts.

6.4.3 PROLIFERATING T CELLS AND JURKAT CELLS CONTAIN MULTIPLE PHOSPHORYLATED H1 SUBTYPES

H1 samples were extracted from cycling, activated T cells. HPCE separation of H1 histones displayed the presence of multiple peaks due to phosphorylation in addition to the unphosphorylated subtypes. Exponentially growing Jurkat cells displayed a somewhat increased level of H1 phosphorylation, compared with any T cell sample. All migration orders coincided exactly with before published data [27]. The differences between T cells and Jurkat cells were also shown by the H1.5 phosphorylation patterns obtained after RP-HPLC separation prior to HPCE (Figure 5, insets).

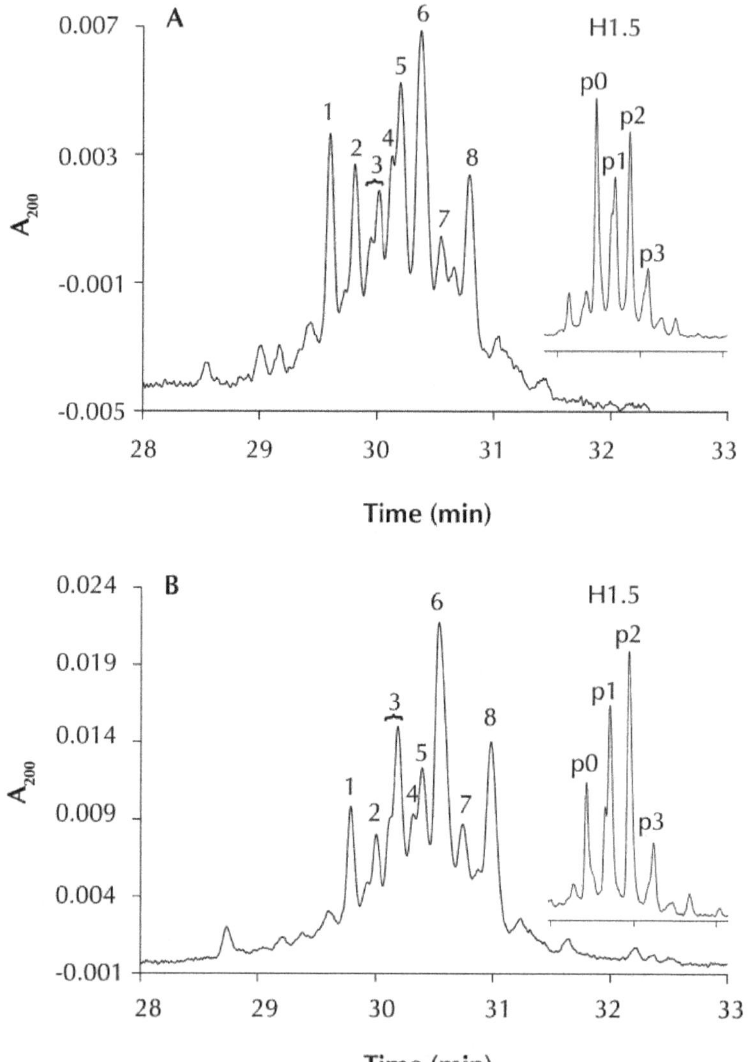

FIGURE 5 The HPCE separations of H1 histones and RP-HPLC-fractionated H1.5 (inset) from **(A)** activated T cells and **(B)** exponentially growing Jurkat cells.
The peaks were identified as: (1), unphosphorylated H1.5; (2) unphosphorylated H1.4; (3) monophosphorylated H1.5; (4) monophosphorylated H1.4; (5) unphosphorylated H1.3; (6) diphosphorylated H1.5, together with unphosphorylated H1.2 and possibly diphosphorylated H1.4; (7) monophosphorylated H1.3; (8) monophosphorylated H1.2 together with triphosphorylated H1.5.

6.4.4 FLOW SORTING OF T CELLS AND JURKAT CELLS IN DIFFERENT CELL CYCLE PHASES

Flow sorting DNA histograms (with sorting gates for G_1, S, and G_2/M populations) of cycling T cells and Jurkat cells are shown in Figure 6. The sorted populations were reanalyzed after sorting to check the purity of the different populations (Figure 6). Flow sorting of Jurkat cells resulted in almost pure cell cycle populations (Table 2). Sorting of cycling T cells resulted in relatively pure G_1 and S populations, but there was some cross-contamination of the G_2/M populations seen during reanalysis, primarily by cells with a measured DNA content corresponding to G_1 cells (Table 2 and Figure 6). In this study, one of the T cell samples (T cells 3) had a higher G_1 cross-contamination of the S phase cells (Table 2) than did the other T cell samples. This can be explained by an increase in the spreading of flow sorting droplets in this particular experiment.

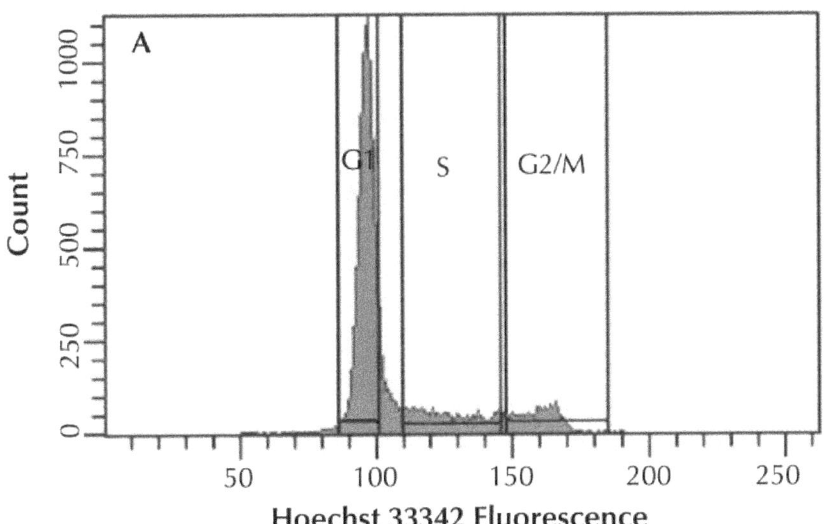

FIGURE 6 *(Continued)*

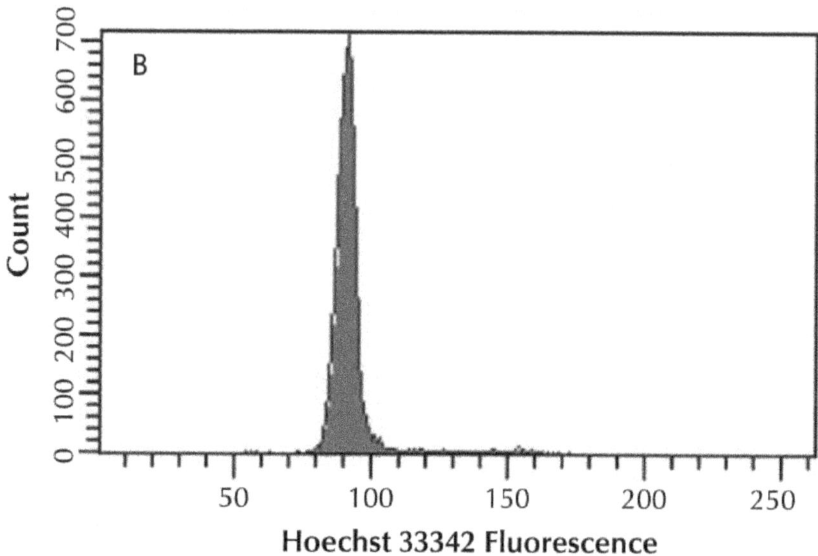

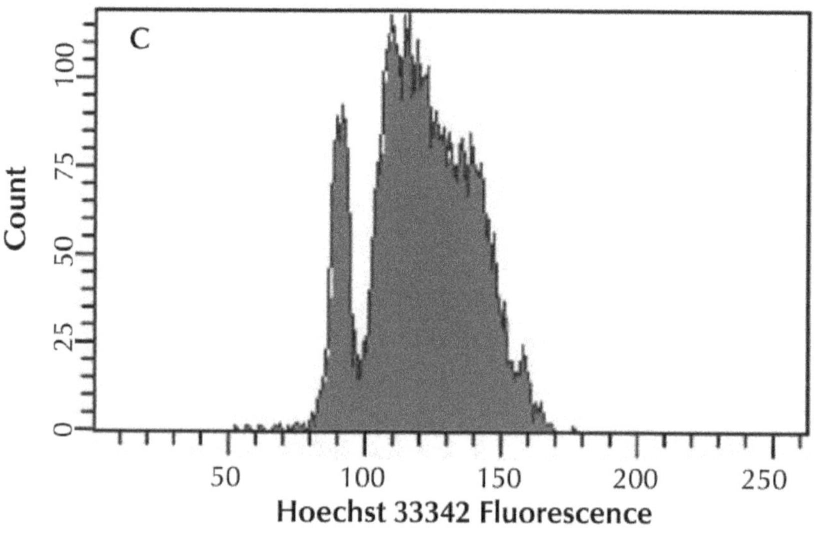

FIGURE 6 *(Continued)*

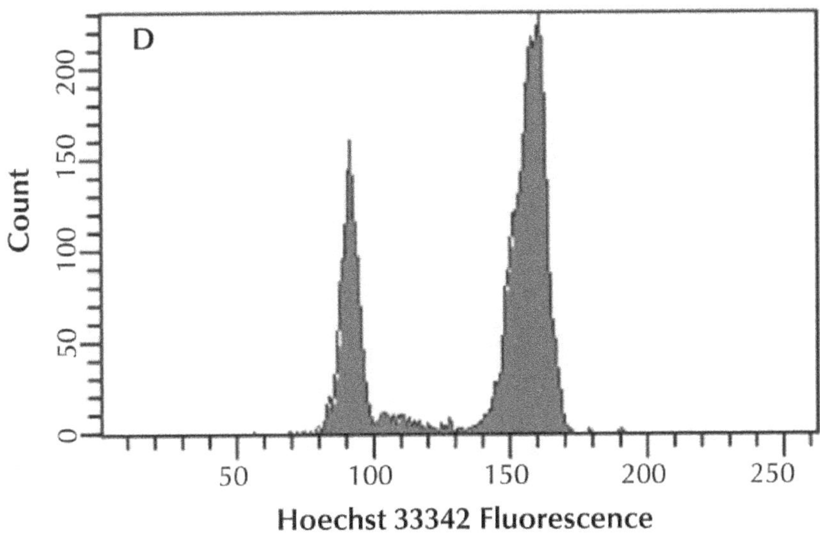

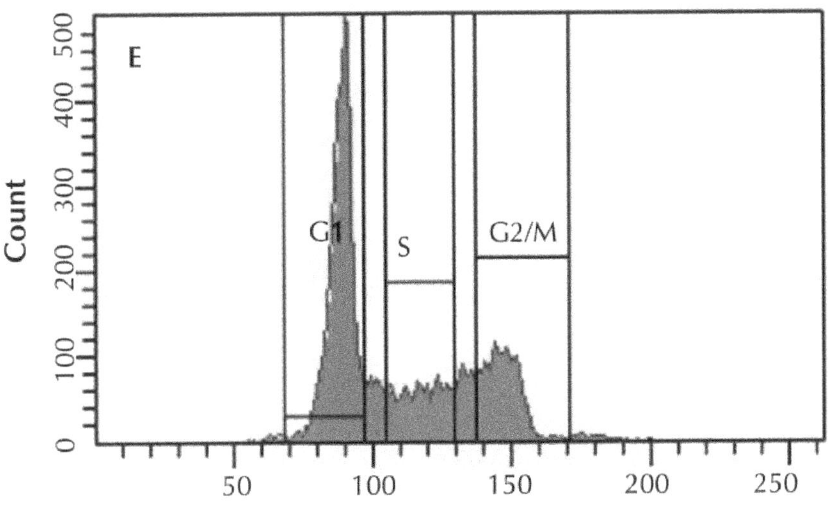

FIGURE 6 *(Continued)*

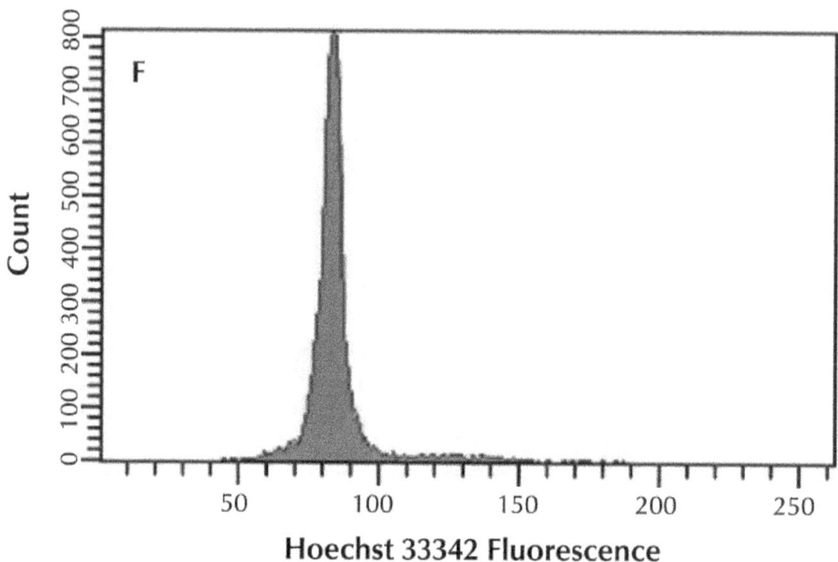

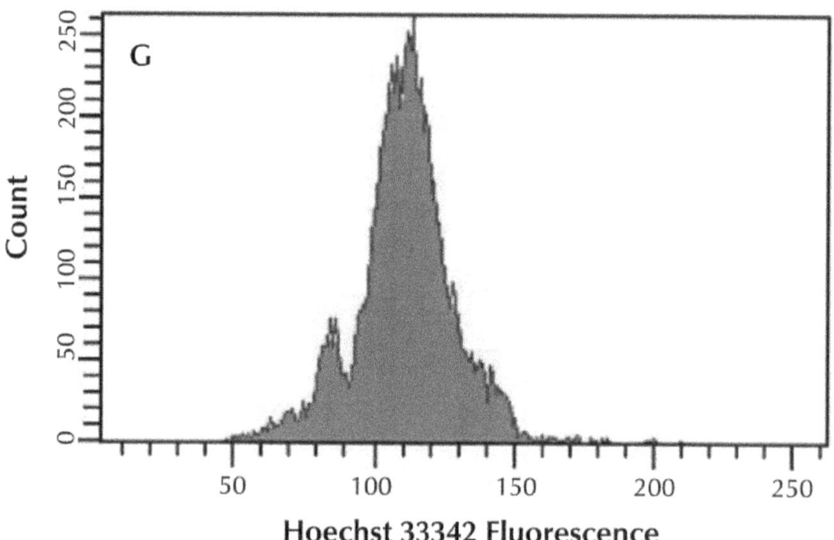

FIGURE 6 *(Continued)*

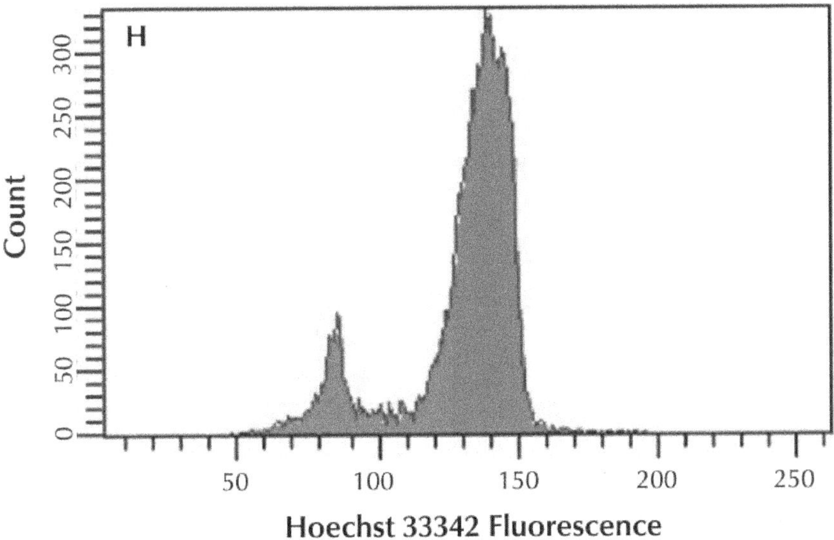

FIGURE 6 The DNA histograms with sorting gates of Hoechst 33342-stained (A) T cells and (E) Jurkat cells.

After cell sorting, the different cell populations were reanalyzed. (B-D) Reanalysis of sorted T cell populations in (B) G_1, (C) S, and (D) G_2/M populations. (F-H) Reanalysis of sorted Jurkat populations in (F) G_1, (G) S, and (H) G_2/M populations.

TABLE 2 Purity of flow sorted populations from activated T cells and Jurkat cells

	Purity, %		
Sample	**G1**	**S**	**G2/M**
T cells 1	96.6	88.4	63.6
T cells 2	95.9	83.4	70.5
T cells 3	92.1	71.0	60.9
Jurkat 1	94.8	88.0	89.3
Jurkat 2	93.1	86.1	87.7
Jurkat 3	93.4	88.5	86.2

Gréen *et al. Epigenetics and Chromatin* 2011 **4**:15 doi:10.1186/1756-8935-4-15

The cell cycle distribution of the DNA histograms from Hoechst 33342-stained cells at flow sorting was determined using Modfit (Figure 7). Cell cycle data are presented in Table 3. From these data, it is evident that there were fewer T cells in G_2/M compared with Jurkat cells. This may be an explanation for the lower purity of the sorted G_2/M populations from T cells.

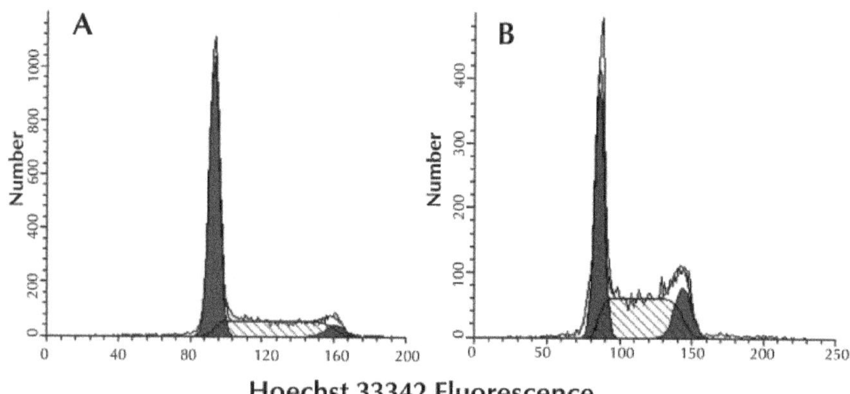

FIGURE 7 Cell cycle analysis of DNA histograms after gating in forward FSC and SSC and doublet discrimination.
(A) Activated T cells, **(B)** Jurkat cells.

TABLE 3 Cell cycle distribution of cell populations stained with Hoechst 33342 selected for sorting (after gating in forward/side scatter (FSC/SSC) and doublet discrimination).

Sample	G1, %	S, %	G2/M, %
T cells 1	58.9	35.3	5.8
T cells 2	64.7	31.0	4.3
T cells 3	69.6	26.4	4.0
Jurkat 1	54.1	33.1	15.5
Jurkat 2	41.2	44.7	14.1
Jurkat 3	50.3	38.2	11.5

Gréen *et al. Epigenetics and Chromatin* 2011 **4**:15 doi:10.1186/1756-8935-4-15

6.4.5 THE PHOSPHORYLATION OF H1 HISTONES STARTS IN THE G_1 PHASE OF THE CELL CYCLE IN NORMAL PROLIFERATING T CELLS

The Histone H1 subtype and phosphorylation pattern was determined using HPCE for G_1, S, and G_2/M T cell populations (Figure 8). Only small variations were detected between the three T cell samples. Furthermore, H1.5 phosphorylation was also examined after RP-HPLC separation followed by HPCE of the isolated H1.5 peak from the RP-HPLC fractionation of H1 histones (insets in Figure 8).

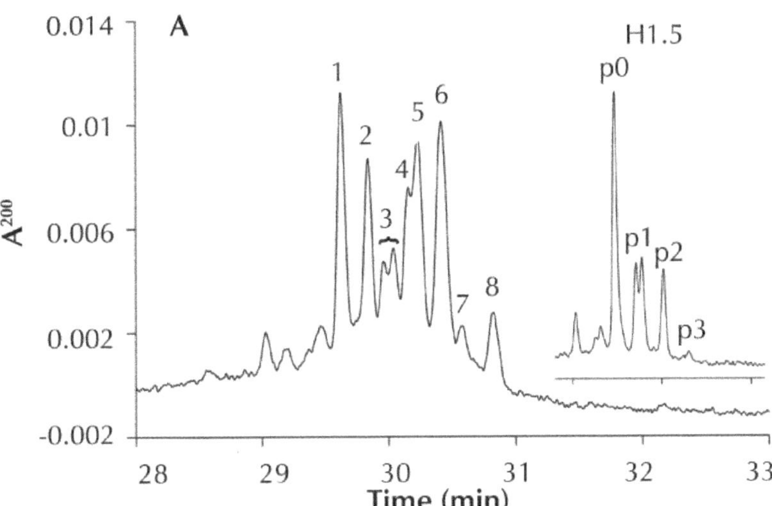

FIGURE 8 *(Continued)*

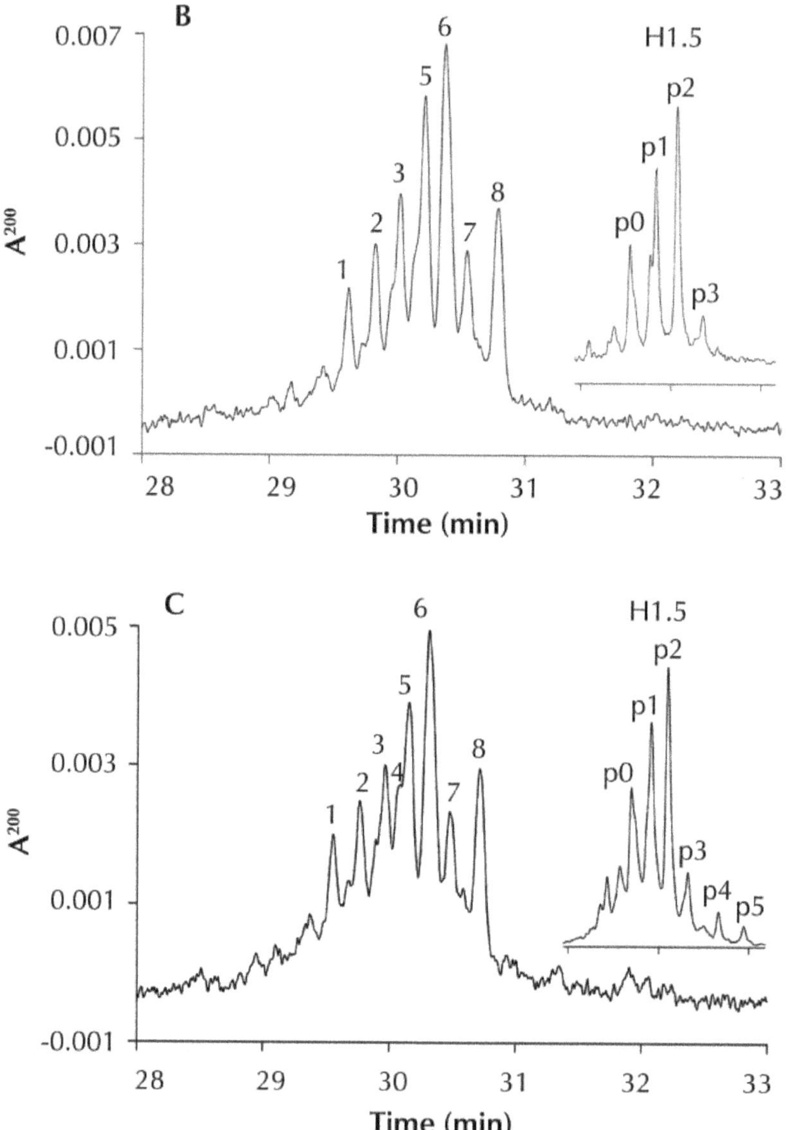

FIGURE 8 The HPCE separations of H1 histones and RP-HPLC-fractionated H1.5 (insets) extracted from activated, flow sorted T cells.
Cells in **(A)** G_1, **(B)** S, and **(C)** G_2/M phase. Peak designations as in Figure 5.

In G_1 T cells, approximately 50% of H1.5 was present in its unphosphorylated form (as determined by peak heights). Most of the remaining H1.5 was either mono- or diphosphorylated. The same pattern is probably to be true also for H1.4, but this cannot be verified due to the co-migration of diphosphorylated H1.4 with unphosphorylated H1.2, and diphosphorylated H1.5 (peak 6 in Figure 8). H1.2 monophosphorylation was evident. The level of H1.3 phosphorylation was low.

Cells in S phase had more extended H1.5 phosphorylation, with a clear increase in mono-, di-, and triphosphorylated H1.5. A clear reduction of unphosphorylated H1.5 was evident (peak 1). Histone H1.4 phosphorylation also increased, which was seen through reduction of the peak containing unphosphorylated H1.4., H1.2, and H1.3 monophosphorylation increased.

The S phase phosphorylation pattern was largely preserved in the sorted G_2/M T cell populations. It was evident that the extent of H1.5 mono- and diphosphorylation was preserved, whereas a small increase in triphosphorylated H1.5 could be detected. In this study, the presence of p4 and p5 hyperphoshorylated forms was indicated during G_2/M. These phosphorylations probably originate from the metaphase cells in this population, because these forms have been detected in mitotic CEM cells [23]. However, we could not detect higher phosphorylation forms of the other subtypes, although they are predicted to be present in metaphase cells. This finding, and that of the low amounts of tetra- and pentaphosphorylated forms of H1.5, can probably be explained by the relatively short time during mitosis when these forms occur. Further studies are needed to address the issue of mitotic phosphorylation.

6.4.6 EXPONENTIALLY GROWING JURKAT CELLS CONTAIN MORE EXTENSIVELY PHOSPHORYLATED H1 SUBTYPES IN THE G_1 PHASE OF THE CELL CYCLE COMPARED WITH ACTIVATED T CELLS

After flow sorting of exponentially growing Jurkat cells, H1 histones from G_1, S, and G_2/M cell populations were extracted and separated by HPCE. The H1 subtype and phosphorylation pattern was reproducible between the Jurkat samples.

In G_1 Jurkat cells, highly phosphorylated H1.5 was detected. Histone H1.4 monophosphorylation was evident, and possibly diphosphorylated H1.4 was present as a part of peak 6. H1.2 monophosphorylation was detected (Figure 9, peak 8). The level of H1.3 phosphorylation was low (Figure 9, peak 7).

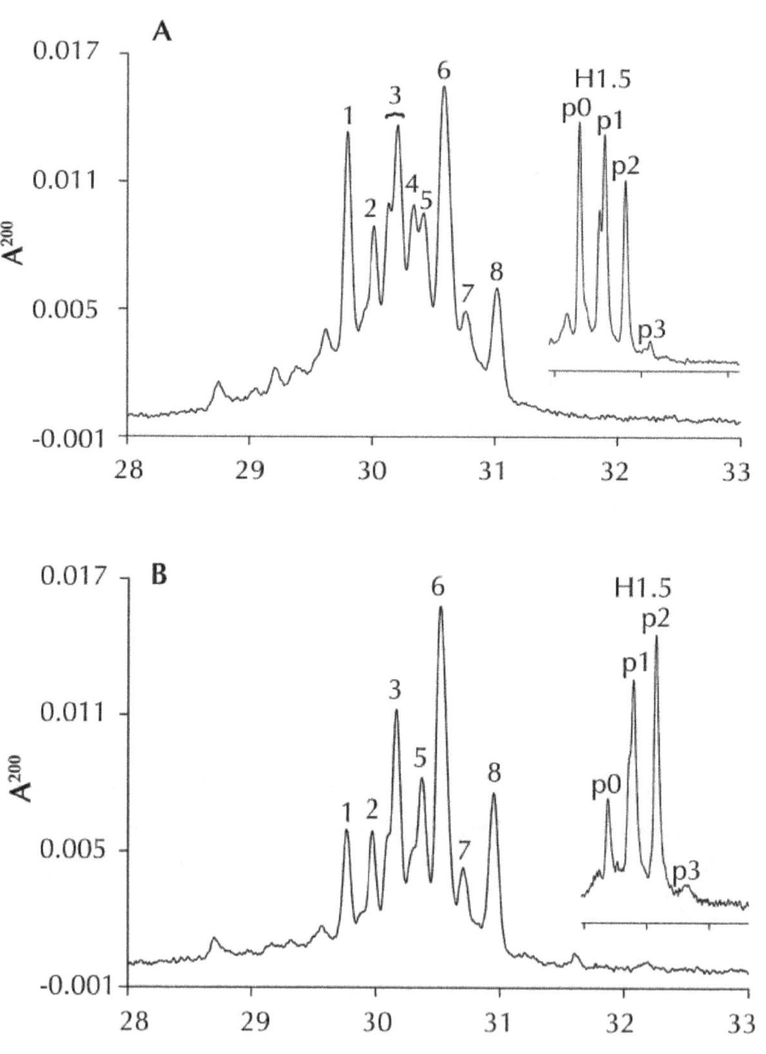

FIGURE 9 *(Continued)*

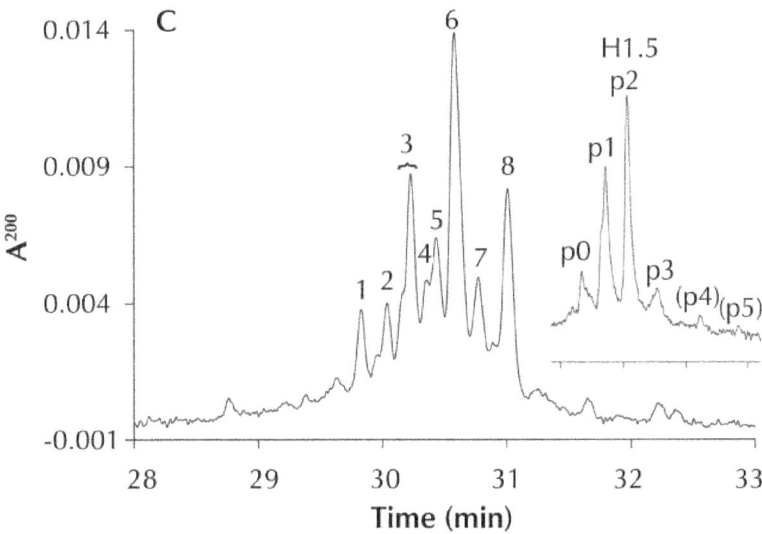

FIGURE 9 The HPCE separations of H1 histones and RP-HPLC-fractionated H1.5 (insets) from flow sorted Jurkat cells.
Cells in **(A)** G_1, **(B)** S, and **(C)** G_2/M phase Peak designations as in Figure 5.

In Jurkat cells sorted from S phase, H1.5 phosphorylation increased substantially. The level of unphosphorylated H1.4 decreased slightly, whereas monophosphorylated H1.4 decreased, probably due to an increase in diphosphorylated H1.4. H1.2 monophosphorylation was increased, whereas H1.3 phosphorylation was virtually unaffected.

In G_2/M, the H1 phosphorylation pattern resembled that in S phase, but the extent of phosphorylation increased somewhat for all subtypes. This is also evident in which unphosphorylated H1.5 decreased and higher phosphorylated forms were detected (p4 and p5). The purity of the sorted G_2/M cells (Table 2) was high, but some late S-phase cells might still have been present in these samples.

The major difference between activated T cells and Jurkat cells was a more extended phosphorylation in G_1 Jurkat cells. In addition, G_2/M Jurkat cells contained a lower level of unphosphorylated H1.5 compared with G_2/M T cells. However, this difference may be explained by a contamination of G_1 cells in the sorted G_2/M T cell populations, resulting in an underestimation of G_2/M phosphorylation. Therefore, we anticipate that T cells and Jurkat cells exhibit an almost similar H1 phosphorylation pattern in S phase and in G_2/M phase.

6.5 CONCLUSION

Increasing evidence indicates that H1 phosphorylation is important in the priming of chromatin for DNA replication. A result indicate that an interphase serine phosphorylation pattern becomes largely established during G_1 or early S phase, and confirm that complementary serine phosphorylation of newly synthesized H1 histones takes place mainly during the S phase of the cell cycle. We also detected a significant increase in the H1.5 content upon activation of T cells, indicating that expression of this subtype may be coupled to proliferative capacity. The T lymphoblastoid cells showed a more extended H1 phosphorylation in G_1 compared with normal T cells, which may be a part or a consequence of aberrant cell cycle control in malignant cells.

KEYWORDS

- **Cell division**
- **Cell-cycle**
- **Histone**
- **Nucleosome**
- **Segregation**

ACKNOWLEDGMENT

We are most grateful to Dr. Marie Larsson for expert help with T cell purification and activation. This work, as part of the European Science Foundation EUROCORES Program EuroDYNA, was supported by funds from the Austrian Science Foundation (I23-B03), the EC 6th framework program under contract number ERAS-CT-2003-980409, and the Swedish Cancer Society.

AUTHORS' CONTRIBUTIONS

Anna Gréen designed the study, performed cell culturing and flow cytometry, and wrote most of the manuscript; Bettina Sarg performed RP-HPLC and HPCE, and wrote parts of the manuscript; Henrik Gréen performed flow sorting and wrote parts of the manuscript; Anita Lönn isolated proteins; Herbert H. Lindner designed RP-HPLC and HPCE analysis, analyzed RP-HPLC and HPCE data, and helped supervise the project; and Ingemar Rundquist conceived and supervised the project, and wrote the final manuscript. All authors read and approved the final manuscript.

REFERENCES

1. Luger, K., Mader, A. W., Richmond, R. K., Sargent, D. F., and Richmond, T. J. Crystal structure of the nucleosome core particle at 2.8 a resolution. *Nature*, **389**, 251–260 (1997).
2. Thomas, J.O. Histone H1: location and role. *Curr Opin Cell Biol*, **11**, 312–317 (1999).
3. Izzo, A., Kamieniarz, K., and Schneider, R. The histone H1 family: specific members, specific functions? *Biol Chem*, **389**, 333–343 (2008).
4. Robinson, P. J. and Rhodes, D. Structure of the "30 nm" chromatin fibre: a key role for the linker histone. *Curr Opin Struct Biol*, **16**, 336–343 (2006).
5. Brown, D. T. Histone H1 and the dynamic regulation of chromatin function. *Biochem Cell Biol*, **81**, 221–227 (2003).
6. Raghuram, N, Carrero, G., Th'ng, J., and Hendzel, M. J. Molecular dynamics of histone H1. *Biochem Cell Biol*, **87**, 189–206 (2009).
7. Woodcock, C. L., Skoultchi, A. I., and Fan, Y. Role of linker histone in chromatin structure and function: H1 stoichiometry and nucleosome repeat length. *Chromosome Res*, **14**, 17–25 (2006).
8. Fan, Y. and Skoultchi, A. I. Genetic analysis of H1 linker histone subtypes and their functions in mice. *Methods Enzymol*, **377**, 85–107 (2004).
9. Sancho, M., Diani, E., Beato, M., and Jordan A. Depletion of human histone H1 variants uncovers specific roles in gene expression and cell growth. *PLOS Genet*, **4**, e1000227 (2008).
10. Konishi, A., Shimizu, S., Hirota, J., Takao, T., Fan, Y., Matsuoka, Y., Zhang, L., Yoneda, Y., Fujii, Y., Skoultchi, A. I., and Tsujimoto, Y. Involvement of histone H1.2 in apoptosis induced by DNA double-strand breaks. *Cell*, **114**, 673–688 (2003).

11. Roth, S. Y. and Allis, C. D. Chromatin condensation: does histone H1 dephosphorylation play a role? *Trends Biochem Sci*, **17**, 93–98 (1992).
12. Happel, N. and Doenecke, D. Histone H1 and its isoforms: contribution to chromatin structure and function. *Gene*, **431**, 112 (2009).
13. Horn, P. J., Carruthers, L. M., Logie, C., Hill, D. A., Solomon, M. J., Wade, P. A., Imbalzano, A. N., Hansen, J. C., and Peterson, C. L. Phosphorylation of linker histones regulates ATP-dependent chromatin remodeling enzymes. *Nat Struct Biol*, **9**, 263–267 (2002).
14. Stavreva, D. A. and McNally, J. G. Role of H1 phosphorylation in rapid GR exchange and function at the MMTV promoter. *Histochem Cell Biol*, **125**, 83–89 (2006).
15. Bradbury, E. M., Inglis, R. J., and Matthews, H. R. Control of cell division by very lysine rich histone (F1) phosphorylation. *Nature*, **247**, 257–261 (1974).
16. Alexandrow, M. G. and Hamlin, J. L. Chromatin decondensation in S-phase involves recruitment of Cdk2 by Cdc45 and histone H1 phosphorylation. *J Cell Biol*, **168**, 875–886 (2005).
17. Herrera, R. E., Chen, F., and Weinberg, R. A. Increased histone H1 phosphorylation and relaxed chromatin structure in Rb-deficient fibroblasts. *Proc Natl Acad Sci USA*, **93**, 11510–11515 (1996).
18. Hale, T. K., Contreras, A., Morrison, A. J., and Herrera, R. E. Phosphorylation of the linker histone H1 by CDK regulates its binding to HP1alpha. *Mol Cell*, **22**, 693–699 (2006).
19. Chadee, D. N., Allis, C. D., Wright, J. A., and Davie, J. R. Histone H1b phosphorylation is dependent upon ongoing transcription and replication in normal and ras-transformed mouse fibroblasts. *J Biol Chem*, **272**, 8113–8116 (1997).
20. Thiriet, C. and Hayes, J. J. Linker histone phosphorylation regulates global timing of replication origin firing. *J Biol Chem*, **284**, 2823–2829 (2009).
21. Gurley, L. R., Walters, R. A., and Tobey, R. A. Sequential phsophorylation of histone subfractions in the Chinese hamster cell cycle. *J Biol Chem*, **250**, 3936–3944 (1975).
22. Talasz, H., Helliger, W, Puschendorf, B., and Lindner, H. *In vivo* phosphorylation of histone H1 variants during the cell cycle. *Biochemistry*, **35**, 1761–1767 (1996).
23. Sarg, B., Helliger, W., Talasz, H., Forg, B., and Lindner, H. H. Histone H1 phosphorylation occurs site-specifically during interphase and mitosis: identification of a novel phosphorylation site on histone H1. *J Biol Chem*, **281**, 6573–6580 (2006).
24. Talasz, H., Sarg, B., and Lindner, H. H. Site-specifically phosphorylated forms of H1.5 and H1.2 localized at distinct regions of the nucleus are related to different processes during the cell cycle. *Chromosoma*, **118**, 693–709 (2009).
25. Chadee, D. N., Taylor, W. R., Hurta, R. A., Allis, C. D., Wright, J. A., and Davie, J. R. Increased phosphorylation of histone H1 in mouse fibroblasts transformed with oncogenes or constitutively active mitogen-activated protein kinase kinase. *J Biol Chem*, **270**, 20098–20105 (1995).
26. Chadee, D. N., Peltier, C. P., and Davie, J. R. Histone H1(S)-3 phosphorylation in Ha-ras oncogene-transformed mouse fibroblasts. *Oncogene*, **21**, 8397–8403 (2002).
27. Green, A., Sarg, B., Koutzamani, E., Genheden, U., Lindner, H. H., and Rundquist, I. Histone H1 dephosphorylation is not a general feature in early apoptosis. *Biochemistry*, **47**, 7539–7547 (2008).

28. Gurley, L. R., Valdez, J. G., and Buchanan, J. S. Characterization of the mitotic specific phosphorylation site of histone H1. Absence of a consensus sequence for the p34cdc2/cyclin B kinase. *J Biol Chem*, **270**, 27653–27660 (1995).
29. Rundquist, I. and Lindner, H. H. Analyzes of linker histone--chromatin interactions in situ. *Biochem Cell Biol*, **84**, 427–436 (2006).
30. Eirin-Lopez, J. M., Gonzalez-Tizon, A. M., Martinez, A., and Mendez, J. Birth-and-death evolution with strong purifying selection in the histone H1 multigene family and the origin of orphon H1 genes. *Mol Biol Evol*, **21**, 1992–2003 (2004).
31. Doenecke, D., Albig, W., Bode, C., Drabent, B., Franke, K., Gavenis, K., and Witt, O. Histones: genetic diversity and tissue-specific gene expression. *Histochem Cell Biol*, **107**, 110 (1997).
32. Khochbin, S. Histone H1 diversity: bridging regulatory signals to linker histone function. *Gene*, **271**, 112 (2001).
33. Lennox, R. W. and Cohen, L. H. The histone H1 complements of dividing and nondividing cells of the mouse. *J Biol Chem*, **258**, 262–268 (1983).
34. Wang, Z. F., Sirotkin, A. M., Buchold, G. M., Skoultchi, A. I., and Marzluff, W. F. The mouse histone H1 genes: gene organization and differential regulation. *J Mol Biol*, **271**, 124–138 (1997).
35. Vindelov, L. L., Christensen, I. J., and Nissen, N. I. A detergent-trypsin method for the preparation of nuclei for flow cytometric DNA analysis. *Cytometry*, **3**, 323–327 (1983).
36. Lindner, H., Sarg, B., and Helliger, W. Application of hydrophilic-interaction liquid chromatography to the separation of phosphorylated H1 histones. *J Chromatogr A*, **782**, 55–62 (1997).
37. Lindner, H., Helliger, W., Dirschlmayer, A., Talasz, H., Wurm, M., Sarg, B., Jaquemar, M., and Puschendorf, B. Separation of phosphorylated histone H1 variants by high-performance capillary electrophoresis. *J Chromatogr*, **608**, 211–216 (1992).
38. Lindner, H., Wurm, M., Dirschlmayer, A., Sarg, B., and Helliger, W. Application of high-performance capillary electrophoresis to the analysis of H1 histones. *Electrophoresis*, **14**, 480–485 (1993).

CHAPTER 7

POST-TRANSLATIONAL MODIFICATIONS

STEPHANIE D. BYRUM, SEAN D. TAVERNA, and ALAN J. TACKETT

CONTENTS

7.1 INTRODUCTION

Eukaryotic genomes are highly organized into transcriptionally active (euchromatic) and silent (heterochromatic) chromatin regions. Conversion of chromatin between the two major forms is regulated in part through interactions between chromatin-modifying enzymes and nucleosomes. Nucleosomes are the fundamental unit of chromatin and consist of approximately 147 base pairs of DNA wrapped around an octameric core of the histones H2A, H2B, H3, and H4 [1]. Chromatin structure plays a key role in the regulation of gene activity and its mis-regulation is a theme characteristic of many types of disease and cancer [1]. The N-terminal tails of histones, which protrude outside of the nucleosome core [2], are subject to many sites and types of post-translational modifications (PTMs), which, in turn, help regulate biological processes through altering nucleosome stability or the function of chromatin-associated complexes [3, 4]. For example, acetylation of histone lysine residues on the N-terminal tail has been correlated to active gene transcription either by countering the negative charge of the DNA backbone, or through the recruitment or stabilization of bromodomain-containing proteins [3, 5, 6].

A major emphasis in the field of chromatin biology is the understanding of how histone PTMs and protein-protein interactions are associated with specific gene loci to regulate gene transcription. Current technologies like ChIP (chromatin immunoprecipitation), affinity purification of protein-histone complexes for proteomic analysis, and more recent technology that allows for the purification of chromosome sections for proteomic analysis are used to study protein interactions on chromosomes [7-10]. One pitfall of these technologies is the challenge of purifying cognate histones (i.e., preserving the *in vivo* associated histones during isolation of chromatin). To overcome this pitfall, it has been previously reported how to monitor and prevent dynamic exchange of histones during chromatin purification [11]. *In vivo* chemical cross-linking reagents, such as formaldehyde, can be used to prevent histone exchange during the purification of chromatin sections [12]. However, there is a balanced level of chemical cross-linking needed to trap protein-protein and protein-DNA interactions, while still allowing for the solubility of chromatin for purification and access of affinity reagents [12].

Recently a quantitative approach using I-DIRT is published, an isotopic labeling technique utilizing affinity purification and mass spectrometry, to measure levels of histone exchange in purified chromatin sections [11]. Here it is described the bioinformatic analysis, which expands on this published chapter, reporting the significance of proper cross-linking to capture histones with transcription activating PTMs during chromatin purification. In this chapter, one is able to gain new insights into the dynamic exchange of histones and post-translationally modified histones.

7.2 EXPERIMENTAL METHODS

Detailed methods are described in Byrum et al. [10]. Briefly, *Saccharomyces cerevisiae HTB1::TAP-HIS3 BY4741* (Open Biosystems) cells grown in isotopically light media and cells from an arginine auxotrophic strain (*arg4::KAN BY4741*, Open Biosystems) cultured in isotopically heavy media ($^{13}C_6$ arginine) were grown to midlog phase (3.0×10^7 cells/mL) and cross-linked using either 0%, 0.05%, 0.25%, or 1.25% formaldehyde (FA). The cells were harvested, mixed 1:1 by cell weight (isotopically light cells: heavy cells), and lysed under cryogenic conditions. The cell powder was resuspended in affinity purification buffer (20 mM HEPES pH 7.4, 300 mM NaCl, 0.1% tween-20, 2 mM $MgCl_2$, and 1% Sigma fungal protease inhibitors) and the DNA sheared to ~1 kb sections. Small chromatin sections containing TAP tagged H2B histones were affinity purified on IgG-coated Dynabeads and the eluted proteins were resolved with a 4–20% Tris-Glycine gel. Following colloidal Coomassie-staining, histone gel bands were excised, trypsin digested, and tryptic peptides were subjected to tandem mass spectrometric analysis with a coupled Eksigent NanoLC-2D and Thermo LTQ-Orbitrap mass spectrometer [12]. The histone purification experiments were performed in triplicate.

The isotopically light and heavy arginine containing histone peptides were identified using a Mascot (version 2.2.03) database search. Peptide identification can be made with mass spectrometric database searching software other than Mascot with equivalent results. The search parameters included: precursor ion tolerance 10 ppm, fragment ion tolerance 0.6 Da, fixed modification of carbamidomethyl on cysteine, variable modification of oxidation on methionine and acetyl on lysine, and 2 missed cleavages possible with trypsin. The Mascot results were uploaded into Scaffold 3 (version 3.00.01) for viewing the proteins and peptide information. A false discovery rate of 1% was used as the cut off value for arginine containing histone peptides. The monoisotopic peak intensity (I) values for each arginine containing peptide were extracted using Qual Browser (version 2.0, Thermo). The percent light for each peptide was calculated as $I_L/(I_L + I_H)$. The average of all peptides identified for each percentage of cross-linking was calculated along with the standard error. The number of unique identified peptides was: bulk H3 (26, 14, 9, and 8), H3K9acK14ac (7, 4, 8, and 8), bulk H4 (25, 8, 8, and 13), and H4K12acK16ac (7, 4, 5, and 3) for 0%, 0.05%, 0.25% and 1.25% FA, respectively. Percent light peptide reported here differs from the Byrum et al. [10] report [11] as it is separated PTM containing and unmodified peptides in the current chapter.

7.3 RESULTS AND DISCUSSION

The potential roles histone modifications play in regulating gene transcription and the recruitment of protein complexes to specific gene loci have made them attractive therapeutic targets for a variety of diseases including cancer. In order to preserve and study histone PTMs that occur on specific sites of chromatin, histone exchange must be prevented during the chromatin purification process. Previously transient I-DIRT technology was utilized to investigate the level of chemical cross-linking with formaldehyde necessary to prevent histone exchange during chromatin purification [11]. Here, new bioinformatic analyzes are performed that reveal differential exchange rates for histones containing PTMs correlated to active gene transcription. As shown in Figure 1 and detailed in the experimental methods section, isotopically light histones were isolated via a TAP tag on H2B in the presence of an equivalent amount of isotopically heavy histones. The exchange of histones (i.e., the incorporation of isotopically heavy histones during the isolation of isotopically light histones) was followed with mass spectrometry.

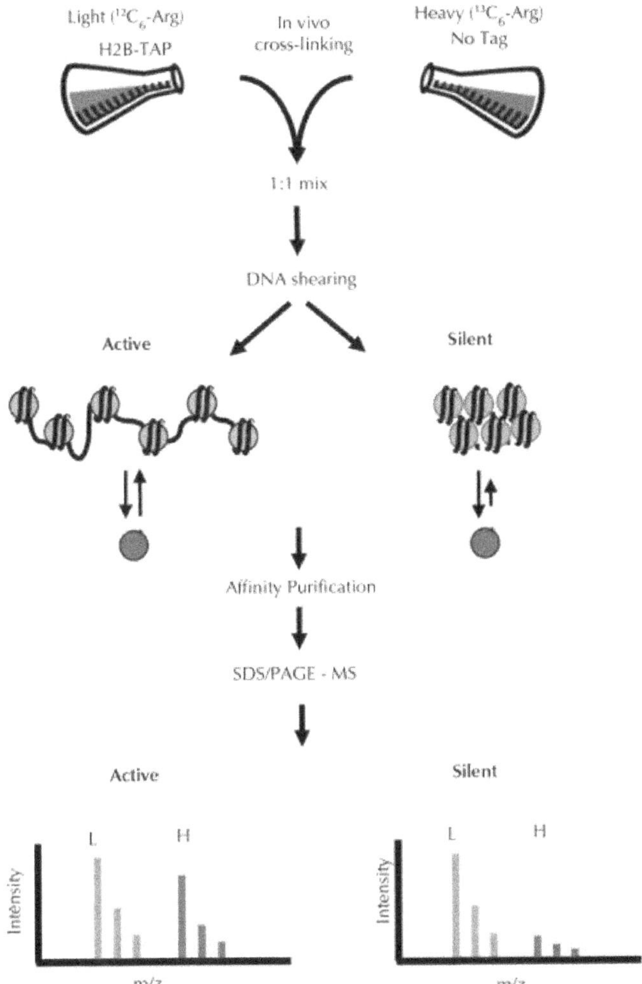

FIGURE 1 Quantitative analysis of histone exchange. *S. cerevisiae* H2B-TAP cells were grown in isotopically light media ($^{12}C_6$-Arg) while an arginine auxotrophic strain was grown in isotopically heavy ($^{13}C_6$-Arg) media. Cultures were chemically cross-linked with formaldehyde, harvested independently, mixed 1:1 by cell weight, and cryogenically co-lysed. Chromatin was sheared to ~1 kb and affinity purified on IgG coated Dynabeads. Histones were resolved by SDS-PAGE and the percent light peptides were measured by mass spectrometry. Depending on the level of *in vivo* cross-linking, histones will dissociate and re-associate with the purified chromatin. This exchange can be monitored by measuring the incorporation of isotopically heavy histones (red circles). Actively transcribing chromatin is more loosely packaged and will undergo histone exchange more readily. Silent chromatin is more densely packaged and is less likely to undergo histone exchange.

Mascot analysis of the mass spectrometric data obtained from H2B-TAP cells treated with increasing amounts of formaldehyde identified lysine acetylation marks on histone H3 lysine 9 and lysine 14 (H3K9acK14ac) and histone H4 lysine 12 and lysine 16 (H4K12acK16ac). H3K9acK14ac and H4K12acK16ac are reported marks of active gene transcription, as is acetylation at many other histone lysines [5,6,13,14]. Representative mass spectra of bulk H3, H3K9acK14ac, bulk H4, and H4K12acK16ac peptides for each percentage of cross-linking are shown in Figure 2. The average percent light of all peptides identified for each histone is plotted in Figure 3. Percent light values approaching 100% light peptides indicate minimal histone exchange during purification while those near 50% light peptides reflect rapid exchange. Peptides from the H2B-TAP control were ~100% light at all formaldehyde concentrations tested. The reason that the H2B-TAP peptides are ~100% light is that the TAP tagged version of H2B is only expressed in the strain grown in isotopically light media. This isotopically light TAP tagged version of H2B migrates slower in SDS-PAGE due to the ~20 kDa molecular mass addition of the TAP tag; thus, excision of this band on the gel is exclusively for isotopically light H2B-TAP as all other histones migrate further in the gel. Non-specific proteins co-enriching with H2B-TAP have ~50% light peptides, reflecting the mixing of isotopically light and heavy cultures prior to purification. Without cross-linking, ~10% histone exchange during purification was observed (Figure 3). As reported previously in Byrum et al. [10], mild cross-linking at 0.05% actually increased the observed level of histone exchange during purification, which was not observed at elevated levels of cross-linking. It is predicted that cross-linking more readily stabilizes densely packaged areas of chromatin like heterochromatin, while leaving less densely packaged regions less stable. In accordance, as densely packaged chromatin becomes more heavily cross-linked, it becomes less represented in the analysis due to less efficient DNA shearing and solubility for purification. At a low level of formaldehyde (0.05%), histone H3K9acK14ac peptides are closer to non-specific percent light indicating rapid histone exchange; however, bulk histone H3 is ~80% light. This reveals that histones modified with activating transcription marks exchange more readily than histones without the transcription activating marks. This likely reflects the less densely packaged euchromatin that is more transcriptionally active. At 0.25% formaldehyde, acetylated histone

H3K9acK14ac showed greater exchange compared with bulk H3; however, they both have increased percent light peptides indicating the minimization of exchange with increasing formaldehyde cross-linking. Bulk histone H4 and H4K12acK16ac had similar percentages of light peptides at 0.05% formaldehyde; however, acetylated H4 showed more exchange than bulk H4 at 0.25% formaldehyde. All bulk and acetylated peptides had ~100% light peptides at 1.25% formaldehyde, which indicated that the histones are minimally exchanged. Therefore, 1.25% formaldehyde is sufficient to prevent exchange of histones containing PTMs correlated to gene transcription during our purification of chromatin sections. The percent of formaldehyde cross-linking is specific for yeast synthetic media as other medias require different levels depending on their amine or cross-linking moiety content.

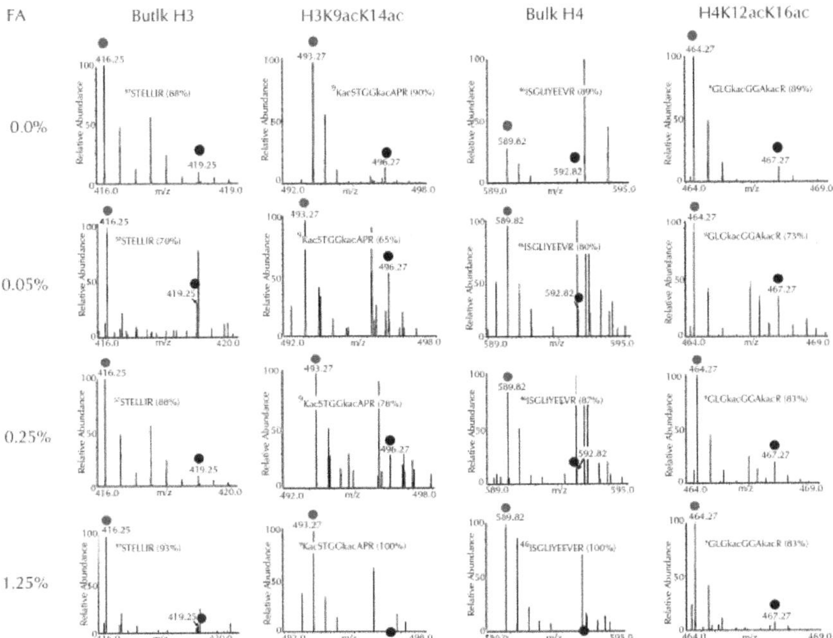

FIGURE 2 Mass spectra of PTM-containing histone peptides. Mass spectra were collected with an Orbitrap mass analyzer for doubly charged peptides from bulk histone H3, H3K9acK14ac, bulk histone H4, and H4K12acK16ac. Blue circles indicate the isotopically light peak while red circles indicate the isotopically heavy peak. The percent isotopically light is shown in parentheses and *in vivo* formaldehyde (FA) cross-linking percentages are listed.

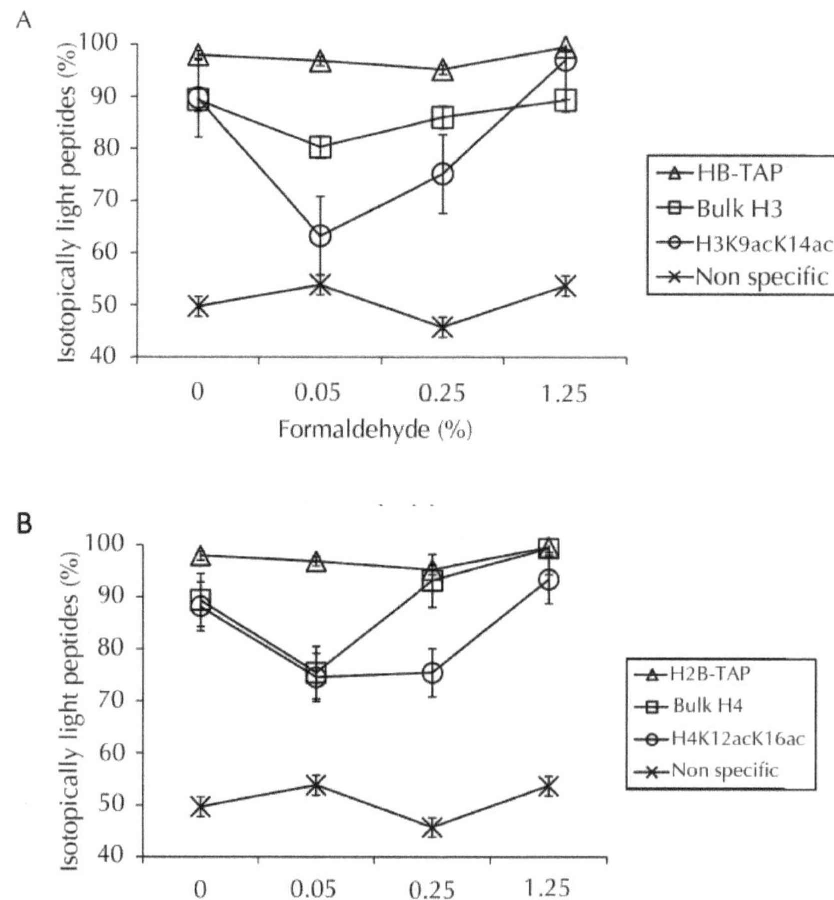

FIGURE 3 Histone exchange occurs more readily in chromatin containing transcription activating PTMs. **(A)** The average and standard error of isotopically light arginine containing peptides for bulk H3, H3K9acK14ac, H2B-TAP, and 15 non-specifically associating proteins are plotted as a function of formaldehyde cross-linking. **(B)** Plot of bulk H4, H4K12acK16ac, H2B-TAP, and 15 non-specific proteins as a function of formaldehyde cross-linking. Levels approaching 100% light peptides indicate minimal histone exchange while levels at ~50% light peptides reflect rapid exchange.

7.4 CONCLUSION

Previously the application of I-DIRT technology has been published to determine the level of histone dissociation/re-association during chromatin purification [11]. In this chapter, additional bioinformatic analyzes have been applied to study the dynamics of exchange for histones containing transcription-activating PTMs. As demonstrated in the histone exchange analysis shown in Figure 3, it is shown that chromatin marked for gene transcription is susceptible to the loss of histones during purification and therefore requires sufficient levels of *in vivo* chemical cross-linking to preserve the native chromatin composition. The technique reported in Byrum et al. [10] and further analyzed here is relevant for a variety of genome-wide studies, and should be considered when preservation of *in vivo* chromatin content is essential for functional analyzes, especially when examining transcriptional processes.

KEYWORDS

- **Arginine Auxotrophic Strain**
- **Eukaryotic Genomes**
- **Histone**
- **I-DIRT Technology**
- **Nucleosome Core**

ACKNOWLEDGMENT

This Chapter was funded by NIH R01DA025755, P20RR015569, P20RR016460 and F32GM093614.

AUTHORS' CONTRIBUTIONS

SDB carried out the experiments, data analysis, and drafted the manuscript. SDT and AJT conceived of the study and participated in its design and coordination. AJT helped to draft the manuscript. All authors read and approved the final manuscript.

REFERENCES

1. Allis, C. D., Jenuwein, T., Reinberg, D., and Caparros, M. L. *Epigenetics*. Cold Spring Harbor Laboratory Press (2006).
2. Suto, R. K., Clarkson, M. J., Tremethick, D. J., and Luger, K. Crystal structure of a nucleosome core particle containing the variant histone H2A.Z. *Nature Structural Biology* **7:**1121–1124 (2000).
1. Taverna, S. D., Li, H., Ruthenburg, A. J., Allis, C. D., and Patel, D. J. How chromatin-binding modules interpret histone modifications: lessons from professional pocket pickers. *Nat Struct Mol Biol* **14,** 1025–1040 (2007).
3. Felsenfeld, G. and Groudine, M. Controlling the double helix. *Nature* **421,** 448–453 (2003).
4. Dion, M. F., Altschuler, S. J., Wu, L. F., and Rando, O. J. Genomic characterization reveals a simple histone H4 acetylation code. *Proceedings of the National Academy of Sciences* **102,** 5501–5506 (2005).
5. Luebben, W. R., Sharma, N., and Nyborg, J. K., Nucleosome eviction and activated transcription require p300 acetylation of histone H3 lysine 14. *Proceedings of the National Academy of Sciences* **107,** 19254–19259 (2010).
6. Tackett, A. J., Dilworth, D. J., Davey, M. J., O'Donnell, M., Aitchison, J. D., Rout, M. P., and Chait, B. T. Proteomic and genomic characterization of chromatin complexes at a boundary. *Journal of Cell Biology* **169,** 35–47 (2005).
7. Dilworth, D. J., Tackett, A. J., Rogers, R. S., Yi, E. C., Christmas, R. H., Smith, J. J., Siegel, A. F., Chait, B. T., Wozniak, R. W., and Aitchison, J. D. The mobile nucleoporin Nup2p and chromatin-bound Prp20p function in endogenous NPC-mediated transcriptional control. *Journal of Cell Biology* **171,** 955–965 (2005).
8. Pokholok, D. K., Harbison, C. T., Levine, S., Cole, M., Hannett, N. M., Lee, T. I., Bell, G. W., Walker, K., Rolfe, P. A., Herbolsheimer, E., Zeitlinger, J., Lewitter, F., Gifford, D. K., and Young, R. A. Genome-wide Map of Nucleosome Acetylation and Methylation in Yeast. *Cell* **122,** 517–527 (2005).
9. Déjardin, J. and Kingston, R. E. Purification of Proteins Associated with Specific Genomic Loci. *Cell* **136,** 175–186 (2009).
10. Byrum, S., Mackintosh, S. G., Edmondson, R. D., Cheung, W. L., Taverna, S. D., Tackett, A. J. Quantitative Analysis of Histone Exchange during Chromatin Purification. *Journal of Integrated OMICS* **1,** 61–65 (2011).

11. Smart, S. K., Mackintosh, S. G., Edmondson, R. D., Taverna, S. D., and Tackett, A. J. Mapping the local protein interactome of the NuA3 histone acetyltransferase. *Protein Science* **18,** 1987–1997 (2009).
12. Shogren-Knaak, M., Ishii, H., Sun, J-M., Pazin, M. J. Davie, J. R., and Peterson, C. L. Histone H4-K16 Acetylation Controls Chromatin Structure and Protein Interactions. *Science* **311,** 844–847 (2006).
13. Jin, Q., Yu, L-R., Wang, L., Zhang, Z., Kasper, L. H., Lee, J-E., Wang, C., Brindle, P. K., Dent, S. Y. R., and Ge, K. Distinct roles of GCN5/PCAF-mediated H3K9ac and CBP/p300-mediated H3K18/27ac in nuclear receptor transactivation. *EMBO J* **30,** 249–262 (2011).

PART IV

CHROMATIN MODIFICATIONS

CHAPTER 8

CHROMATIN SIGNATURE

BRANDON J. THOMAS, ERIC D. RUBIO, NIKLAS KRUMM, PILIB Y. BROIN, KAROL BOMSZTYK, PIRI WELCSH, J OHN M. GREALLY, AARON A. GOLDEN, and ANTON KRUMM

CONTENTS

8.1 INTRODUCTION

Allele-specific gene expression is an integral component of cellular programming and development and contributes to the diversity of cellular phenotypes [1, 2]. Allelic differences in gene expression are mediated by either parent-of-origin-specific selection (imprinting) or stochastic selection of alleles for activation and/or silencing. The importance of genomic imprinting has recently been highlighted by RNA sequencing studies that demonstrated widespread allelic differences in gene expression in mouse brain affecting more than 1,300 genes [3]. The extent of sex- and stage-specific expression of individual alleles emphasizes the essential role of allelic transcriptional regulation in development. In addition to the extensive occurrence of imprinted parent-of-origin-specific expression, gene expression patterns of clonal cell populations are also modified by random or stochastic silencing of either the maternal or paternal allele. Well-known loci displaying allele-specific expression include odorant receptor genes, immunoglobulins and various receptor proteins [4-6]. Additionally, previous large-scale studies have provided new data demonstrating that parent-of-origin-specific expression is employed much more frequently than previously thought [7]. These new findings illustrate the scale and complexity of genomic allele-specific expression. However, the precise molecular mechanism underlying the allelic bias in gene expression is not very well understood.

The best-characterized locus with strict monoallelic imprinted gene expression is the region containing the insulin-like growth factor 2 (*IGF2*) and *H19* genes [8]. The regulation of this locus relies on the imprinting control region (ICR), which acquires DNA methylation on the paternal allele during normal development of the male germline. Methylation of cytosines at the ICR inhibits binding of the zinc finger protein CTCF to the paternal allele, preventing its role as an insulator and allowing long-range interactions of the *IGF2* promoter with enhancer elements downstream of the *H19* gene [9-11]. In contrast, the unmethylated ICR on the maternal allele recruits CTCF, effectively preventing promoter-enhancer interactions and maintaining repression of the maternal *IGF2* gene.

The well-documented requirement of CTCF for imprinted expression at the *IGF2*/*H19* gene locus is thought to result from its role in establishing and/or maintaining long-distance interactions between regulatory elements [12]. Allele-specific binding of CTCF to the ICR has long been known to be essential for the formation of chromatin loops. While the precise mechanism of CTCF's role in long-distance chromatin interactions remains unknown, several studies have provided a rationale for the differential expression of the maternal and paternal *IGF2* gene by revealing an interaction of CTCF with cohesin, a protein complex known for its requirement during sister chromatid cohesion in mitosis [13-16]. Chromosome conformation capture experiments in combination with RNA interference assays recently confirmed the CTCF and cohesin-dependent formation of higher-order chromatin structures at the *IGF2*/*H19* and other gene loci [17-19].

In addition to DNA methylation, histone modifications also contribute to the maintenance of allele-specific expression. DNA methylation of ICRs is accompanied by repressive histone markers, including histone H3 trimethylated at lysine 9 (H3K9me3). In contrast, the unmethylated allele is characterized by permissive histone markers, including histone H3 trimethylated at lysine 4 [20]. Colocalization of epigenetic markers including DNA methylation and histone H3 dimethylated at lysine 9 has been exploited to identify epigenetically distinct parental alleles. Chromosomal regions displaying overlaps of euchromatin and heterochromatin-specific markers have been enriched for known imprinted genes [21].

Despite the importance of monoallelic expression in cellular development and differentiation, little is known about the establishment and maintenance of random monoallelic expression. The link between allele-specific binding of CTCF and monoallelic expression of the *IGF2* gene prompted us to test whether the presence of CTCF and H3K9me3 specifies a chromatin arrangement, which demarcates random monoallelically expressed alleles. Using array-based chromatin immunoprecipitation (ChIP-chip), 293 loci displaying these chromatin markers were identified. The *IGF2BP1* gene locus was selected to further examine whether the presence of CTCF and H3K9me3 comprises a necessary chromatin arrangement for a specific expression profile analogous to the monoallelic behavior observed at the *IGF2/H19* locus. Surprisingly, colocalization of CTCF and H3K9me3 does not provide a reliable measure of monoallelic binding of CTCF at the *IGF2BP1* gene. Our studies included allele-specific sequencing of immunoprecipitated chromatin to demonstrate that chromatin at each *IGF2BP1* allele is bivalent. Importantly, both alleles recruit RNA polymerase II, suggesting that silencing of one *IGF2BP1* allele occurs after transcription initiation. By establishing which epigenetic configurations are involved in governing monoallelic gene expression, the understanding of epigenetic mechanisms will be broaden as they relate to cancer progression and cellular differentiation.

8.2 METHODS

8.2.1 CHIP-CHIP ANALYSIS

The amplification and preparation of immunoprecipitated DNA derived from HBL100 cells for hybridization to ENCODE arrays (Roche NimbleGen Inc., Madison, WI, USA) was performed essentially as described previously [53]. Sample labeling and array hybridization were performed at NimbleGen Systems Inc. Genomic control DNA was labeled with Cy3, and sample DNA was labeled with Cy5. Both Cy3- and Cy5-labeled DNA were hybridized to high-density arrays tiling through ENCODE regions with 50-mer oligonucleotides across nonrepetitive genomic regions. The ratios of the Cy3 and Cy5 intensities of each probe were calculated using NimbleGen Systems' proprietary software.

8.2.2 PEAK DETECTION AND FALSE-POSITIVE RATE CALCULATION

A genomic sequence was considered a possible CTCF-binding site if there were at least four probes among the sequence probe and the flanking probes within a window covering 250 bp on both sides of the probe had $\log_2$ ratio values above a specified cutoff value. The cutoff value was calculated separately for each chromosome. The cutoff value is a given percentage of the value (mean + 6 × standard deviation) of the $\log_2$ ratio values of all the probes covering the chromosome. The possible binding sites thus detected are called peaks. To calculate the false-positive rate (FPR) by data permutation, the $\log_2$ ratio values among probes were scrambled to generate a randomized data set for each individual chromosome. Multiple repetitions of this process generated 20 randomized data sets for each chromosome. Subsequently, the peak detection algorithm described above was applied to count the average number of peaks in the 20 randomized data sets using the same cutoff. The ratio of that number to the number of peaks from the nonrandomized data set is the FPR. The FPR is associated with the threshold setting, which is indicated by the value of cutoff P. Peak detection and randomization of data sets were repeated for different threshold settings of P. The corresponding FPRs were calculated and assigned to peaks. The FPR value assigned to the individual peaks is the value associated with the cutoff P at which the peak is first detected.

Peak discovery was performed using chromatin immunoprecipitate: input ratios combined from adjacent oligonucleotides within 250-bp regions. The FPR of detection was estimated by permutation analyzes in which the experimentally determined $\log_2$ ratio values were reassigned to probes in a random fashion, allowing selection of stringency and specificity levels. To define sites of CTCF interaction with high confidence, peaks were required to be present in all three biological replicates and to be generated at a FPR <0.05.

8.2.3 CHROMATIN IMMUNOPRECIPITATION

Chromatin was prepared for immunoprecipitation as described previously [54] by cross-linking the cells in 1% formaldehyde for 5 min and subjecting them to subsequent sonication until the bulk of DNA was 300–600 bp in size. Chromatin corresponding to 2×10^7 cells was immunoprecipitated with anti-CTCF antibody (D31H2; Cell Signaling Technology, Danvers, MA, USA), anti-H3K9me3 antibody (ab8898; Abcam, Cambridge, MA, USA), anti-trimethyl K4-histone H3 antibody (ab8580; Abcam), anti-trimethyl K27-histone H3 antibody (Millipore 07-449, Billerica MA, USA) or anti-RNA polymerase II antibody (sc899; Santa Cruz Biotechnology, Santa Cruz, CA, USA). Immunoprecipitates were washed, the DNA protein cross-links were reversed and the recovered DNA was tested by performing conventional quantitative PCR as described previously [54]. RNA polymerase II ChIP experiments were performed using the Matrix ChIP protocol [55]. Sequences of primers specific for the gene loci under study as well as the reference primers are available upon request.

8.2.4 RNA EXTRACTION AND RT-PCR

Synthesis of cDNA was carried out according to the manufacturer's instructions (Qiagen, Valencia, CA, USA) using 1 μg of total RNA. For detection of pre-mRNA, RNA preparations were pretreated with TURBO DNase I (Ambion/Applied Biosystems) as described in the manufacturer's protocol. RT was carried out at 37°C for 1 hr.

8.2.5 CELL CULTURE

Cell lines were cultured in RPMI 1640 medium supplemented with 10% FCS, 2 mM L-glutamine and the antibiotics penicillin (50 U/mL) and streptomycin.

8.2.6 SODIUM BISULFITE CONVERSIONS

gDNA was treated with sodium bisulfite using the EZ DNA Methylation Kit (Zymo Research, Orange, CA, USA) according to the manufacturer's instructions. PCR amplification of bisulfite-treated DNA was performed using ZymoTaq DNA Polymerase (Zymo Research Corporation, Irvine, CA, USA) and conversion-specific primers targeted to the *IGF2BP1* CTCF region (forward primer: 5'-TATTTTTTAGTTGGGTTAAT-TGGTG-3', reverse primer: 5'-ATACTACCTCTCCTTCCAAAATCTC-3'). The amplified products were purified by gel electrophoresis and sequenced. Each case was scored as methylated or unmethylated, and the percentage of methylation was calculated using BiQ Analyzer software [33].

8.2.7 TAQMAN ALLELIC DISCRIMINATION ASSAYS

TaqMan allelic discrimination assays were performed according to the manufacturer's instructions with the following adjustments: cDNA from B lymphoblasts was preamplified for 14 cycles. PCR products were gel-purified and subsequently used as templates in the genotyping of samples. The specific primer sequences used are avaliable upon request.

8.2.8 IN VITRO CTCF BINDING ANALYSIS USING IMMOBILIZED TEMPLATES

Crude nuclear extract was prepared from 1×10^9 Jurkat cells grown in growth media (RPMI 1640 with 10% fetal bovine serum) according to methods described previously [56]. Biotinylated template DNA was generated by PCR amplification of the *IGF2BP1* intronic region using a biotinylated/nonbiotinylated primer combination. The specific primer sequences are available upon request. For each binding reaction, 1 pM biotinylated DNA template was coupled to 50-µg streptavidin-linked magnetic beads (Dynabeads M-280 Streptavidin; Invitrogen, Carlsbad, CA, USA). Templates immobilized to magnetic beads were washed three times in B&W buffer (5 mM Tris, pH 7.5, 0.5 mM ethylenediaminetetraacetic acid (EDTA), 1 M NaCl) and resuspended in Jurkat nuclear extract. After a 2-hr incubation at 4°C, immobilized templates were washed three times in Dignam buffer D (20 mM 4-(2-hydroxyethyl)-1-piperazineethanesulfonic acid, pH 7.9, 20% glycerol, 0.1 M KCl, 1 mM EDTA, 0.1 mM ethylene glycol tetraacetic acid, 1% Nonidet P-40, 1 mM dithiothreitol) containing protease inhibitor (P8340; Sigma, St Louis, MO, USA). To recover template-bound proteins, beads were incubated in elution buffer (5 mM Tris, pH 7.5, 0.5 mM EDTA, 1 M $NaHCO_3$) including protease inhibitors. After a 5-min incubation, the eluate was removed and transferred into a fresh tube. The presence of CTCF in the eluate was determined using standard Western blot analysis protocols.

8.3 DISCUSSION

Allele-specific expression in which one parental allele is stochastically or parent-of-origin-specifically silenced is widespread in mammalian organisms. Large-scale, allele-specific gene expression analyzes have revealed that 5% to 10% of autosomal genes show random monoallelic transcription [7]. The stability of allele-specific expression through many cell passages suggests that epigenetic modifications maintain this specific type of gene regulation throughout generations of cells. Analogously to the regulation at the imprinted *IGF2/H19* locus, the hypothesis was tested whether monoallelic binding of CTCF, a characteristic marker for the *IGF2/H19* ICR, also underlies random monoallelic expression. Using ChIP-chip analyzes, chromosomal loci is identified that are enriched in both CTCF and H3K9me3 and cross-correlated their positions with previously published lists of monoallelically expressed genes. Our data indicate that genomic loci enriched for both CTCF and H3K9me3 do not significantly correlate with monoallelically expressed genes. While this lack of correlation could be formally attributed to variations in monoallelic expression between different cell lines and types, it should be noted that the genome-wide pattern of CTCF binding is very consistent between different cell lineages [30, 38, 39]. Thus, if CTCF and H3K9me3 contribute to allele-specific expression, it should be detectable through allele-specific association of CTCF and H3K9me3. Focusing on the

IGF2BP1 gene, it was tested whether monoallelic expression in a pedigree of LCLs correlates with monoallelic binding of CTCF. Although binding of CTCF to its targets is thought to be sensitive to DNA methylation, surprisingly the cytosine residue closely flanking the CTCF target was found motif at the *IGF2BP1* gene to be consistently methylated without any effect on CTCF recruitment. Indeed, our *in vitro* analyzes of the binding requirements using immobilized templates confirmed that methylation of cytosine residues within the *IGF2BP1* sequence does not affect CTCF binding. These data are consistent with those in previous studies in which researchers found that cytosine methylation outside the CTCF core motif did not affect the binding affinity of bacterially expressed wild-type and mutant CTCF proteins [40]. This information is useful for the identification of the genomic subset of CTCF sites that might contribute to differential cell- and stage-specific expression due to their sensitivity to cytosine methylation, potentially mediating changes in large-scale chromatin organization during development and disease.

A number of studies have examined the correlation of allele-specific expression with allele-specific association of epigenetic markers [21, 41-45]. The data produced by these studies have established common signatures of imprinted alleles, including H3K9me3 and H3K4me3, providing a powerful means by which to identify novel imprinted or monoallelically expressed loci [46-48]. In contrast to the strict allele-specific association of DNA methylation and chromatin markers at imprinted genes, histone modifications at the nonimprinted, monoallelically expressed *IGF2BP1* gene do not predict the active allele. Both H3K4me3 and H3K27me3, markers characteristic of active and inactive loci, are associated with each allele, as both sequence variants of SNP rs4794017 are present in the DNA of heterozygous individuals recovered from ChIP experiments. Moreover, loading of RNA polymerase II also does not provide a reliable marker for identifying the transcribed allele. Our ChIP experiments identified both sequence variants at SNP rs4794017 within the promoter proximal region of anti-RNA polymerase II immunoprecipitated DNA. Because only one LCL in this chapter was informative for determining an association of RNA polymerase II at the *IGF2BP1* alleles, it could not be defined how frequently this type of regulation occurs within cell lineages and throughout the genome. However, other investigators have reported similar results at the *PCNA* gene. Maynard *et al.* [44] found that both *PCNA* alleles in IMR90 cells are bound by RNA polymerase II, although only one allele generates full-length mRNA. Together, these data suggest that transcription elongation not only is a general rate-limiting step in the transcription of the vast majority of genes [34, 35, 37] but also regulates the expression of a subset of monoallelically expressed genes.

The expression of *IGF2BP1* in differentiated cell types, including LCLs, is significantly lower than in embryonic stem cells. In an attempt to determine whether allele-specific expression also contributes to *IGF2BP1* regulation early in development, both gDNA and cDNA were genotyped in 11 human embryonic stem cell (hESC) lines. However, while only three hESC lines were informative (heterozygous at SNP rs11655950), all three expressed *IGF2BP1* in a biallelic manner. Although the number of available and informative hESC lines is not sufficient to clearly define a role for allele-specific elongation in early developmental stages, it is believed that it is unlikely that this mechanism is restricted to cell types with low levels of *IGF2BP1* expression. Control of transcriptional activity through promoter proximal pausing or premature termination of transcription is not restricted to specific gene classes characterized by low levels of transcriptional activity [35]. It is speculated that distinct positioning of the homologous alleles within the nuclear space and association with distinct "transcription factories" may contribute to monoallelic transcription elongation.

The *IGF2BP1* gene is highly expressed during embryonic development and is required for the regulation of mRNA stability of several genes involved in growth regulation, including the *IGF2*, β-catenin and *MYC* genes [23-25]. Consistent with its role in early developmental stages, the *IGF2BP1* gene is downregulated in differentiated cell types, and overexpression of *IGF2BP1* is known to occur in multiple human cancers, including breast, lung and colon [49-52]. Thus, changes in the level of *IGF2BP1* expression through silencing of only one allele could provide a safeguard against pathogenesis and disease.

8.4 RESULTS

8.4.1 COLOCALIZATION OF CTCF AND H3K9ME3 IN THE HUMAN GENOME

Allele-specific binding of CTCF to the ICR regulates parent-of-origin-specific expression of the *IGF2* gene and correlates with differential cytosine methylation and the presence of H3K9me3 [9-11]. A large-scale survey was carried out to identify genomic sites with chromatin markers similar to those at the ICR of the *IGF2/H19* locus. Using ChIP-chip, CTCF binding sites were identified by tiling through the nonrepetitive portion of the genome in 100-bp intervals. Genomic sites bound by CTCF were assembled on a condensed array set that tiled through 9,823 sites using overlapping probes, and replicate ChIP experiments were performed. By using conservative criteria (positive signal in three replicates; P <0.05) in this analysis, it was identified that the 8,462 loci that interact with CTCF. To identify the subset of sites that associate with both CTCF and H3K9me3, the association of these 8,462 loci with H3K9me3 was tested using the condensed DNA array set. These analyzes revealed 293 loci that are both bound by CTCF and marked by H3K9me3 (distances of CTCF and H3K9me3 peaks <500 bp). Of the 293 loci, 115 directly mapped to coding regions. Of the remaining loci (174 of 293), the majority (147 loci) were located in intergenic regions at a distance >10 kb to the nearest 5' end of known genes. Only 27 loci mapped to promoter regions. Overall, 40% of the CTCF/H3K9me3 loci mapped to intergenic regions, 51% mapped to intragenic domains and 9% mapped to promoter regions, a distribution similar to that of the 8,462 CTCF loci (44%, 51% and 10% respectively). Notably, the CTCF-regulated *IGF2/H19* locus is included in the subset of 293 loci, suggesting that our experimental approach may be useful for the identification of similarly expressed genes.

8.4.2 *IGF2BP1 ALLELES ARE STOCHASTICALLY EXPRESSED IN HUMAN B CELLS*

Genes classified as "monoallelically expressed" encompass both imprinted genes, such as the *IGF2* gene, where monoallelic expression is regulated in a parent-of-origin-specific manner, and stochastic loci, where individual alleles are randomly selected for expression independent of parental origin. In recent studies in which allele-specific transcription was assessed in several human cell lines, more than 300 (7.5%) of 4,000 human genes examined were subject to random monoallelic expression, with a majority of the latter being capable of biallelic expression [7].

To examine whether CTCF binding at sites marked by H3K9me3 is indicative of monoallelic expression, first the genomic positions of our 293 loci was compared with the list of genes expressed in a random allele-specific manner. Only a small number of genes (8 of 293 loci) were common to both the monoallelically expressed cohort described by Gimelbrant *et al.* [7] and our CTCF/H3K9me3 set of ChIP-chip binding loci.

To further examine the correlation between CTCF/H3K9me3 and monoallelic expression, 12 genes located near one of the 293 CTCF/H3K9me3 sites were selected (*DIAPH1*, *FUS1*, *PKP1*, *ARFGAP2*, *PCDHGA*, *MTHFR*, *LAIR1*, *GPR3*, *ARMET*, *NPR1*, *NHLRC1* and *IGF2BP1*) to search lymphoblastoid cell lines (LCLs) derived from a pedigree from the Center d'Etude du Polymorphisme Humaine (CEPH) for SNPs in exonic and 3'-UTR regions. The monoclonality of LCLs was confirmed by analysis of their immunoglobulin heavy chain (IgH) gene rearrangement [22]. Sequencing of genomic DNA (gDNA) and cDNA of LCLs identified the insulin-like growth factor binding protein gene *IGF2BP1* as the only candidate gene expressed from only one allele. *IGF2BP1* is an RNA-binding protein that regulates transcript stability and translation of the imprinted *IGF2* gene [23]. In addition, *IGF2BP1* binds to H19, MYC and β-TrCP1 mRNA to regulate message half-life, localization and translation of RNA, suggesting that the regulation of *IGF2BP1* expression may affect disease and development [24, 25]. It was focused on *IGF2BP1* to examine the contribution of CTCF and H3K9me3 markers colocalized at intron 5 to allele-specific expression (Figure 1).

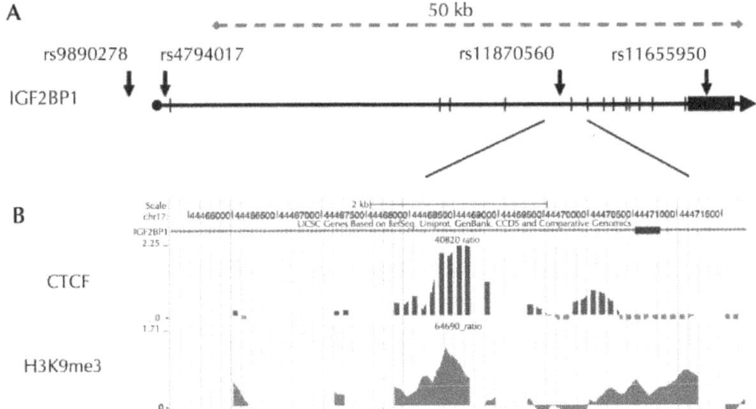

FIGURE 1 Colocalization of CTCF and H3K9me3 at the *IGF2BP1* locus**.** Array-based chromatin immunoprecipitation (ChIP-chip) data for both CTCF and histone H3 trimethylated at lysine 9 (H3K9me3) identify candidate loci for analysis of monoallelic expression. **(A)** Depiction of the *IGF2BP1* gene with specific SNPs examined in this chapter (arrows). **(B)** Close-up portion of the locus with tracks for CTCF enrichment (top track) and H3K9me3 association (bottom track) near SNP site rs11870560. The ChIP-chip data are displayed using the UCSC Genome Browser. DNA derived from CTCF ChIP experiments was analyzed by using microarrays with hybridization probes spaced 100 bp apart. The higher resolution of the H3K9me3 ChIP-chip data is due to the use of condensed array sets that tiled through all of the CTCF-positive regions with probes overlapping each other by 12 nt.

Sequencing of gDNA identified 10 individuals that were heterozygous at SNP rs11655950 in the 3'-UTR of *IGF2BP1*. All heterozygous SNPs were subsequently typed in cDNA. A comparison of the transcriptome-derived genotypes to genomic genotypes indicated that six individuals expressed *IGF2BP1* primarily from only one allele. In contrast, four individuals were found to express both *IGF2BP1* alleles. SNP determination for genomic and cDNA for CEPH family 1331 was confirmed by allelic discrimination assays based on fluorogenic probes (TaqMan allelic discrimination assay; Applied Biosystems, Foster City, CA, USA), which yielded identical results. The TaqMan allelic discrimination assay, a real-time PCR based approach, yields a scatterplot of genotypes capable of quantitatively detecting a range of 1:1 and 1:5 ratios of individual alleles in DNA mixtures at SNP rs11655950. Individuals GM7033 and GM6989 were found to express the paternally inherited *IGF2BP1* allele, while GM7030 and GM7005 were found to express the maternally inherited allele (Figure 2). Individuals GM7007 and GM7016 also exhibited monoallelic expression of *IGF2BP1*, but could not identify the mode of expression because of the limited pedigree. These data indicate that monoallelic expression at the *IGF2BP1* gene locus is not determined by parent-of-origin markings; instead, it is defined by stochastic choice.

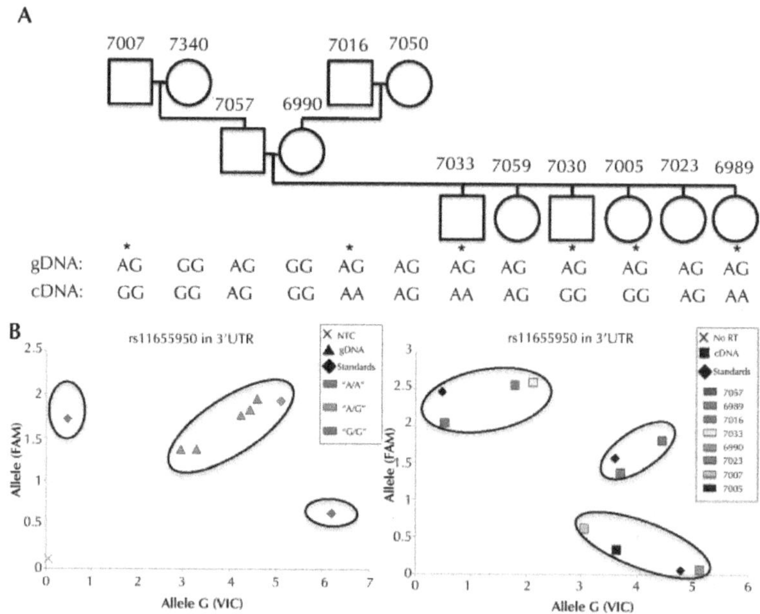

FIGURE 2 Analysis of allele-specific *IGF2BP1* expression. Comparative analysis of sequence variations in B lymphoblasts of the CEPH pedigree family 1331 reveals monoallelic expression of the *IGF2BP1* gene. **(A)** Pedigree analysis was carried out for the SNP site rs11655950 located in the 3'-UTR of the *IGF2BP1* gene. Each individual is shown with CEPH family identification, sample identification and genetic information (SNP genomic DNA (gDNA) genotype- or transcript-derived genotype). Individuals with monoallelic *IGF2BP1* gene expression are indicated by asterisks. If the individual is homozygous at the SNP, allele-specific expression cannot be defined. **(B)** Left: Genotyping results at rs11655950 with gDNA from members of CEPH family 1331. gDNA was analyzed using the TaqMan SNP Genotyping Assay. This assay discriminates between sequence variants using two allele-specific probes carrying two different fluorophores, VIC and FAM. Individuals coded in red and green represent cell lines that are homozygous for alleles A and G, respectively. Orange-labeled individuals contain both A and G alleles at SNP rs11655950 and represent informative cell lines used for further analysis of monoallelic expression. Diamonds indicate cDNA samples, and black × indicates averaged triplicates of a no-template control (NTC) near the origin of the graph. Right: Genotyping results of transcript-derived cDNA from heterozygous B lymphoblasts. Individuals are color-coded in the figure key. No-RT controls (No RT) from cDNA synthesis are shown near the origin of the graph and are indicated by a black X. Control samples (standards) of stem cell lines previously genotyped as homozygous AA, heterozygous AG and homozygous GG were plotted and are indicated by diamonds.

8.4.3 CTCF BINDS TO ITS TARGET MOTIF AT THE *IGF2BP1* LOCUS INDEPENDENTLY OF DNA METHYLATION

Binding of CTCF to its target motifs at both the human and mouse ICR of the *IGF2/H19* locus is sensitive to DNA methylation [10, 26]. To test whether monoallelic expression of *IGF2BP1* in some individuals is also regulated by monoallelic DNA methylation of CTCF binding motifs, a role for CpG methylation and allele-specific binding of CTCF was examined at this locus.

To precisely determine the DNA sequence required for CTCF binding at the *IGF2BP1* locus, it was searched for potential motifs using SOMBRERO [27], a *de novo* motif-finding algorithm that uses multiple self-organizing maps (SOM) to cluster sequences of a specific length (reads) from a set of input sequences (such as enriched genomic loci identified by ChIP-chip experiments). Motif alignment using STAMP [28] and comparison to the JASPAR transcription factor database [29] identified a distinct cohort of 68 motif models, all of which were identical to the canonical CTCF motif previously reported [30]. The clustered reads associated with all 68 motif models were mapped back to sequences enriched in our ChIP-chip analysis and were displayed using the UCSC Genome Browser. Using this approach, 28,713 peaks were identified, each composed of multiple overlapping reads, within the original 8,462 ChIP-chip loci. Using a strategy similar to that used to study ChIP-seq clustering [31], our frequency analysis of these peak heights yielded a bimodal distribution with an evident power law at low peak heights deviating to a clear excess in the numbers of peaks with heights >10. Consequently the peak populations partitioned into low-confidence and high-confidence groups using the peak height threshold of 10.

Using this approach, three potential motifs were identified (X, Y and Z) (Figure 3) within the 350-bp region of the *IGF2BP1* gene locus enriched in our ChIP-chip experiments. Two of the putative binding sites, Y and Z, accumulated a significant number of matches to motif models. However, only one of the three putative CTCF binding sites belongs to the group of high-confidence binding sites (site Y) (Figure 3). In support of our *in silico* analysis of CTCF binding, previously published high-resolution ChIP-seq data on CTCF binding revealed enrichment of sequences surrounding motifs Y and Z, suggesting that either one or both motifs is required for CTCF recruitment.

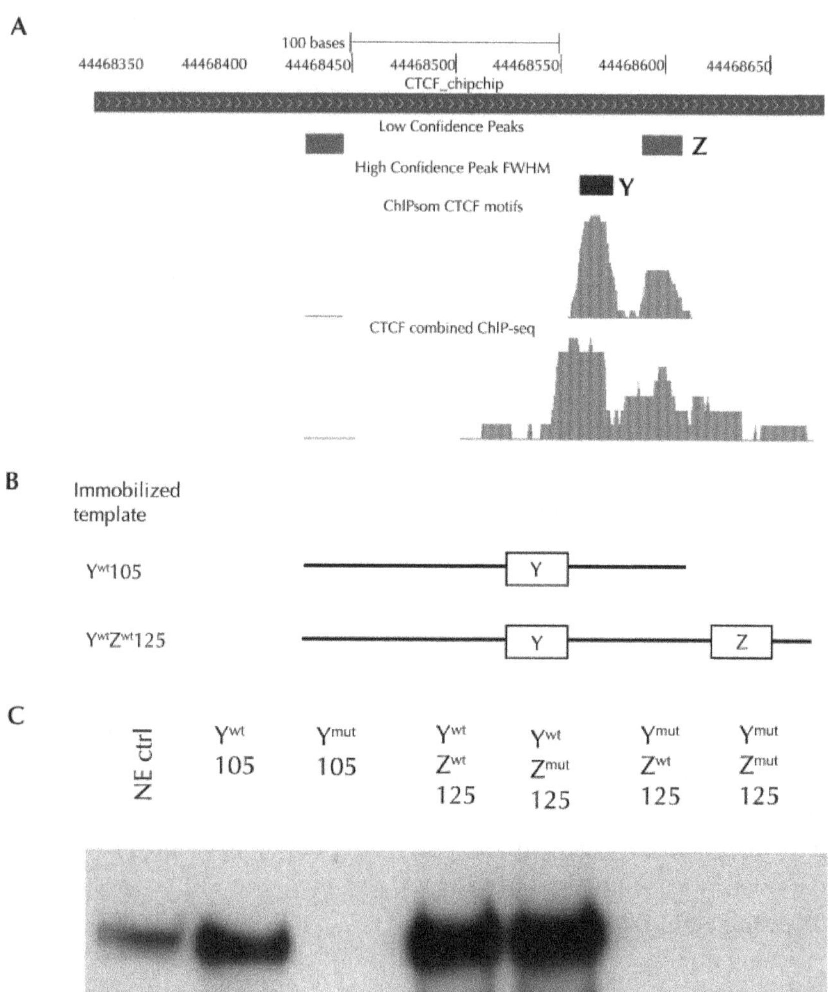

FIGURE 3 Functional CTCF sequence motifs at the intronic region of the *IGF2BP1* gene. **(A)** UCSC Genome Browser display of relative positions of high- and low-confidence CTCF target motifs, ChIP-chip, ChIP sequencing (ChIP-seq) and ChIP self-organizing maps results. **(B)** Y^{wt} 105-bp and $Y^{wt}Z^{wt}$ 125-bp templates employed in the immobilized template assay. **(C)** Western blot analysis of CTCF recruitment to Y^{wt} 105-bp and $Y^{wt}Z^{wt}$ 125-bp templates containing combinations of wild-type and mutated CTCF target sequences. Motif Y is sufficient for recruitment of CTCF.

To further define the contribution of motifs Y and Z to CTCF binding, their ability was measured to recruit CTCF *in vitro* using immobilized template assays. Wild-type and mutant DNA templates containing either one or both motifs were linked to magnetic beads, incubated with nuclear extract, washed and tested for association with CTCF by performing Western blot analysis. A 105-bp template containing the wild-type *IGF2BP1* intronic sequence efficiently recruited CTCF (Y^{wt} 105-bp template). In contrast, CTCF binding was severely reduced when the putative CTCF motif Y was mutated by four base substitutions. To test the contribution of the adjacent motif Z to CTCF binding at the *IGF2BP1* locus, several 125-bp DNA templates were generated that encompassed both CTCF target motifs. Targeted mutations at specific positions of motif Y and/or motif Z were introduced to test the contribution of each motif to recruitment of CTCF. The 125-bp template recruited CTCF more efficiently than the 105-bp template. However, motif Z does not contribute to CTCF recruitment, since targeted mutations in motif Z do not influence the level of CTCF binding. Consistent with this notion, CTCF binding is undetectable in the absence of a wild-type motif Y.

CTCF binding site Y at the *IGF2BP1* gene contains a single CpG residue adjacent to the 14-bp core sequence of CTCF (Figure 4A). To establish whether binding of CTCF to Y^{wt} is inhibited by cytosine methylation, Y^{wt} 105-bp immobilized templates was tested after *in vitro* methylation of cytosine residues by CpG methyltransferase M.Sssl. For comparison, CTCF motifs containing a higher CpG content were examined, including site A of the MYC gene [32] as well as the B1 sequence of the ICR of the human *IGF2/H19* locus [10]. Cytosine methylation at the human B1 sequence is known to inhibit binding of CTCF. Consistent with this, recruitment of CTCF *in vitro* to immobilized templates containing the B1 sequence or the MYC site A is highly sensitive to DNA methylation (Figure 4B). In contrast, CpG methylation of the Y^{wt} motif has no effect on CTCF recruitment. Replacement of the Y^{wt} core motif by the CTCF-binding sites of the chicken FII insulator element yields similar results. However, CTCF binding becomes sensitive to CpG methylation upon modification of the core motif to the mouse R3 sequence, a homologue of the human B1 sequence. In combination, despite the presence of a methylable CpG residue, binding of CTCF to the Y^{wt} sequence of the *IGF2BP1* gene *in vitro* is not sensitive to CpG methylation.

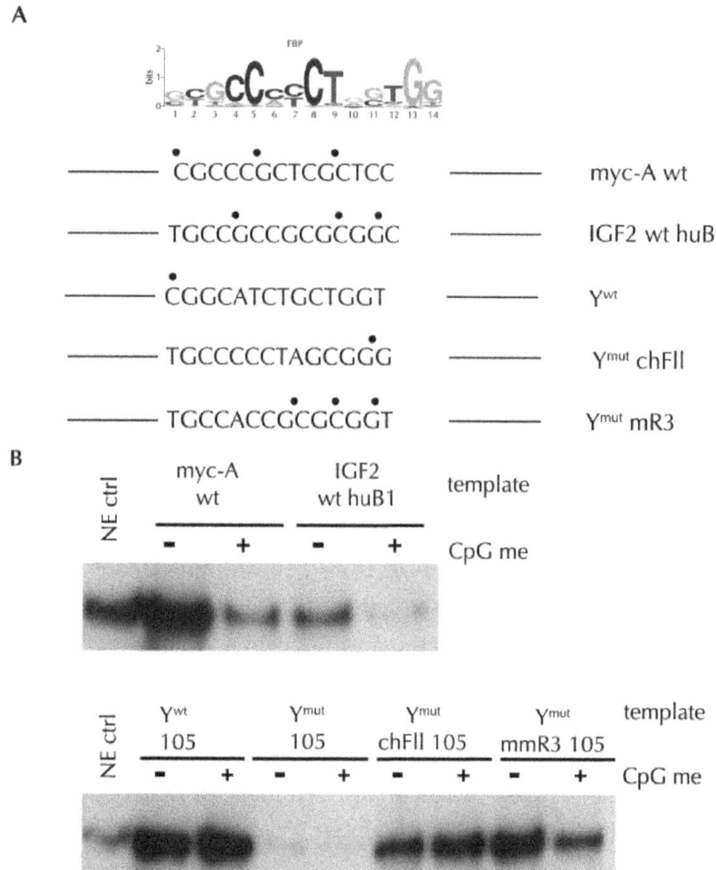

FIGURE 4 Cytosine methylation of the CTCF core motif Y does not influence binding of CTCF. **(A)** CTCF motifs used in the context of the 105-bp immobilized template derived from the intronic region of the *IGF2BP1* gene are shown. The position frequency matrix of the CTCF target motif is shown at the top. Only the sense strand of the motifs is shown. CpG residues are indicated by filled black circles. Myc-A, IGF2 huB1 and Y^{wt} are CTCF target sequences derived from *MYC*, *IGF2* and *IGF2BP1* gene loci. Y^{mut} chFII and Y^{mut} mmR3 contain the CTCF target sequence of the chicken HS4 insulator [57] and the CTCF target region of the mouse imprinting control region R3 [10]. **(B)** Top: control experiments revealed the sensitivity of CTCF binding to DNA methylation (CpG me) at the myc-A and IGF2 huB1 templates. Bottom: methylation of the 105-bp Y^{wt} template did not affect the recruitment of CTCF. While methylated chicken FII CTCF target sites efficiently recruited CTCF, CpG methylation of the mouse R3 sequence decreased the binding of CTCF.

To confirm that our *in vitro* characterization of CTCF binding accurately reflected the *in vivo* association of CTCF with the *IGF2BP1* locus, the methylation status of the CTCF motif and adjacent CpG residues in the *IGF2BP1* intronic region in both biallelically (GM7057) were evaluated and monoallelically (GM6989) expressing cells by using bisulfite sequencing (Figure 5). The methylation levels were calculated using BiQ Analyzer software [33]. Our data reveal that the CpG residue at the 5' end of the CTCF binding motif Y is invariably methylated. In addition, other methylable residues in this region exhibited some degree of DNA methylation. To further confirm binding of CTCF to methylated *IGF2BP1* intronic sequences, DNA derived from immunoprecipitates of ChIP experiments were bisulfite-sequenced with CTCF antibodies. As a control, the *IGF2BP1* region derived from anti-H3K9me3 ChIP experiments was bisulfite-sequenced. The results confirmed our *in vitro* finding that demonstrated an association of CTCF with a methylated motif.

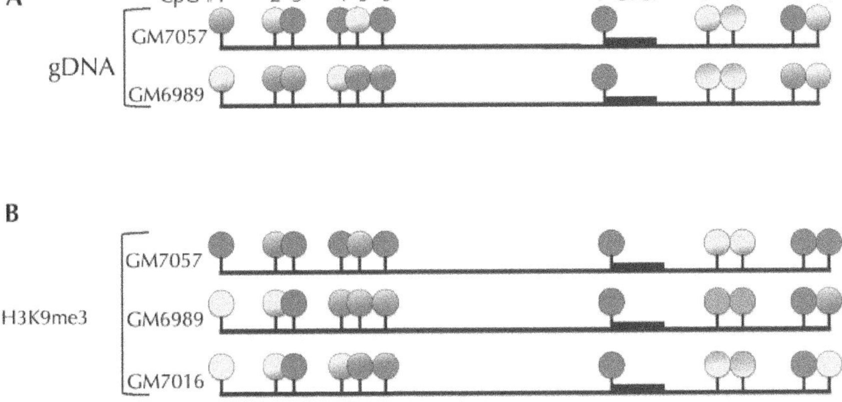

FIGURE 5 *(Continued)*

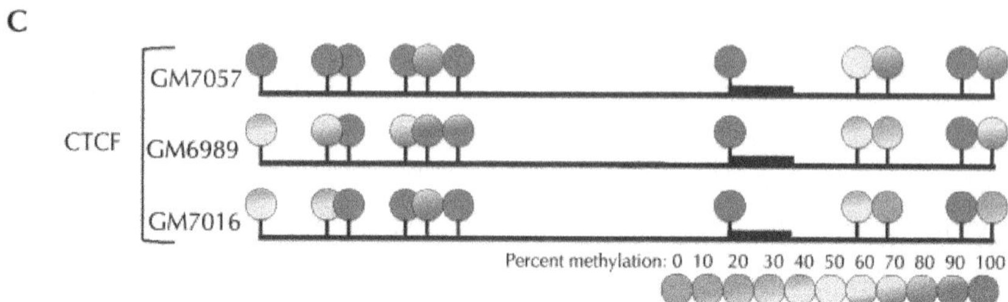

FIGURE 5 DNA methylation analysis of the *IGF2BP1* CTCF binding region. Analysis of DNA methylation with bisulfite sequencing at the intronic CTCF binding region of the *IGF2BP1* gene is shown. **(A)** The percentage of methylation of CpG sites in gDNA derived from cell lines that express *IGF2BP1* from only one allele (GM7016, GM6989) or from both alleles (GM7057) is shown. The CpG residue located within the CTCF binding motif is invariably methylated and is indicated by the thick black bar located adjacent to CpG site 7 (indicated by asterisks). **(B)** The percentage of methylation at each CpG site of the *IGF2BP1* CTCF site in DNA samples recovered from anti-H3K9me3 ChIP. **(C)** The percentage of methylation at each CpG site of the *IGF2BP1* CTCF site in DNA samples recovered from anti-CTCF ChIP experiments. The level of DNA methylation is represented according to the heat map keys located at the bottom of the figure.

8.4.4 CTCF AND H3K9ME3 COLOCALIZE AT BOTH THE MATERNAL AND PATERNAL IGF2BP1 ALLELES

Consistent methylation of the CTCF-binding motif in *IGF2BP1* indicated that DNA methylation is not allele-specific. To directly determine whether CTCF is bound monoallelically, the allele-specific association of both CTCF and H3K9me3 was determined by sequencing DNA recovered from ChIP experiments. First the informative cell lines were identified by genotyping individuals from CEPH pedigree 1331 at SNP sites located close to the CTCF binding site. Cell lines derived from both monoallelically (GM7016 and GM6989) and biallelically (GM7057) expressing individuals were heterozygous at SNP site rs11870560 at the CTCF site (Figure 6A). First the allelic discrimination assay was applied to serial dilutions of known homozygous of the two possible alleles to test its ability to quantitatively assess the contribution of each allele in a DNA mixture. This assay provides quantitative results with high sensitivity and reproducibility within a ten-fold range of DNA concentrations, thus making it a useful tool for allelic discrimination of immunoprecipitated DNA. Two monoallelically (GM7016 and GM6989) were used and one biallelically (GM7057) expressing cell lines to genotype DNA recovered from ChIP assays using either anti-CTCF or anti-H3K9me3 antibodies. Each analysis was performed in triplicate. Equal proportions of the two sequence variants were detected in DNA derived from ChIP assays with either H3K9me3 or CTCF antibodies, indicating that CTCF associates with both the maternal and paternal alleles (Figure 6B). Thus, monoallelic expression of the *IGF2BP1* gene is not mediated through monoallelic binding of CTCF.

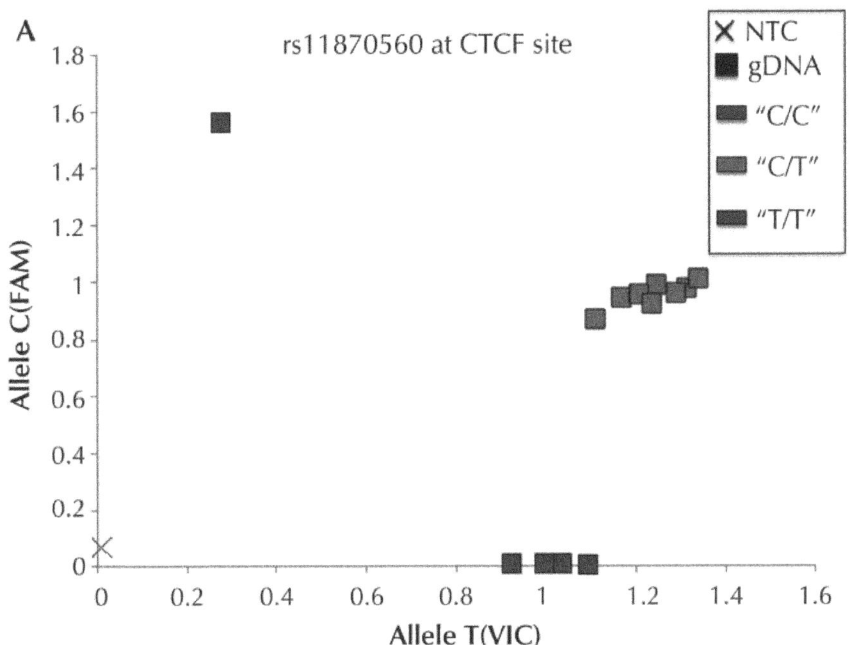

FIGURE 6 *(Continued)*

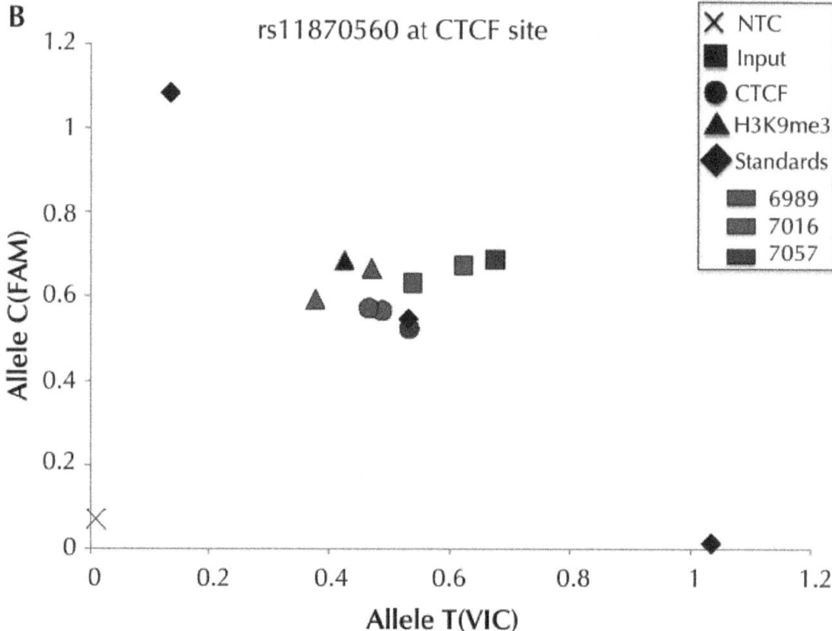

FIGURE 6 Allelic Specificity of CTCF and H3K9me3. Informative ChIP templates were analyzed using the TaqMan allelic discrimination assay to address the allelic association of CTCF and H3K9me3. **(A)** Genotyping results at rs11870560 identify informative cell lines useful for the detection of allele-specific association of CTCF and H3K9me3. gDNA obtained from monoallelic and biallelic cell lines were genotyped using the TaqMan allelic discrimination assay. Squares represent gDNA samples and are coded in red and green to represent cell lines that are homozygous for allele C and allele T, respectively. Orange indicates heterozygous individuals. Averaged triplicate of a no-template control (NTC) is shown near the origin of the graph. **(B)** Genotyping at SNP rs11870560 with DNA templates recovered from ChIP experiments was used to identify the enrichment of the two alleles with either CTCF (circle) or H3K9me3 (triangle). Each color shown in the figure key represents a lymphoblastoid cell line (LCL) derived from an individual of the pedigree, while the shape represents the source of each sample (for example, squares signify input samples, while circles and triangles indicate ChIP samples obtained with CTCF and H3K9me3 antibodies, respectively). Immunoprecipitated templates were generated using the ChIP protocol described in Materials and Methods. Both monoallelic and biallelic cell lines indicate biallelic distribution of both CTCF and H3K9me3. Diamonds indicate control LCL samples (standards) previously genotyped as homozygous CC, heterozygous CT and homozygous TT.

8.4.5 THE IGF2BP1 PROMOTER ASSOCIATES WITH BOTH ACTIVE AND SILENT HISTONE MODIFICATIONS IN B CELLS

To define alternative mechanisms responsible for random monoallelic expression of *IGF2BP1*, it was sought to identify markers that distinguish the active and inactive alleles. K27-trimethylated and K4-trimethylated histone H3, respectively, mark transcriptionally silent and active chromatin. The relative enrichment of these two histone markers were determined at the *IGF2BP1* promoter for each allele in both monoallelically and biallelically expressing cell lines using ChIP with anti-H3K-4me3 and anti-H3K27me3 antibodies. Both H3K4me3 and H3K27me3 were detected at the *IGF2BP1* gene promoter. To determine whether any of the histone modifications selectively associates with either allele, it was again searched for informative sequence SNPs at the *IGF2BP1* promoter region in the CEPH pedigree. Cell lines derived from individuals GM6989 (monoallelically expressing cell line) and 7057 (biallelically expressing cell line) were heterozygous at SNP rs9890278 located upstream of the transcription initiation site, whereas GM7007 (monoallelically expressing cell line) was heterozygous for SNP rs4794017 located 1 kb downstream of the transcription initiation site. To address whether active and silent alleles in these cell lines are distinguished by specific histone markers, SNPs rs9890278 and rs4794017 were sequenced in gDNA recovered from ChIP experiments using anti-H3K4me3 and anti-H3K27me3 antibodies. The results revealed that both H3K4me3 and H3K27me3 are detected on both alleles in a bivalent fashion (Figure 7). In combination, our results indicate that both active and silent histone markers (H3K4me3 and H3K27me3) coexist in the promoter region of both *IGF2BP1* alleles in monoallelically as well as biallelically expressing cell lines. These data indicate that allele-specific expression of *IGF2BP1* cannot be explained by differential association of active and silent histone markers.

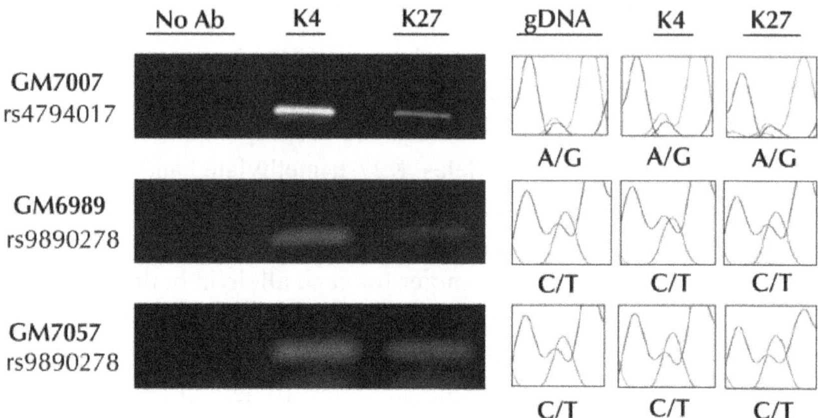

FIGURE 7 *IGF2BP1* promoter region is enriched with activating and silencing chromatin modifications. DNA recovered from ChIP experiments using anti-H3K4me3, anti-H3K27me3 and RNA polymerase II ChIP templates was genotyped by sequencing the *IGF2BP1* promoter region containing sequence variant rs4794017 or rs9890278. Left: Enrichment of H3K4me3 (K4) and H3K27me3 (K27) in monoallelically (GM7007, GM6989), and biallelically (GM7057) expressing cell lines. The positions of informative SNPs rs479017 and rs9890278 are shown in Figure 1. Both activating and silencing marks are significantly enriched. Right: Sequences enriched by ChIP were excised and sequenced. The results show an association of both alleles with active and silent histone modifications at the *IGF2BP1* promoter region independent of transcriptional status.

8.4.6 SILENCING OF THE INACTIVE IGF2BP1 ALLELE BY INHIBITION OF RNA POLYMERASE II ELONGATION

Monoallelic expression of *IGF2BP1* cannot be attributed solely to selective activation or silencing of one allele through histone modifications, since H3K4me3 as well as H3K27me3 are detected at both alleles. H3K-4me3 is typically associated with transcriptionally active alleles, raising the question whether allele-specific transcription elongation or RNA processing accounts for monoallelic expression of the *IGF2BP1* gene. To address this hypothesis, it was again searched mono- and biallelically expressing cell lines for sequence SNPs near the site of transcription initiation at the *IGF2BP1* promoter. Within CEPH pedigree 1331, only line GM7007 contained a heterozygous genotype at SNP site rs4794017 located within intron 1, 1 kb downstream of the transcription initiation site. RNA polymerase II ChIP were performed on chromatin prepared from this monoallelically expressing line. Quantitative real-time PCR analyzes revealed enrichment of *IGF2BP1* promoter sequences similar to the enrichment observed at the MYC promoter. Immunoprecipitated DNA was PCR-amplified and sequenced (Figure 8A). Identification of both sequence variants at rs4794017 in DNA recovered from ChIP experiments indicates that RNA polymerase II is associated with both *IGF2BP1* alleles, which is consistent with the presence of H3K4me3 at the promoter of both alleles.

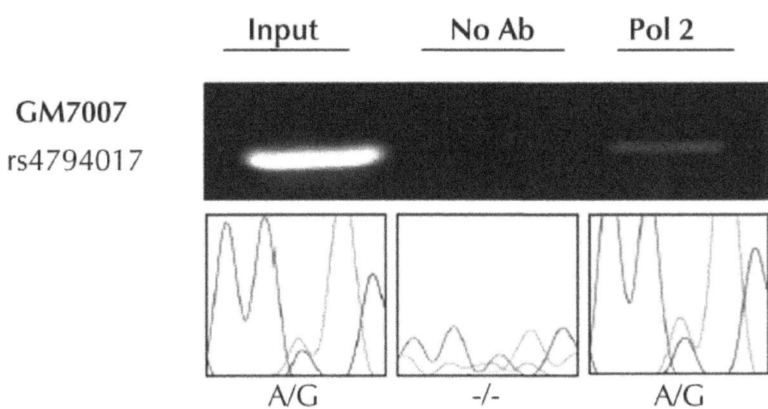

FIGURE 8 *(Continued)*

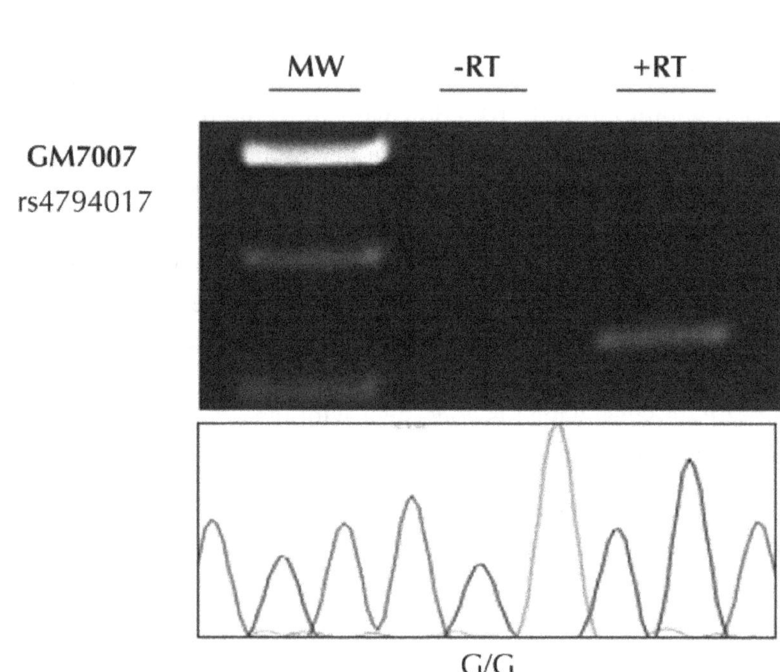

FIGURE 8 RNA polymerase II associates with both alleles in a monoallelically expressing cell line. **(A)** Recruitment of RNA polymerase II to the *IGF2BP1* promoter was examined by ChIP in monoallelically expressing GM7007 cells. DNA recovered from chromatin that had been immunoprecipitated with anti-RNA polymerase II antibodies (Pol2) was amplified and sequenced for allelic association. Sequencing results (bottom) reveal that both alleles of the monoallelically expressing cell line GM7007 associate with RNA polymerase II near SNP site rs4794017. In contrast, sequencing of DNA from "no antibody" ChIP reactions failed to produce sequence reads. **(B)** Allele specificity of precursor mRNA was determined by sequencing of cDNA prepared from total RNA of GM7007 cells. RNA had been extensively pretreated with DNase I to eliminate gDNA prior to reverse transcription by RT. Subsequently, cDNA samples were amplified using primers flanking rs4794017. In the absence of RT (-RT), no amplification products were oberved. +RT amplicons were gel-purified and sequenced. Bottom: Sequence traces at the heterozygous SNP site rs4794017 located 1 kb downstream of the transcription initiation site in cDNA of GM7007 indicate a single allele.

These data suggest that allele specificity of transcription is achieved after recruitment of RNA polymerase to both alleles, such as through transcriptional pausing and/or selective RNA processing. A major rate-limiting step in transcription elongation is pausing of RNA polymerase

II in the promoter proximal region immediately downstream of the transcription initiation site [34-37]. The 5' portion of the *IGF2BP1* gene were sequenced of all monoallelically expressing cell lines to identify sequence variants that would be useful for allelic identification of promoter proximal regions occupied by RNA polymerase II or for the determination of the allelic origin of unspliced, precursor pre-mRNA transcripts. Since no additional informative sequence variants were identified, it was focused on the detection and sequencing of pre-mRNA transcripts about 1 kb downstream of the transcription initiation site in GM7007. Using the informative SNPs located within intron 1 of this gene, nascent unspliced RNA with primers designed were targeted to amplify a region containing SNP site rs4794017. To avoid detection of gDNA in RNA samples, DNA was efficiently removed by treatment with an engineered, highly active form of DNase I (TURBO DNase I; Applied Biosystems/Ambion, Austin, TX, USA). This protocol allowed detection of pre-mRNA free of gDNA contamination (Figure 8B). Sequencing of amplified *IGF2BP1* pre-cDNA revealed only one of the two sequence variants at SNP rs4794017, indicating that pre-mRNA transcripts are transcribed from only one allele despite the presence of RNA polymerase II on both alleles. Thus, our data indicate that monoallelic expression of the *IGF2BP1* gene is regulated through allele-specific transcriptional elongation prior to SNP site rs4794017, located approximately 600 bp downstream of the first intron splice site.

8.5 CONCLUSION

Allele-specific gene expression is common in the human genome and is thought to contribute to phenotypic variation. The allele-specific association of CTCF, H3K9me3 and DNA methylation is a characteristic marker of imprinted gene expression at the *IGF2/H19* locus, raising the question whether these epigenetic markers are useful for identifying both imprinted and random monoallelically expressed genes throughout the genome. In this chapter, it has been demonstrated that colocalization of CTCF and H3K9me3 does not represent a reliable chromatin signature indicative of monoallelic expression. In addition, it is concluded that allele-specific binding of CTCF requires methylation of very specific cytosine residues within the target motif, effectively limiting the number of CTCF binding sites potentially affected by allele-specific binding. In addition, the active and inactive alleles of random monoallelically expressed genes do not necessarily correlate with active or inactive histone markers. Remarkably, the selection of individual alleles for expression at the *IGF2BP1* locus occurs during early stages of transcription elongation.

KEYWORDS

- **Alleles**
- **Homozygous**
- **Immunoprecipitation**
- **Lysine 9**
- **Oligonucleotide**

ACKNOWLEDGMENT

We thank Carol Ware, Angel Nelson, Jennifer Hesson and Chris Cavanaugh at the Institute for Stem Cell and Regenerative Medicine for providing us with the stem cells used in this chapter. This chapter was supported by grants from the National Institutes of Health (National Cancer Institute grant CA109597), the US Department of Defense (grant W81X-WH-08-1-0636) and the John H. Tietze Foundation (to AK) and by a Mary Gates Endowment scholarship (to BJT).

COMPETING INTERESTS

The authors declare that they have no competing interests.

AUTHORS' CONTRIBUTIONS

AK conceived of and designed the chapter. BJT, EDR and AK performed the experiments. PÓB, AAG, JMG and NK provided bioinformatics support and carried out the statistical analyzes. PW and KB contributed the samples. BJT, PW, AAG and AK drafted the chapter. All authors read and approved the final manuscript.

REFERENCES

1. Delaval, K. and Feil, R. Epigenetic regulation of mammalian genomic imprinting. *Curr Opin Genet Dev* **14,** 188–195 (2004).
2. Ferguson-Smith, A. C. and Surani, M. A. Imprinting and the epigenetic asymmetry between parental genomes. *Science* **293,** 1086–1089 (2001).
3. Gregg, C., Zhang, J., Weissbourd, B., Luo, S., Schroth, G. P., Haig, D., and Dulac, C. High-resolution analysis of parent-of-origin allelic expression in the mouse brain. *Science* **329,** 643–648 (2010).
4. Chess, A., Simon, I., Cedar, H., and Axel, R. Allelic inactivation regulates olfactory receptor gene expression. *Cell* **78,** 823–834 (1994).
5. Bix, M. and Locksley, R. M. Independent and epigenetic regulation of the interleukin-4 alleles in $CD4^+$ T cells. *Science* **281,** 1352–1354 (1998).
6. Holländer, G. A., Zuklys, S., Morel, C., Mizoguchi, E., Mobisson, K., Simpson, S., Terhorst, C., Wishart, W., Golan, D. E., Bhan, A. K., and Burakoff, S. J. Monoallelic expression of the interleukin-2 locus. *Science* **279,** 2118–2121 (1998).
7. Gimelbrant, A., Hutchinson, J. N., Thompson, B. R., and Chess, A. Widespread monoallelic expression on human autosomes. *Science* **318,** 1136–1140 (2007).
8. Reik, W. and Walter, J. Genomic imprinting: parental influence on the genome. *Nat Rev Genet* **2,** 21–32 (2001).
9. Bell, A. C. and Felsenfeld, G. Methylation of a CTCF-dependent boundary controls imprinted expression of the *Igf2* gene. *Nature* **405,** 482–485 (2000).
10. Hark, A. T., Schoenherr, C. J., Katz, D. J., Ingram, R. S., Levorse, J. M., and Tilghman, S. M. CTCF mediates methylation-sensitive enhancer-blocking activity at the *H19/Igf2* locus. *Nature* **405,** 486–489 (2000).
11. Kanduri, C., Pant, V., Loukinov, D., Pugacheva, E, Qi, C. F., Wolffe, A., Ohlsson, R., and Lobanenkov, V. V. Functional association of CTCF with the insulator upstream of the *H19* gene is parent of origin-specific and methylation-sensitive. *Curr Biol* **10,** 853–856 (2000).
12. Phillips, J. E. and Corces, V. G. CTCF: master weaver of the genome. *Cell* **137,** 1194–1211 (2009).
13. Parelho, V., Hadjur, S., Spivakov, M., Leleu, M., Sauer, S., Gregson, H. C., Jarmuz, A., Canzonetta, C., Webster, Z., Nesterova, T., Cobb, B. S., Yokomori, K., Dillon, N.,

Aragon, L., Fisher, A. G., and Merkenschlager, M. Cohesins functionally associate with CTCF on mammalian chromosome arms. *Cell* **132,** 422–433 (2008).

14. Rubio, E. D., Reiss, D. J., Welcsh, P. L., Disteche, C. M., Filippova, G. N., Baliga, N. S., Aebersold, R., Ranish, J. A., and Krumm, A. CTCF physically links cohesin to chromatin. *Proc Natl Acad Sci USA* **105,** 8309–8314 (2008).
15. Stedman, W., Kang, H., Lin, S., Kissil, J. L., Bartolomei, M. S., and Lieberman, P. M. Cohesins localize with CTCF at the KSHV latency control region and at cellular c-myc and *H19/Igf2* insulators. *EMBO J* **27,** 654–666 (2008).
16. Wendt, K. S., Yoshida, K., Itoh, T., Bando, M., Koch, B., Schirghuber, E., Tsutsumi, S., Nagae, G., Ishihara, K., Mishiro, T., Yahata, K., Imamoto, F., Aburatani, H., Nakao, M., Imamoto, N., Maeshima, K., Shirahige, K., and Peters, J. M. Cohesin mediates transcriptional insulation by CCCTC-binding factor. *Nature* **451,** 796–801 (2008).
17. Hadjur, S., Williams, L. M., Ryan, N. K., Cobb, B. S., Sexton, T., Fraser, P., Fisher, A. G., and Merkenschlager, M. Cohesins form chromosomal *cis*-interactions at the developmentally regulated *IFNG* locus. *Nature* **460,** 410–413 (2009).
18. Hou, C., Dale, R., Dean, A. Cell type specificity of chromatin organization mediated by CTCF and cohesin. *Proc Natl Acad Sci USA* **107,** 3651–3656 (2010).
19. Nativio, R., Wendt, K. S., Ito, Y., Huddleston, J. E., Uribe-Lewis, S., Woodfine, K., Krueger, C., Reik, W., Peters, J. M., and Murrell, A. Cohesin is required for higher-order chromatin conformation at the imprinted *IGF2-H19* locus. *PLoS Genet* **5,** e1000739 (2009).
20. Kacem, S. and Feil, R. Chromatin mechanisms in genomic imprinting. *Mamm Genome* **20,** 544–556 (2009).
21. Wen, B., Wu, H., Bjornsson, H., Green, R. D., Irizarry, R., and Feinberg, A. P. Overlapping euchromatin/heterochromatin-associated marks are enriched in imprinted gene regions and predict allele-specific modification. *Genome Res* **18,** 1806–1813 (2008).
22. Diss, T. C., Pan, L., Peng, H., Wotherspoon, A. C., and Isaacson, P. G. Sources of DNA for detecting B cell monoclonality using PCR. *J Clin Pathol* **47,** 493–496 (1994).
23. Nielsen, J., Christiansen, J., Lykke-Andersen, J., Johnsen, A. H., Wewer, U. M., and Nielsen, F. C. A family of insulin-like growth factor II mRNA-binding proteins represses translation in late development. *Mol Cell Biol* **19,** 1262–1270 (1999).
24. Noubissi, F. K., Elcheva, I., Bhatia, N., Shakoori, A., Ougolkov, A., Liu, J., Minamoto, T., Ross, J., Fuchs, S. Y., and Spiegelman, V. S. CRD-BP mediates stabilization of *βTrCP1* and c-*myc* mRNA in response to β-catenin signalling. *Nature* **441,** 898–901 (2006).
25. Runge, S., Nielsen, F. C., Nielsen, J., Lykke-Andersen, J., Wewer, U. M., and Christiansen, J H19 RNA binds four molecules of insulin-like growth factor II mRNA-binding protein. *J Biol Chem* **275,** 29562–29569 (2000).
26. Engel, N., Thorvaldsen, J. L., and Bartolomei, M. S. CTCF binding sites promote transcription initiation and prevent DNA methylation on the maternal allele at the imprinted *H19/Igf2* locus. *Hum Mol Genet* **15,** 2945–2954 (2006).
27. Mahony, S., Hendrix, D., Golden, A., Smith, T. J., and Rokhsar, D. S. Transcription factor binding site identification using the self-organizing map. *Bioinformatics* **21,** 1807–1814 (2005).
28. Mahony, S. and Benos, P. V. STAMP: a web tool for exploring DNA-binding motif similarities. *Nucleic Acids Res* (35 Web server), W253–W258 (2007).

29. Sandelin, A., Alkema, W., Engström, P., Wasserman, W. W., and Lenhard, B. JASPAR: an open-access database for eukaryotic transcription factor binding profiles. *Nucleic Acids Res* (32 Database), D91–D94 (2004).
30. Kim, T. H., Abdullaev, Z. K., Smith, A. D., Ching, K. A., Loukinov, D. I., Green, R. D., Zhang, M. Q., Lobanenkov, V. V., and Ren, B. Analysis of the vertebrate insulator protein CTCF-binding sites in the human genome. *Cell* **128,** 1231–1245 (2007).
31. Zhang, Z. D., Rozowsky, J., Snyder, M., Chang, J, and Gerstein, M. Modeling ChIP sequencing in silico with applications. *PLoS Comput Biol* **4,** e1000158 (2008).
32. Gombert, W. M. and Krumm, A. Targeted deletion of multiple CTCF-binding elements in the human C-MYC gene reveals a requirement for CTCF in C-MYC expression. *PLoS One* **4,** e6109 (2009).
33. Bock, C., Reither, S., Mikeska, T., Paulsen, M., Walter, J., and Lengauer, T. BiQ Analyzer: visualization and quality control for DNA methylation data from bisulfite sequencing. *Bioinformatics* **21,** 4067–4068 (2005).
34. Guenther, M. G., Levine, S. S., Boyer, L. A., Jaenisch, R., and Young, R. A. A chromatin landmark and transcription initiation at most promoters in human cells. *Cell* **130,** 77–88 (2007).
35. Krumm, A., Hickey, L. B., and Groudine, M. Promoter-proximal pausing of RNA polymerase II defines a general rate-limiting step after transcription initiation. *Genes Dev* **9,** 559–572 (1995).
36. O'Brien, T., Lis, J. T. RNA polymerase II pauses at the 5' end of the transcriptionally induced *Drosophila hsp70* gene. *Mol Cell Biol* **11,** 5285–5290 (1991).
37. Zeitlinger, J., Stark, A., Kellis, M., Hong, J. W., Nechaev, S., Adelman, K., Levine, M., and Young, R. A. RNA polymerase stalling at developmental control genes in the *Drosophila melanogaster* embryo. *Nat Genet* **39,** 1512–1516 (2007).
38. Heintzman, N. D., Hon, G. C., Hawkins, R. D., Kheradpour, P., Stark, A., Harp, L. F., Ye, Z., Lee, L. K., Stuart, R. K., Ching, C. W., Ching, K. A., Antosiewicz-Bourget, J. E., Liu, H., Zhang, X., Green, R. D., Lobanenkov, V. V., Stewart, R., Thomson, J. A., Crawford, G. E., Kellis, M., and Ren, B. Histone modifications at human enhancers reflect global cell-type-specific gene expression. *Nature* **459,** 108–112 (2009).
39. Mikkelsen, T. S., Xu, Z., Zhang, X., Wang, L., Gimble, J. M., Lander, E. S., and Rosen, E. D. Comparative epigenomic analysis of murine and human adipogenesis. *Cell* **143,** 156–169 (2010).
40. Renda, M., Baglivo, I., Burgess-Beusse, B., Esposito, S., Fattorusso, R., Felsenfeld, G., and Pedone, P. V. Critical DNA binding interactions of the insulator protein CTCF: a small number of zinc fingers mediate strong binding, and a single finger-DNA interaction controls binding at imprinted loci. *J Biol Chem* **282,** 33336–33345 (2007).
41. Kadota, M., Yang, H. H., Hu, N., Wang, C., Hu, Y., Taylor, P. R., Buetow, K. H., and Lee, M. P. Allele-specific chromatin immunoprecipitation studies show genetic influence on chromatin state in human genome. *PLoS Genet* **3,** e81 (2007).
42. Kerkel, K., Spadola, A., Yuan, E., Kosek, J., Jiang, L., Hod, E., Li, K., Murty, V. V., Schupf, N., Vilain, E., Morris, M., Haghighi, F., and Tycko, B. Genomic surveys by methylation-sensitive SNP analysis identify sequence-dependent allele-specific DNA methylation. *Nat Genet* **40,** 904–908 (2008).

43. Knight, J. C., Keating, B. J., Rockett, K. A., Kwiatkowski, D. P. *In vivo* characterization of regulatory polymorphisms by allele-specific quantification of RNA polymerase loading. *Nat Genet* **33,** 469–475 (2003).
44. Maynard, N. D., Chen, J., Stuart, R. K., Fan, J. B., Ren, B. Genome-wide mapping of allele-specific protein-DNA interactions in human cells. *Nat Methods* **5,** 307–309 (2008).
45. McCann, J. A., Muro, E. M., Palmer, C., Palidwor, G., Porter, C. J., rade-Navarro, M. A., and Rudnicki, M. A. ChIP on SNP-chip for genome-wide analysis of human histone H4 hyperacetylation. *BMC Genomics* **8,** 322 (2007).
46. Delaval, K., Govin, J., Cerqueira, F., Rousseaux, S., Khochbin, S., and Feil, R. Differential histone modifications mark mouse imprinting control regions during spermatogenesis. *EMBO J* **26,** 720–729 (2007).
47. Fournier, C., Goto, Y., Ballestar, E., Delaval, K., Hever, A. M., Esteller, M., and Feil, R. Allele-specific histone lysine methylation marks regulatory regions at imprinted mouse genes. *EMBO J* **21,** 6560–6570 (2002).
48. Mikkelsen, T. S., Ku, M., Jaffe, D. B., Issac, B., Lieberman, E., Giannoukos, G., Alvarez, P., Brockman, W., Kim, T. K., Koche, R. P., Lee, W., Mendenhall, E., O'Donovan, A., Presser, A., Russ, C., Xie, X., Meissner, A., Wernig, M., Jaenisch, R., Nusbaum, C., Lander, E. S., Bernstein, B. E. Genome-wide maps of chromatin state in pluripotent and lineage-committed cells. *Nature* **448,** 553–560 (2007).
49. Ioannidis, P., Kottaridi, C., Dimitriadis, E., Courtis, N., Mahaira, L., Talieri, M, Giannopoulos, A., Iliadis, K., Papaioannou, D., Nasioulas, G., Trangas, T. Expression of the RNA-binding protein CRD-BP in brain and non-small cell lung tumors. *Cancer Lett* **209,** 245–250 (2004).
50. Ioannidis, P., Mahaira, L., Papadopoulou, A., Teixeira, M. R., Heim, S., Andersen, J. A., Evangelou, E., Dafni, U., Pandis, N., and Trangas, T. CRD-BP: a c-Myc mRNA stabilizing protein with an oncofetal pattern of expression. *Anticancer Res* **23,** 2179–2183 (2003).
51. Ioannidis, P., Mahaira, L., Papadopoulou, A., Teixeira, M. R., Heim, S., Andersen, J. A., Evangelou, E., Dafni, U., Pandis, N., and Trangas, T. 8q24 copy number gains and expression of the c-myc mRNA stabilizing protein *CRD-BP* in primary breast carcinomas. *Int J Cancer* **104,** 54–59 (2003).
52. Ioannidis, P., Trangas, T., Dimitriadis, E., Samiotaki, M., Kyriazoglou, I., Tsiapalis, C. M., Kittas, C., Agnantis, N., Nielsen, F. C., Nielsen, J., Christiansen, J., and Pandis, N. C-*MYC* and IGF-II mRNA-binding protein (CRD-BP/IMP-1) in benign and malignant mesenchymal tumors. *Int J Cancer* **94,** 480–484 (2001).
53. Bieda, M., Xu, X., Singer, M. A., Green, R., and Farnham, P. J. Unbiased location analysis of E2F1-binding sites suggests a widespread role for E2F1 in the human genome. *Genome Res* **16,** 595–605 (2006).
54. Gombert, W. M., Farris, S. D., Rubio, E. D., Morey-Rosler, K. M., Schubach, W. H., and Krumm, A. The c-*myc* insulator element and matrix attachment regions define the c-*myc* chromosomal domain. *Mol Cell Biol* **23,** 9338–9348 (2003).
55. Flanagin, S., Nelson, J. D., Castner, D. G., Denisenko, O., and Bomsztyk, K. Microplate-based chromatin immunoprecipitation method, Matrix ChIP: a platform to study signaling of complex genomic events. *Nucleic Acids Res* **36,** e17 (2008).

56. Dignam, J. D., Lebovitz, R. M., and Roeder, R. G. Accurate transcription initiation by RNA polymerase II in a soluble extract from isolated mammalian nuclei. *Nucleic Acids Res* **11,** 1475–1489 (1983).
57. Chung, J. H., Bell, A. C., and Felsenfeld, G. Characterization of the chicken β-globin insulator. *Proc Natl Acad Sci USA* **94,** 575–580 (1997).

CHAPTER 9

BIVALENT CHROMATIN MODIFICATION

MARCO DE GOBBI, DAVID GARRICK, MAGNUS LYNCH, DOUGLAS VERNIMMEN, JIM R. HUGHES, NICOLAS GOARDON, SIDINH LUC, KAREN M. LOWER, JACQUELINE A. SLOANE-STANLEY, CRISTINA PINA, SHAMIT SONEJI, RAFFAELE RENELLA, TARIQ ENVER, STEPHEN TAYLOR, STEN EIRIK W. JACOBSEN, PARESH VYAS, RICHARD J. GIBBONS, and DOUGLAS R. HIGGS

CONTENTS

9.1 INTRODUCTION

In recent years it has been suggested that the epigenetic program may play a key role in determining cell fate, including the decision to undergo self-renewal or commitment. Based on genome-wide chromatin immunoprecipitation (ChIP) studies combined with expression analysis, it has been suggested that the chromatin associated with many genes controlling lineage fate decisions is uniquely marked in stem cells. Their histone signature is referred to as bivalent as it includes modifications associated both with repression (H3K27me3) imposed by the polycomb group proteins (PcG), and activation (H3K4me3) encoded by the Set/MLL histone methyltransferase, the mammalian homologue of the trithorax group proteins (trxG) [1-6]. Despite having both "active" and 'repressive' chromatin marks, such genes were thought not to be expressed. Taken together, these observations led to an attractive model suggesting that a preimposed epigenetic signature suppresses expression of lineage control genes in stem cells (maintaining a pluripotent state) while at the same time 'poising' such genes for subsequent activation (reviewed in [7]). In favor of this, many lineage-control genes have a bivalent signature [1-5]. However, as the model has evolved, more recently it has been shown that RNA polymerase II (PolII) may be present but stalled at the promoters of bivalent genes [8, 9] and that short (abortive) transcripts may be detected at their promoters [10]. Furthermore, although embryonic stem cells (ES cells) lacking the PcG repressive complex 2 (PRC2) aberrantly express developmental regulators [11] they maintain pluripotency [12]. Similarly, two recent experiments in which components of the SET1/MLL core subunit (Dpy-30, RbBP5 and WDR5) were reduced to similar levels showed opposite phenotypes. In one chapter there was maintenance of self-renewal with a defect in differentiation [13], and in another there was a loss of self-renewal [14]. Together these observations suggest that the current models explaining the significance bivalently marked chromatin may require revision.

An understated problem in testing the prevailing bivalent chromatin hypothesis is that the criteria for identifying such signatures are poorly defined. Closer analysis of publicly available chromatin datasets from human embryonic stem (ES) cells shows that even contiguous bivalent chromatin domains can be modified in widely different ways with respect to the relative levels and the distributions of H3K4me3 and H3K27me3 across the locus. Sequential ChIP analyzes of a few developmental genes have shown that H3K4me3 and H3K27me3 may colocalise and, by extrapolation, it has been implied that all genes whose promoters are marked (to any degree) by both modifications are truly bivalent. According to this paradigm, many specialised, lineage-specific genes are bivalent [3, 4, 15] and as cells differentiate, chromatin modifications resolve into active or repressed states. However, since the original observations, it has become clear that bivalent chromatin modifications (indistinguishable from those seen in pluripotent cells) can also be newly established and/or maintained in differentiating cells [16-19]. Therefore, the functional significance of such bivalently marked genes has been questioned [20] and more studies have been urged to determine the mechanisms underlying these chromatin structures [19].

The human α globin genes are located within a well characterized multigene cluster whose analysis has elucidated many of the general principles underlying the transcriptional and epigenetic regulation of mammalian gene expression. The α globin cluster (5'-HBZ-HBM-HBA2-HBA1-HBQ-3') provides unequivocal examples of specialized, tissue-specific genes consistently scored as bivalent in ES cells [4, 5] (Figure 1). Their fully activated expression depends on one or more of four remote conserved regulatory elements (MCS-R1 to R4) [21], which interact with their promoters in erythroid cells via a looping mechanism [22]. Although expressed in a strictly tissue-specific and developmental-stage-specific manner during erythropoiesis, the α globin promoters and much of the body of the associated genes lie within unmethylated CpG islands [23]. Previously, many aspects of the transcription factor and epigenetic programs associated with hematopoiesis have been described and how they are played out on the α globin cluster [24-27]. Here, this model has been used to investigate in detail the relationship between chromatin marks and mRNA expression during commitment and differentiation into erythroid cells. It appears that, rather than carrying a preimposed bivalent epigenetic signature which silences them, in pluripotent ES cells the α globin genes are repressed by PcG and comodified at readily detectable levels by H3K-4me3 when expressed, even at basal levels. To ensure that characterization of the α globin genes is revealing a general principle that could be relevant to other bivalent genes, global analysis of H3K4me3/H3K27me3 modification and gene expression was performed. The results suggest that the understanding of chromatin bivalency at the α globin locus may explain similar marks found at many other bivalent genes both in pluripotent and differentiating cells and highlights an alternative mechanism for generating bivalent domains.

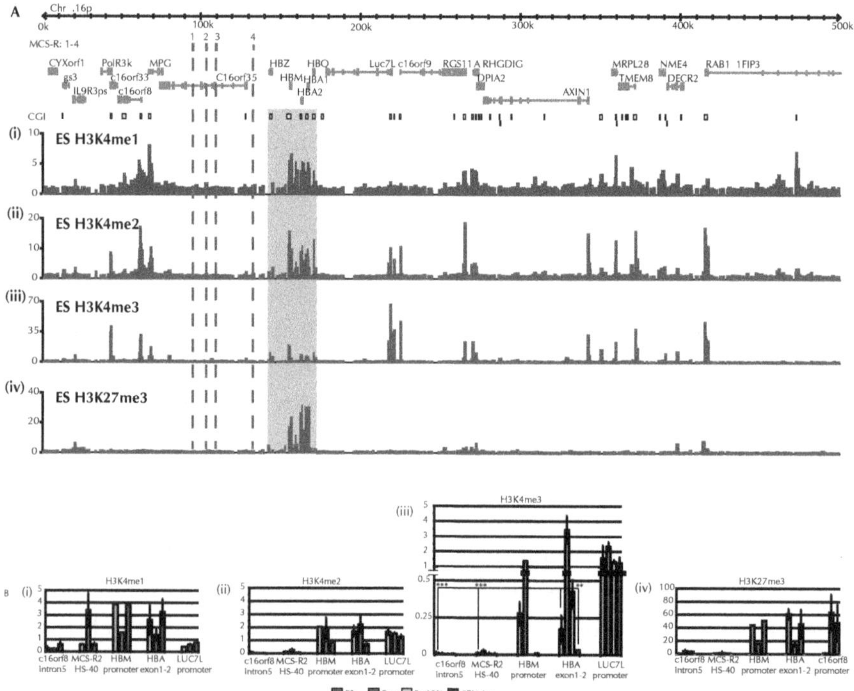

FIGURE 1 Chromatin marks at the telomeric region of chromosome 16p in embryonic stem (ES) cells. (a) The genes, multispecies conserved regulatory sequences (MCS-R1-4) and CpG islands (CGI) are shown at the top. The red shaded box represents the α globin cluster. The y axis represents the enrichment of chromatin immunoprecipitation (ChIP) DNA over input DNA. (b) ChIP quantitative PCR (qPCR) data at selected amplicons. Ery = primary human erythroblasts; Ery10% = mixed population consisting of 10% erythroblasts and 90% Ly; ES = human ES cells; Ly = Epstein-Barr virus (EBV)-transformed lymphoblastoid cell line. The fold enrichment has been calculated relative to a control sequence in the ACTB promoter. The error bars show the standard deviation of three independent experiments. $^{**}P < 0.05$; $^{***}P < 0.01$ (both by Student's t test).

9.2 METHODS

9.2.1 ETHICS

The part of the chapter involving human participants was approved by the Institutional Ethics Committee (approval number 06/Q1606/110). A written informed consent was obtained from all patients.

9.2.2 PRIMARY CELLS AND CELL CULTURE

The EBV-Ly cell lines were cultured in RPMI 1640 supplemented with 10% (v/v) fetal calf serum, 2 mM L-glutamine, 50 U/ml penicillin and 50 μg/ml streptomycin. Isolation and culture of primary human erythroblasts was carried out as described previously [35]. The human embryonic stem cell line, H1 (WiCell, Madison, WI, USA), was grown on irradiated mouse embryonic fibroblasts in medium containing Dulbecco's modified Eagle medium (DMEM):F12, serum replacer (Invitrogen, Paisley, UK), L-glutamine, β-mercaptoethanol, non-essential amino acids and basic fibroblast growth factor (8 ng/ml). Cells were passaged every 5–7 days with collagenase IV to maintain undifferentiated human embryonic stem cells.

9.2.3 FLOW CYTOMETRIC ANALYSIS AND SORTING

Normal human bone marrow samples were collected from individuals undergoing total hip replacement for osteoarthritis. CD34+ cells were enriched using MACS (Miltenyi Biotech, Bergisch Gladbach, Germany) immunomagnetic beads and cryopreserved in 90% fetal bovine serum (FBS)/10% dimethylsulfoxide (DMSO). All experiments were carried out using cryopreserved CD34+ cells that were thawed, washed in Iscove's modified Dulbecco's medium (IMDM)/10% FBS and cultured overnight in StemSpanSFEM (StemCell Technologies, Grenoble, France) in the presence of recombinant human stem cell factor (100 ng/ml), Flt3-ligand

(100 ng/ml) and thrombopoietin (100 ng/ml) (Peprotech, Rock Hill, NJ, USA).

The CD34+ cells were first stained with purified anti-CD2, RPA-2.10; CD3, HIT3a; CD4, RPA-T4; CD7,124-1D1; CD8, RPA-T8; CD10, CB-CALLA; CD11b, ICRF44; CD14, 61D3; CD19, HIB19; CD20, 2H7; CD56, MEM188; GPA, GA-R2 (eBioscience, San Diego, CA, USA). Subsequently cells were stained with Pacific Blue conjugated goat F(ab')2 anti-mouse IgG conjugates (H+L) (Invitrogen). Finally, cells were stained with fluorescein isothiocyanate (FITC)-conjugated anti-CD38 (HIT2), phycoerythrin (PE)-conjugated anti-CD45RA (HI100), PE-Cy7-conjugated anti-CD123 (6H6) (eBioscience), PE-Cy5-conjugated anti-CD34 (581) (Beckman Coulter, High Wycombe, UK) and allophycocyanin (APC)-conjugated anti-CD110 (BAH-1) (Becton Dickinson, Franklin Lakes, NJ, USA). HSCs were isolated as Lin- CD34+ CD38-, CMPs as Lin- CD34+ CD38+ CD123low/+ CD45RA- CD110-, GMPs as Lin- CD34+ CD38+ CD123+ CD45RA+ CD110- and MEPs as Lin- CD34+ CD38+ CD123-/low CD45RA- CD110-. Dead cells were excluded by Hoechst 33258 (Invitrogen) staining or by 7-aminoactinomycin D (Sigma-Aldrich, Gillingham, UK). Appropriate unstained, single stained and Fluorescence Minus One controls were used to determine the background staining level and compensation in each channel. All sorting and analyzes were performed on three laser-equipped MoFlo (Dako Cytomation, Ely, UK) or BD FACS AriaII (Becton Dickinson) machines. Single ES cells were sorted with an automated cell deposition unit into 96-well plates. FACS data were analyzed with Summit software (Dako Cytomation).

9.2.4 CHIP ASSAYS

ChIP analyzes were performed according to the Millipore ChIP protocol (Millipore 17-295, Billerica, MA, USA), as described previously [26]. Input and immunoprecipitated material were analyzed by real time PCR using a series of PCR amplicons and 5'FAM-3'TAMRA probes across the α globin locus [26]. ChIP-chip experiments were carried out on the custom α globin tiling path microarray, as described previously [22]. The enrichment of ChIP DNA over input DNA was calculated as ratio of the background corrected ChIP signal divided by the background corrected input signal (both globally normalised). The antibodies used in the experiments were: anti-mono/di/trimethyl Lys4 histone H3 (07-436, 07-030, and 07-473, respectively), anti-trimethyl Lys27 histone H3 (07-449) and anti-trimethyl Lys9 histone H3 (07-442 lot 24416) from Millipore; anti-trimethyl Lys36 histone H3 (ab9050) from Abcam (Cambridge, UK); anti-RNA-PolII (H224) from Santa Cruz Biotechnology (Santa Cruz, CA, USA), anti-EZH2 (36-6300) from Zymed (San Francisco, CA, USA).

For ChIP in the artificially mixed population of cells, 10% of primary erythroblasts were mixed with 90% of EBV-ly cells. At any one time, about 80% of erythroid cells are positive for nascent α globin RNA transcripts [36] and 100% of these cells accumulate high levels of globin RNA during the late stages of erythropoiesis. Therefore these cells are presumed to be 100% modified by H3K4me3. By contrast, in lymphocytes, the α globin genes are repressed and the H3K4me3 signal at the α globin promoter is at background level.

Sequential ChIP analyzes were performed as follows: 5 μg of H3K4me3 or H3K27me3 antibody were immobilised to 50 μl of protein A agarose beads (Millipore) by crosslinking with 2.5 mM BS3 (Thermo Fisher Scientific, Waltham, MA, USA). Chromatin precipitated using the first antibody was eluted in 500 μl of 0.1 M NaHCO3 and 1% SDS. This solution was then diluted tenfold in ChIP Dilution buffer (Millipore) and subjected to immunoprecipitation using the second antibody. The second antibody was not crosslinked to protein A.

9.2.5 GENE EXPRESSION ANALYSIS

Total RNA was extracted using TriReagent (Sigma). Contaminating DNA was removed from RNA preps with the DNA-free kit (Ambion/Applied Biosystems, Austin, TX, USA) according to the manufacturer's instructions. cDNA was generated using 1–5 μg of total RNA and random primers using the Prostar RT-PCR kit (Stratagene/Agilent Technologies, Santa Clara, CA, USA). Negative control cDNA samples generated without reverse transcriptase were analyzed in all experiments. Real time qPCR experiments were carried out on ABI Prism 7000 Sequence Detection System

(Applied Biosystems) using a set of primers/probe detecting HBA2 cDNAs. The results were normalised to a control sequence in the 18S ribosomal RNA (RNRI) gene (Eurogentec, Southampton, UK). To detect short abortive transcripts at the 5' region of HBA cDNA, two different sets of primers were used in a SybrGreen qPCR reaction in expression analysis.

Multiplex single-cell RT-PCR analysis was performed as previously described [33, 37]. Single cells were deposited into 96-well PCR plates using a single cell depositor unit coupled to a fluorescence-activated cell sorting (FACS) ARIAII cell sorter (providing single cells in >99% of the wells, and no wells with more than 1 cell as assessed by routinely sorting fluorescent beads or cells prior to and after single cell sorting). Each well-contained 4 μl of lysis buffer (0.4% NP40, 2.3 mM dithiothreitol (DTT), 0.07 mM dNTP, 0.5 U/μl Rnase Inhibitor). Cell lysates were reverse transcribed in a 10 μl reaction with Superscript III Reverse Transcriptase (Invitrogen) and gene-specific primers. A first round of PCR (40 cycles) was performed by the addition of a PCR mix containing PCR buffer, 1.25 U of Taq polymerase (Invitrogen) and gene-specific forward primers. A total of 1 μl of 1:10 diluted first-round PCR products were amplified in a second-round PCR, which was carried out using fully nested gene-specific primers. PCR products were gel electrophoresed and visualised by ethidium bromide staining. Only control-positive (OCT4 in ES cells, HPRT in EBV-Ly) wells were considered as informative and scored. A total of 270 individual human ES cells and 110 EBV-Ly were analyzed.

9.2.6 STATISTICAL ANALYSIS

In order to correlate genes expression and chromatin state in human ES cells, expression data (accession number GSE8439 [4]) and single cell transcript detection data [29] were cross-referenced with ChIP-Seq data (accession number GSE13084 [6]). The promoter chromatin state was calculated as relative ratio of the signal derived from the number of H3K4me3 and H3K27me3 sequence reads across a window between -3 kb and +3 kb of the annotated TSS. The relationship between H3K4me3/H3K27me3 ratio and expression was calculated by averaging of the H3K4me3/H3K27me3 ratio within a sliding window 100 observations wide, incrementing by 1, using a Spearman rank correlation. Considering the different source of the two ES cell datasets (H1, male cell line, for expression data and H9, female cell line, for ChIP-Seq data), X-chromosome linked genes were excluded from the analysis.

9.3 RESULTS AND DISCUSSION

The H3K4me3 at the α globin genes occurs at a low but significant level in ES cells and increases during erythroid differentiation. To dissect the mechanism(s) underlying epigenetic "bivalency" in ES cells, histone modifications across the telomeric region of chromosome 16 containing the human α globin locus was studied. It is found that, whereas H3K4me1 and H3K4me2 were both enriched at the α globin locus (Figure 1a, i and 1ii), H3K4me3 (considered to be a sensitive mark of recent or ongoing transcriptional activity) was relatively low (Figure 1a, iii). However, quantitative real time PCR (qPCR) (Figure 1b, iii) showed that the level of H3K4me3 at the α globin promoter in ES cells was significantly higher than in lymphocytes in which the α globin genes are considered to be fully repressed (see below). Similar results were obtained at the promoter of the HBM gene, a minor α globin-like gene whose promoter is also associated with a large CpG island [28].

To quantify this low level of H3K4me3, the degree of enrichment seen in pluripotent ES cells were compared to that seen in an artificially mixed population of erythroid and non-erythroid cells. This showed (Figure 1b, iii) that the H3K4me3 enrichment seen in ES cells at the α globin gene was even less than that obtained in a mixed population of cells consisting of 10% erythroid cells (presumably fully modified by H3K4me3) (Figure 1b, iii) and 90% lymphocytes (unmodified by H3K4me3, Figure 1b, iii).

These results show that the chromatin associated with the α globin genes is modified at a significant but low level by H3K4me3 in chromatin derived from a population of pluripotent stem cells. However, the level of H3K4me3 modification increases dramatically (15-fold) as cells differentiate into erythroid cells (Figure 1b, iii). This phenomenon is common to many other genes (previously noted to be bivalent in ES cells), which are expressed at high levels late in erythroid differentiation (for example,

BLVRB, FAM83F, MTSS1, TNXB) [18]. Conversely, H3K4me3 is barely detectable in lymphocytes in which the α globin genes are repressed.

H3K27me3 at the α globin genes occurs at high levels in ES cells and decreases during erythroid differentiation. It has been previously shown that α globin expression is repressed in non-erythroid cells by PcG and its associated silencing mark H3K27me3. This repression is put in place early in development and then is either reduced in the erythroid lineage or maintained in non-globin-expressing cell types [27].

Here, these observations were extended to determine the pattern of H3K27me3 in pluripotent cells across the telomeric 500 kb of chromosome 16p. As noted in several non-erythroid differentiated cells [27], H3K27me3 enrichment extended across a broad region of the α globin cluster in ES cells (Figure 1a, iv).

Of importance, next the relative levels of H3K27me3 enrichment were determined at the HBM and α globin promoters in ES cells and lymphocytes in which the α globin genes are repressed. ChIP-chip results (Figure 1, iv and [27]), in accordance with qPCR data (Figure 1b, iv), indicate that the locus is modified to a similar extent in both cell types. These findings suggest that the α globin genes are highly (possibly maximally) modified by H3K27me3 in pluripotent cells and that this level of modification is maintained when cells differentiate into non-erythroid lineages. By contrast, in the erythroid population, as the PcG is completely cleared [27], H3K27me3 is reduced fourfold compared to that seen in non-erythroid cells (Figure 1b, iv) but not totally removed.

The α globin locus is bivalently modified in ES cells. Chromatin modification at the α globin locus thus resembles that seen at other bivalent domains (for example, CDX2) at which there is a high level of H3K27me3, and a low level of H3K4me3 which increases (with expression) or decreases (with silencing) in specific lineages as cells differentiate. To determine if the observed bivalent architecture results from colocalisation of H3K27me3 and H3K4me3 rather than simply reflecting the presence of two distinct subpopulations of active and silent cells, sequential ChIP analyzes was performed.

At α globin and HBM promoters, chromatin precipitated with an antibody against H3K27me3 was sequentially precipitated by an antibody against H3K4me3 (Figure 2a). Similar results were seen by the "reverse" sequential ChIP (Figure 2b). This shows that at least some chromatin at the α globin genes is truly comodified by H3K27me3 and H3K4me3.

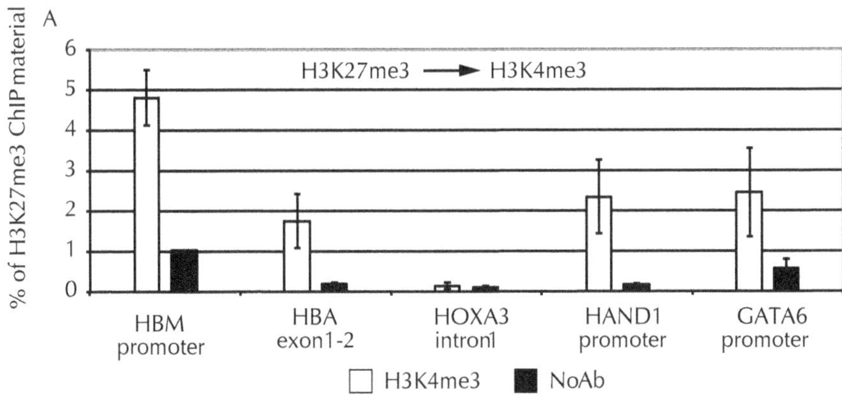

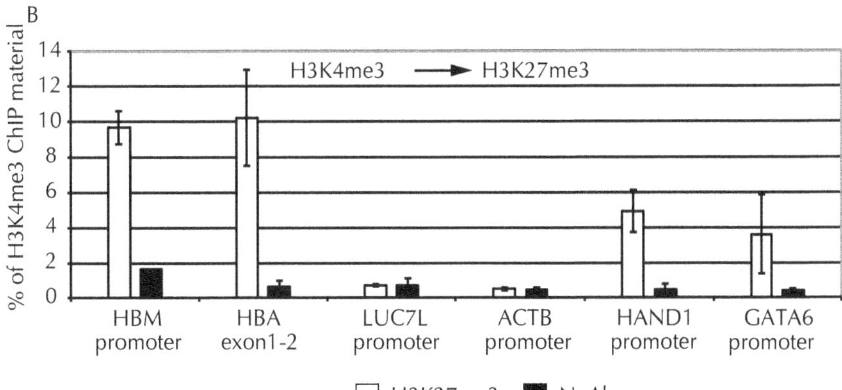

FIGURE 2 Chromatin immunoprecipitation quantitative PCR (ChIP-qPCR) characterisation of the α globin chromatin bivalency. (a) Sequential ChIP with H3K27me3 followed by H3K4me3 antibody. HOXA3 intron represents an H3K27me3 genomic region; HAND1 and GATA6 promoters are control bivalent promoters. (b) Sequential ChIP with H3K4me3 followed by H3K27me3 antibody. LUC7L and ACTB promoter are H3K4me3 modified promoters. The error bars show the standard deviation of three independent ChIP-qPCR experiments.

The α globin locus is expressed at basal levels in a significant proportion of pluripotent cells. Chromatin modification may influence the probability that a locus is transcribed and/or reflect its recent transcriptional state. As for many bivalent genes, the assertion that the α globin genes are not expressed in pluripotent cells is based on the failure to identify binding of PolII in ChIP experiments and the very low (background) signals obtained on RNA microarray analyzes. However, using more sensitive assays it has become increasingly clear that in pluripotent cells many bivalent genes are being transcribed either to produce a variety of short, abortive transcripts [8, 10] or to produce very low levels of full length RNA transcripts [15, 29].

To determine the transcriptional status of the α globin bivalent domain, firstly it was looked for, but could not detect, high levels of 5' abortive transcripts. Then, expression of spliced transcripts were analyzed at the cell population level. It was found that in ES cells the level of α globin mRNA, although very low compared to erythroblasts (approximately 30,000 to 40,000 times less), was at least 10 times higher than that measured in Epstein-Barr virus (EBV)-transformed lymphoblastoid (EBV-Ly) cell lines (Figure 3a). Similarly, HBM mRNA could be detected in ES cells (although 50 times less than α globin) but was not detectable in lymphoblastoid cells. The lower expression of HBM than α globin might be due to differences in promoter sequences or in mRNA stability of this minor globin gene [28]. By comparison, no β globin RNA transcripts were detected in ES and lymphoblastoid cells. It was next estimated that how many ES cells within each population express α globin using single-cell RT-PCR with primers that detect full-length mRNA transcripts. Using a multiplex analysis, this showed a detectable level of α globin in 12% of OCT4 positive ES cells (Figure 3b). Since a single normal erythroid cell contains approximately 20,000 (α + β) globin RNA molecules [30], it can be estimated that pluripotent ES cells express α globin only at basal (two or three copies per cell) levels. This is consistent with observations that chromatin modifications (H3K36me3 and H3K9me3) associated with high rates of transcription were not detected in ES cells. By contrast, no α globin expression was detected in more than 100 EBV-Ly cells (Figure 3c), confirming that in these cells, unlike ES cells, the state of chromatin at the α globin locus is in a completely repressed transcriptional configuration.

A

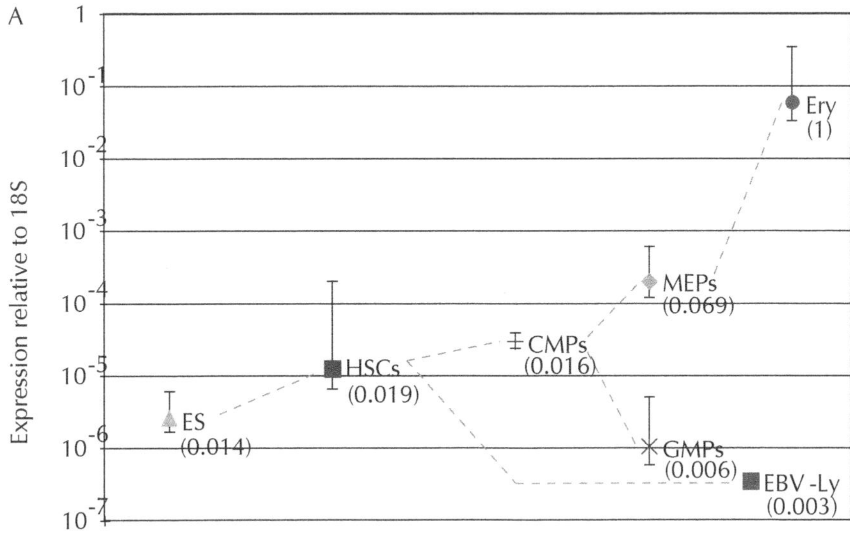

B

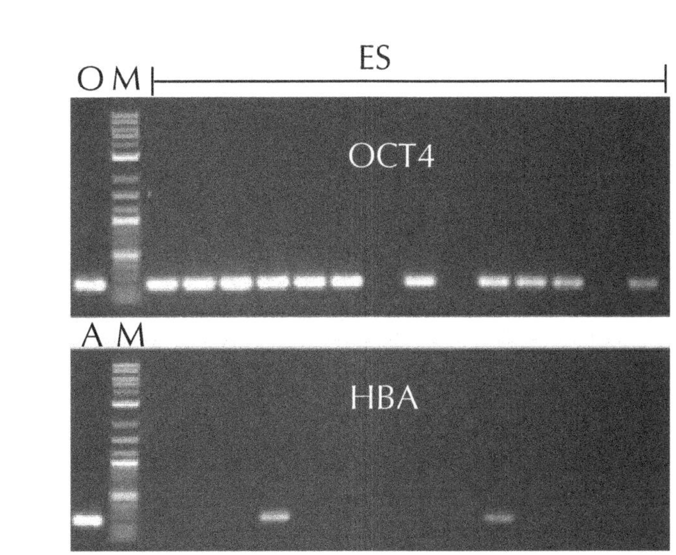

FIGURE 3 *(Continued)*

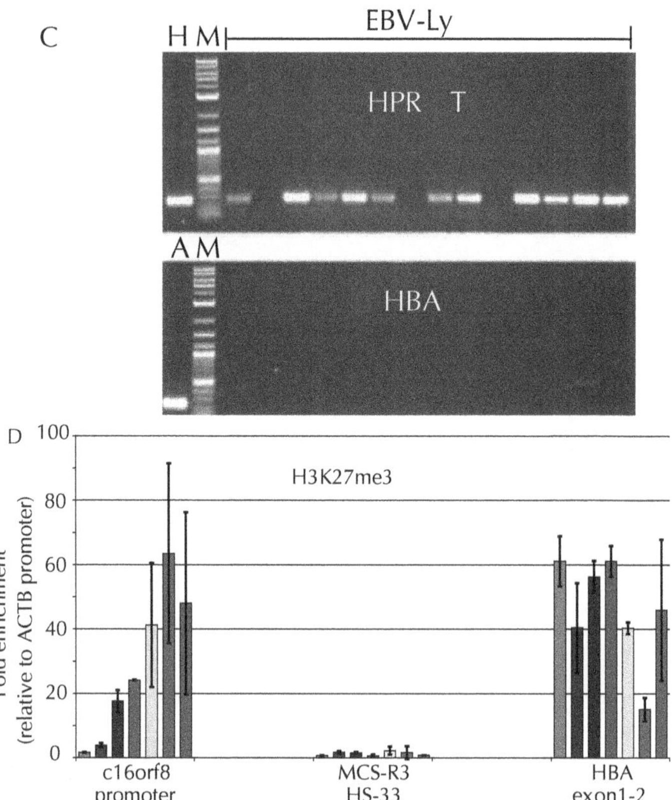

FIGURE 3 α Globin expression and chromatin structure in embryonic stem (ES) cells, hematopoietic progenitors and differentiated cells. **(a)** Relative expression level of α globin. The error bars represent the standard deviation of three independent experiments. CMPs = common myeloid progenitors; EBV-Ly = Epstein-Barr virus (EBV)-transformed lymphoblastoid cell line; ES = ES cells; Ery = primary erythroblasts; GMPs = granulocyte-monocyte progenitors; HSCs = hematopoietic stem cells; MEPs = megakaryocyte-erythroid progenitors. The ratio between the level of H3K4me3 and H3K27me3, as measured by quantitative PCR (qPCR) at the α globin gene (see below), is shown by each of the cell types analyzed. The value obtained in primary erythroblasts was set to 1. **(b)** Examples of OCT4 and α globin expression analysis carried out in 270 single ES cells. Each lane corresponds to the same ES single cell. A = α globin control; O = OCT4 control; M = molecular weight marker. **(c)** Example of HPRT and α globin expression analysis carried out in 110 single EBV-Ly cells. Each lane corresponds to the same EBV-Ly single cell. H = HPRT control. Only OCT4 or HPRT positive cells were considered as informative. **(d)** H3K27me3 chromatin immunoprecipitation (ChIP)-qPCR in hematopoietic progenitors. The fold enrichment is relative to a control sequence in the ACTB promoter. The error bars show the standard deviation of two independent experiments. **(e)** H3K4me3 ChIP-qPCR in hematopoietic progenitors. The fold enrichment is relative to a control sequence in the ACTB promoter. The error bars show the standard deviation of two independent experiments.

Changes in expression and chromatin modification in hematopoietic stem cells and progenitors. To examine how α globin expression changes during hematopoiesis, primary hematopoietic progenitors was studied at various stages of commitment and differentiation (Figure 3a). Mixed populations enriched for hemopoietic stem cells (HSCs) and common myeloid progenitors (CMPs) exhibited a level of α globin gene expression higher than in ES cells. Expression then increased further in fractions enriched for megakaryocyte-erythroid progenitors (MEPs) but decreased in granulocyte-monocyte progenitors (GMPs), confirming that a gradual restriction in the differentiation potential is associated with an upregulation of lineage-specific genes and silencing of lineage-inappropriate genes.

In HSCs high levels of H3K27me3 was detected and relatively low levels of H3K4me3 (Figure 3d); a pattern consistent with the accepted criteria for a bivalent chromatin signature (and as defined for this gene in pluripotent ES cells). The levels of H3K27me3 (and its associated methyltransferase EZH2) in HSCs, CMPs, GMPs, MEPs and EBV-Ly cells were similar and decreased as MEPs differentiated into erythroid cells (Figure 3d). By contrast the levels of H3K4me3, associated with the parallel increase of expression, started to increase as CMPs differentiated into MEPs, prior to the clearance of PcG and H3K27me3. Therefore, at each of these stages of differentiation, the ratio of H3K4me3/H3K27me3 changed, reflecting the level of α globin expression (Figure 3a).

Bivalent genes with a higher H3K4me3 occupancy at the promoter are more often transcribed in ES cells. In genome-wide analyzes it has been shown that, in general, the level of H3K4me3 modification correlates with gene expression and H3K27me3 correlates with silencing [16, 17, 31, 32]. But does this correlation also apply to bivalent genes? The detailed analysis of a single gene presented here shows that, at the bivalent α globin locus, changes in the levels of basal gene expression are reflected by changes in the H3K4me3 mark. Is this behavior an exception? Here the relationship between H3K4me3/H3K27me3 modifications and gene expression at other was evaluated, previously identified bivalent genes.

On a global scale, it was determined how the H3K4me3/H3K27me3 ratio differed at bivalent genes known to be expressed at different levels in ES cells. Data from human ES cell studies [4, 6] were crossreferenced and the bivalent genes subdivided into three “bins” according to their expression. It was found that in a population of ES cells, the chromatin associated with genes with the highest expression had the highest H3K4me3/H3K27me3 ratios, while those with the lowest relative expression had the lowest ratios (Figure 4a). In addition, there was a significant positive association between the absolute level of RNA expression and the H3K4me3/H3K27me3 ratio at the associated transcription start sites (TSSs) ($P = 2.2 \times 10–16$) (Figure 4b). The same data set plotting for each individual bivalent gene was also analyzed the ratio of H3K4me3/H3K27me3 against expression. Clearly they follow the same trend but, due both to biological variation and insensitivity of microarray data to low levels of transcription, there is a considerable scatter.

In a cell population, the relative levels of H3K4me3/H3K27me3 at previously identified bivalent genes could reflect the relative proportions of cells in which the gene in question is marked exclusively by H3K4me3 or H3K27me3. To address this it was looked at a subset of the previously studied bivalent loci with the highest levels of H3K27me3 (top 20%). It seemed likely that, for this set, the genes in question would be modified by H3K27me3 in most, if not all, cells. So it was of interest that, even in this subgroup, a significant positive association between the absolute levels of RNA expression and the levels of associated H3K4me3 ($P = 2.25 \times 10\text{-}8$) was found. Therefore, in comparison to other more stringent forms of silencing (for example DNA methylation), PcG silencing may be incomplete and allow stochastic gene expression.

Finally, the relationship between the ratio of H3K4me3/H3K27me3 at specific bivalent genes and the ability to detect single-cell transcripts expressed from these genes was examined. It has been reported that, as for α globin, expression of other bivalent genes can be sporadically detected in single ES cells [29]. The bivalent genes studied by Gibson et al. [29] were crossreferenced with ChIP data from Ku et al. [6] and subdivided into three classes according to the proportion of cells in which the transcript was detected. Although the number of genes analyzed by single-cell RT-PCR was small, it was found that there was a clear tendency for the genes whose transcripts were never detected, to have the lowest H3K4me3/H3K27me3 ratios (Figure 4c).

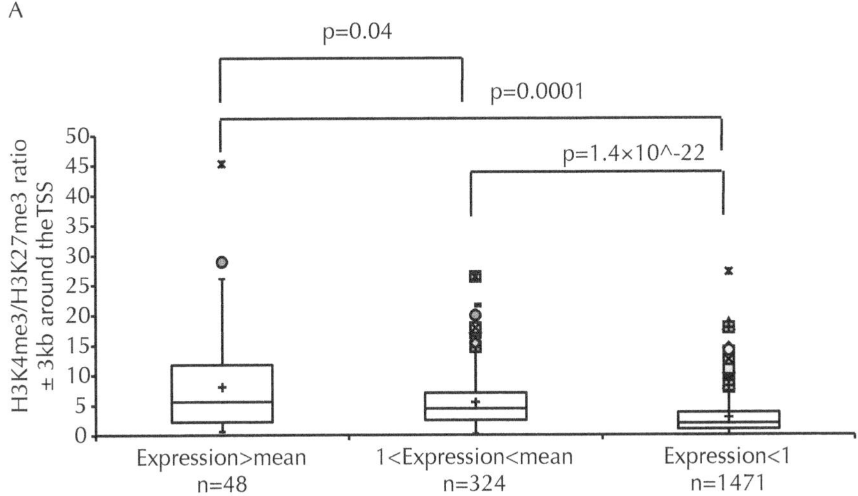

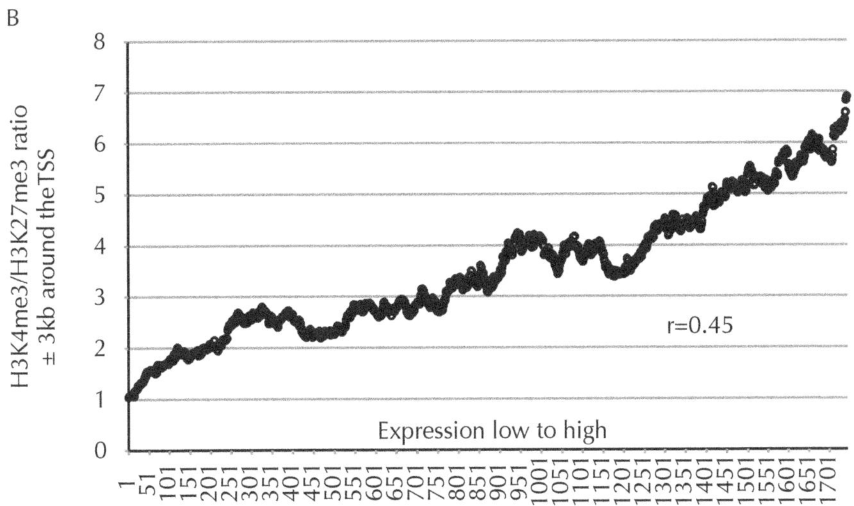

FIGURE 4 *(Continued)*

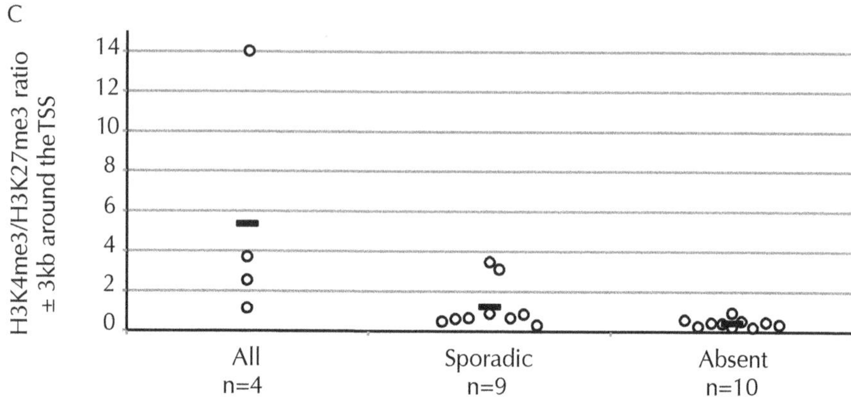

FIGURE 4 Correlation between expression level and H3K4me3/H3K27me3 occupancy at bivalent genes in embryonic stem (ES) cells. **(a)** Box plot showing 25th, 50th and 75th percentiles of H3K4me3/H3K27me3 ratio for 1,843 bivalent genes associated with 3 different expression levels, as calculated by Pan et al. [4] (above the mean value for all expressed genes; between the mean value and 1; below 1). The error bars represent 1.5 times the interquartile range above and below the median. Outliers are plotted individually. The mean of each population is shown as a black cross. Statistical differences were determined using a two-tailed unpaired Student's t test. **(b)** A total of 1,844 bivalent genes were ranked according to their levels of expression (low to high, x axis) and plotted against the moving average (window size 100 genes, step 1) of the ratios of H3K4me3/H3K27me3 (y axis). Spearman's r value is shown in the graph. **(c)** Scatterplot showing the H3K4me3/H3K27me3 ratio for 23 bivalent genes associated with 3 different expression patterns as shown by Gibson et al. [29] (present in all the cells analyzed, present sporadically in some but not all cells, and absent). The mean of each population is shown as a black bar. The H3K4me3/H3K27me3 ratio has been calculated from the signal derived from the number of H3K4me3 and H3K27me3 sequence reads in an area spanning from -3 kb to +3 kb around the transcription start site (TSS). Statistical analysis made with different windows (from -4 kb to +1 kb and from -0.5 kb to +2 kb) gave similar results.

9.4 CONCLUSION

The central question raised here is whether or not bivalent domains, marked both by repressive (H3K27me3) and active (H3K4me3) histone modifications, represent a preprogrammed epigenetic signature of silent but poised chromatin or more simply reflect different degrees of silencing (H3K27me3) and transcription (H3K4me3) within a population of cells. The observation that most clearly supports the prevailing concept of a preprogrammed epigenetic mark in pluripotent cells is the presence of bivalent domains in which, despite the presence of both activating (H3K-4me3) and repressive (H3K27me3) marks, no PolII elongation and full length transcription are detected, leading to the conclusion that all of such genes are poised [8, 9]. However through evaluation of public datasets it is readily notable that the levels and patterns of histone modifications at bivalent genes are very variable. Similarly, a wide range of H3K4me3 levels within bivalent domains of embryonic and haematopoietic stem and progenitor cells has been recently identified [13, 19].

Here, the detailed analysis of a single well-characterized bivalent gene (α globin) expressed during haematopoiesis and the global analysis of the relationship between H3K4me3/H3K27me3 ratio and gene expression suggest another explanation for bivalent chromatin domains that might be considered. It is possible that some genes with bivalent signatures may be poised in pluripotent cells by a common but as yet undefined molecular mechanism. If so, then it is not clear why the associated chromatin modifications are so variable. An alternative explanation, proposed here, is that bivalently marked genes are regulated by PcG and marked by H3K27me3; the different levels of associated H3K4me3 may simply be a sensitive marker of different levels of transcriptional activity. The latter scenario is consistent with previous observations showing that many lineage-specific genes can be stochastically expressed at low but variable levels in multipotent cells (so called multilineage priming) [33].

Although both situations could be considered to be poised the underlying mechanisms and their implications are different. In fact, the first model (repressed but poised for later activation) implies that most of the cells harbour coexisting repressive and activating marks and that both PcG and trxG have to be maintained at the given locus through cell divisions to keep the repressed-but-poised status until further changes in the balance of their activity. By contrast, the second (repressed and marked by variable degrees of stochastic transcription) implies that PcG is not fully efficient at repressing stochastic transcriptional noise. This would provide fertile ground for subsequent activation and changes in cell fate decisions as a tissue-specific transcription factor program emerges [34]. Definitive experiments to discriminate between these two possibilities will require the development of assays to correlate chromatin modification and gene expression within single cells.

KEYWORDS

- α Globin Genes
- Bivalent Domains
- Embryonic Stem Cells
- Haematopoiesis
- Pluripotent Cells

ACKNOWLEDGMENT

We thank the CBRG for bioinformatic support, I Dunham, C Langford and the Microarray Facility of the Wellcome Trust Sanger Institute for assistance with microarrays, K Clark for flow sorting, and T Milne for critical reading of the manuscript. This chapter was supported by the National Institute for Health Research Biomedical Research Centre Program.

COMPETING INTERESTS

The authors declare that they have no competing interests.

AUTHORS' CONTRIBUTIONS

MDG participated in the design of the chapter, carried out the molecular studies and drafted the manuscript; DG, ML, DV, NG, SL, KML, JAS-S, CP and RR carried out the molecular studies; DG, ML and RJG helped to draft the manuscript; JRH and ST participated in the bioinformatic and statistical analyzes; SS performed the statistical analysis; SEWJ, PV, RJG and TE participated in the design of the chapter and its coordination; DHR conceived the chapter and participated in its design and coordination and drafted the manuscript. All authors read and approved the final manuscript.

REFERENCES

1. Azuara, V., Perry, P., Sauer, S., Spivakov, M., Jorgensen, H. F., John, R. M., Gouti, M., Casanova, M., Warnes, G., Merkenschlager, M., and Fisher, A. G. Chromatin signatures of pluripotent cell lines. *Nat Cell Biol* **8,** 532–538 (2006).
2. Bernstein, B. E., Mikkelsen, T. S., Xie, X., Kamal, M., Huebert, D. J., Cuff, J., Fry, B., Meissner, A., Wernig, M., Plath, K., Jaenisch, R., Wagschal, A., Feil, R., Schreiber, S. L., and Lander, E. S. A bivalent chromatin structure marks key developmental genes in embryonic stem cells. *Cell* **125,** 315–326 (2006).
3. Tikkelsen, T. S., Ku, M., Jaffe, D. B., Issac, B., Lieberman, E., Giannoukos, G., Alvarez, P., Brockman, W., Kim, T. K., Koche, R. P., Lee, W., Mendenhall, E., O'Donovan, A., Presser, A., Russ, C., Xie, X., Meissner, A., Wernig, M., Jaenisch, R., Nusbaum, C., Lander, E. S., and Bernstein, B. E. Genome-wide maps of chromatin state in pluripotent and lineage-committed cells. *Nature* **448,** 553–560 (2007).
4. Pan, G., Tian, S., Nie, J., Yang, C., Ruotti, V., Wei, H., Jonsdottir, G. A., Stewart, R., and Thomson, J. A. Whole-genome analysis of histone H3 lysine 4 and lysine 27 methylation in human embryonic stem cells. *Cell Stem Cell* **1,** 299–312 (2007).
5. Zhao, X. D., Han, X., Chew, J. L., Liu, J., Chiu, K. P., Choo, A., Orlov, Y. L., Sung, W. K., Shahab, A., Kuznetsov, V. A., Bourque, G., Oh, S., Ruan, Y., Ng, H. H., and Wei, C. L. Whole-genome mapping of histone H3 Lys4 and 27 trimethylations reveals distinct genomic compartments in human embryonic stem cells. *Cell Stem Cell* **1,** 286–298 (2007).
6. Ku, M., Koche, R. P., Rheinbay, E., Mendenhall, E. M., Endoh, M., Mikkelsen, T. S., Presser, A., Nusbaum, C., Xie, X., Chi, A. S., Adli, M., Kasif, S., Ptaszek, L. M., Cowan, C. A., Lander, E. S., Koseki, H., and Bernstein, B. E. Genomewide analysis of PRC1 and PRC2 occupancy identifies two classes of bivalent domains. *PLoS Genet* **4,** e1000242 (2008).
7. Pietersen, A. M. and van Lohuizen, M. Stem cell regulation by polycomb repressors: postponing commitment. *Curr Opin Cell Biol* **20,** 201–207 (2008).
8. Guenther, M. G., Levine, S. S., Boyer, L. A., Jaenisch, R., and Young, R. A. A chromatin landmark and transcription initiation at most promoters in human cells. *Cell* **130,** 77–88 (2007).
9. Brookes, E. and Pombo, A. Modifications of RNA polymerase II are pivotal in regulating gene expression states. *EMBO Rep* **10,** 1213–1219 (2009).
10. Kanhere, A., Viiri, K., Araujo, C. C., Rasaiyaah, J., Bouwman, R. D., Whyte, W. A., Pereira, C. F., Brookes, E., Walker, K., Bell, G. W., Pombo, A., Fisher, A. G., Young, R. A., and Jenner, R. G. Short RNAs are transcribed from repressed polycomb target genes and interact with polycomb repressive complex-2. *Mol Cell* **38,** 675–688 (2010).
11. Boyer, L. A., Plath, K., Zeitlinger, J., Brambrink, T., Medeiros, L. A., Lee, T. I., Levine, S. S., Wernig, M., Tajonar, A., Ray, M. K., Bell, G. W., Otte, A. P., Vidal, M., Gifford,

D. K., Young, R. A., and Jaenisch, R. Polycomb complexes repress developmental regulators in murine embryonic stem cells. *Nature* **441,** 349–353 (2006).

12. Chamberlain, S. J., Yee, D., and Magnuson, T. Polycomb repressive complex 2 is dispensable for maintenance of embryonic stem cell pluripotency. *Stem Cells* **26,** 1496–1505 (2008).
13. Jiang, H., Skukla, A., Wang, X., Chen, W. Y., Bernstein, B. E., and Roeder, R. G. Role for Dpy-30 in ES cell-fate specification by regulation of H3K4 methylation within bivalent domains. *Cell* **144,** 513–525 (2011).
14. Ang, Y. S., Tsai, S. Y., Lee, D. F., Monk, J., Su, J., Ratnakumar, K., Ding, J., Ge, Y., Darr, H., Chang, B., Wang, J., Rendl, M., Bernstein, E., Schaniel, C., and Leminschka, I. R. Wdr5 mediates self-renewal and reprogramming via the embryonic stem cell core transcriptional network. *Cell* **145,** 183–197 (2011).
15. Efroni, S., Duttagupta, R., Cheng, J., Dehghani, H., Hoeppner, D. J., Dash, C., Bazett-Jones, D. P., Le Grice, S., McKay, R. D., Buetow, K. H., Gingeras, T. R., Misteli, T., and Meshorer, E. Global transcription in pluripotent embryonic stem cells. *Cell Stem Cell* **2,** 437–447 (2008).
16. Roh, T. Y., Cuddapah, S., Cui, K., and Zhao, K. The genomic landscape of histone modifications in human T cells. *Proc Natl Acad Sci USA* **103,** 15782–15787 (2006).
17. Barski, A., Cuddapah, S., Cui, K., Roh, T. Y., Schones, D. E., Wang, Z., Wei, G., Chepelev, I., and Zhao, K. High-resolution profiling of histone methylations in the human genome. *Cell* **129,** 823–837 (2007).
18. Cui, K., Zang, C., Roh, T. Y., Schones, D. E., Childs, R. W., Peng, W., and Zhao, K. Chromatin signatures in multipotent human hematopoietic stem cells indicate the fate of bivalent genes during differentiation. *Cell Stem Cell* **4,** 80–93 (2009).
19. Adli, M., Zhu, J., and Bernstein, B. E. Genome-wide chromatin maps derived from limited numbers of hematopoietic progenitors. *Nat Methods* **7,** 615–618 (2009).
20. Christophersen, N. S. and Helin, K. Epigenetic control of embryonic stem cell fate. *J Exp Med* **207,** 2287–2295 (2009).
21. Hughes, J. R., Cheng, J. F., Ventress, N., Prabhakar, S., Clark, K., Anguita, E., De Gobbi, M., de Jong, P., Rubin, E., and Higgs, D. R. Annotation of cis-regulatory elements by identification, subclassification, and functional assessment of multispecies conserved sequences. *Proc Natl Acad Sci USA* **102,** 9830–9835 (2005).
22. Vernimmen, D., Marques-Kranc, F., Sharpe, J. A., Sloane-Stanley, J. A., Wood, W. G., Wallace, H. A., Smith, A. J., and Higgs, D. R. Chromosome looping at the human alpha-globin locus is mediated via the major upstream regulatory element (HS-40). *Blood* **114,** 4253–4260 (2009).
23. Bird, A. P., Taggart, M. H., Nicholls, R. D., and Higgs, D. R. Non-methylated CpG-rich islands at the human alpha-globin locus: implications for evolution of the alpha-globin pseudogene. *EMBO J* **6,** 999–1004 (1987).
24. Anguita, E., Hughes, J., Heyworth, C., Blobel, G. A., Wood, W. G., and Higgs, D. R. Globin gene activation during haemopoiesis is driven by protein complexes nucleated by GATA-1 and GATA-2. *EMBO J* **23,** 2841–2852 (2004).
25. Vernimmen, D., De Gobbi, M., Sloane-Stanley, J. A., Wood, W. G., and Higgs, D. R. Long-range chromosomal interactions regulate the timing of the transition between poised and active gene expression. *EMBO J* **26,** 2041–2051 (2007).

26. De Gobbi, M., Anguita, E., Hughes, J., Sloane-Stanley, J. A., Sharpe, J. A., Koch, C. M., Dunham, I., Gibbons, R. J., Wood, W. G., and Higgs, D. R. Tissue-specific histone modification and transcription factor binding in alpha globin gene expression. *Blood* **110,** 4503–4510 (2007).
27. Garrick, D., De Gobbi, M., Samara, V., Rugless, M., Holland, M., Ayyub, H., Lower, K., Sloane-Stanley, J., Gray, N., Koch, C., Dunham, I., and Higgs, D. R. The role of the polycomb complex in silencing alpha-globin gene expression in nonerythroid cells. *Blood* **112,** 3889–3899 (2008).
28. Goh, S. H., Terry Lee, Y., Bhanu, N. V., Cam, M. C., Desper, R., Martin, B. M., Moharram, R., Gherman, R. B., and Miller, J. L. A newly discovered human α-globin gene. *Blood* **106,** 1466–1472 (2005).
29. Gibson, J. D., Jakuba, C. M., Boucher, N., Holbrook, K. A., Carter, M. G., and Nelson, C. E. Single-cell transcript analysis of human embryonic stem cells. *Integr Biol (Camb)* **1,** 540–551 (2009).
30. Bhanu, N. V., Trice, T. A., Lee, Y. T., and Miller, J. L. A signaling mechanism for growth-related expression of fetal hemoglobin. *Blood* **103,** 1929–1933 (2004).
31. Schubeler, D., MacAlpine, D. M., Scalzo, D., Wirbelauer, C., Kooperberg, C., van Leeuwen, F., Gottschling, D. E., O'Neill, L. P., Turner, B. M., Delrow, J., Bell, S. P., and Groudine, M. The histone modification pattern of active genes revealed through genome-wide chromatin analysis of a higher eukaryote. *Genes Dev* **18,** 1263–1271 (2004).
32. Weishaupt, H., Sigvardsson, M., and Attema, J. L. Epigenetic chromatin states uniquely define the developmental plasticity of murine hematopoietic stem cells. *Blood* **115,** 247–256 (2010).
33. Hu, M., Krause, D., Greaves, M., Sharkis, S., Dexter, M., Heyworth, C., and Enver, T. Multilineage gene expression precedes commitment in the hemopoietic system. *Genes Dev* **11,** 774–785 (1997).
34. Rase, J. M. and O'Shea, E. K. Control of stochasticity in eukatyotic gene expression. *Science* **304,** 1811–1814 (2004).
35. Pope, S. H., Fibach, E., Sun, J., Chin, K., and Rodgers, G. P. Two-phase liquid culture system models normal human adult erythropoiesis at the molecular level. *Eur J Haematol* **64,** 292–303 (2000).
36. Brown, J. M., Leach, J., Reittie, J. E., Atzberger, A., Lee-Prudhoe, J., Wood, W. G., Higgs, D. R., Iborra, F. J., and Buckle, V. J. Coregulated human globin genes are frequently in spatial proximity when active. *J Cell Biol* **172,** 177–187 (2006).
37. Mansson, R., Hultquist, A., Luc, S., Yang, L., Anderson, K., Kharazi, S., Al-Hashmi, S., Liuba, K., Thoren, L., Adolfsson, J., Buza-Vidas, N., Qian, H., Soneji, S., Enver, T., Sigvardsson, M., and Jacobsen, S. E. Molecular evidence for hierarchical transcriptional lineage priming in fetal and adult stem cells and multipotent progenitors. *Immunity* **26,** 407–419 (2007).

PART V

THE ROLE OF ENVIRONMENT DURING EVOLUTION

CHAPTER 10

ADAPTIVE DIVERGENCE

NORA KHALDI and DENIS C. SHIELDS

CONTENTS

10.1 INTRODUCTION

The isoelectric point (*pI*) and charge of a protein is important for solubility, subcellular localization, and interaction. There is a correlation between subcellular location and protein *pI* [1, 2]. Proteins in the cytoplasm possess an acidic *pI* ($pI < 7.4$), while those in the nucleus have a more neutral *pI* ($7.4 < pI < 8.1$) [1, 2]. It has also been shown that the *pI* can vary greatly, depending on both insertion and deletions between orthologs, and the ecology of the organism [3]. Kirga et al. [3] have shown that the *pI* of membrane proteins of bacteria correlates with their ecological niche, and changes dramatically from acidic to basic. For example, some prokaryotes that infect human have a *pI* that reflects their localization in the human body, compensating for the *pH* change. *E. coli* that resides in the intestines has more acidic proteins, and *H. pylori* that infects the acidic stomach has more negatively charged proteins [3].

For highly abundant proteins, shifts in their *pI* can impact on the function of organs that interact with them. Purtell et al. [4] examined the effects of change in isoelectric point (*pI*) on renal handling of albumin molecules. The authors showed that the increase of the *pI* caused an increase in heterologous albumin secretion and increased nephron permeability.

Milk proteins travel through the various mammalian digestive systems with their different compartments and *pH* levels. For example, carnivorous species possess very acidic stomachs compared to herbivores, and orthologous milk proteins need to travel and perform their function in all these systems. Because of these differences we might expect to observe adaptation of the milk proteins in order to perform orthologous functions, or an adaptation of the *pI* to a new acquired functionality. Large differences in the *pI* of milk proteins might have important consequences on the structure, properties, functionality and interaction of these proteins.

In this chapter the evolutionary changes in the *pI* values of the milk proteins are investigated (Table 1) as a one-dimensional indicator of critical shifts between orthologous milk proteins that might reflect responses to environmental and functional changes between the different mammalian species.

TABLE 1 Function of Milk Proteins.

Protein	Role	Milk fraction
α-S1-casein, β-caseinm and κ- casein	~80% of bovine and 20–45% of human milk protein. Phosphoprotein carriers of minerals and trace elements	Casein micelles
a-lactalburnin	Calcium and other carrier, lactose synthesis [26]	Whey
Lactoferrin	Iron and other metal binding [27], antimicrobial, antiviral [28], antioxidative, cell growth	Whey
Lactadherin	Also known as Milk Fat globule factor 8 (Mfge8); bactericidal and apoptotic properties [29].	Milk fat globule; digestion resistant
Mucin 1	Modulates bacterial adhesion [29]	Milk fat globule; digestion resistant
Xanthine oxidase/dehydrogenase	Fat globule secretion [30] Innate immunity/oxidation [31]	Milk fat globule
Butyrophilin	~40% of protein in Milk Fat Fat Globule Membrane; fat globule secretion [29]	Milk fat globule; rapidly degraded

It is shown that the shifts do not simply reflect differences in sequence lengths between the milk orthologous proteins, and are likely driven by selection. Both sequence length and selection have been recently shown to explain the observed differences in *pI* between mammalian orthologs [5]. It is argued that the differences in the digestive systems due to *pH* and compartmentalization of the different mammals is not the sole driver of major changes in *pI*, and that these selective changes might be due to functional divergence of the protein.

10.2 METHODS

10.2.1 DATA

The human, chimp, monkey macaque, mouse, rat, guinea pig, rabbit, cat, dog, horse, cow, opossum, and platypus protein sequences were downloaded by FTP from the ENSEMBL database at: ftp://ftp.ensembl.org/pub/release-63/fasta/ website.

Out of seventeen identified major milk proteins [18] a subset for analysis on the basis of their belonging were picked to at least 8 mammalian species out of the 13 (Table 2). In addition the eight species needed to include human, chimp, cow, and mouse. These proteins represent the three parts of milk (Table 1): whey, casein, and milk fat globule. The nine major milk proteins defined in human and cow were used to detect their orthologs in the 13 other genomes, defined by reciprocal hits.

10.2.2 ORTHOLOGS AND SEQUENCE EVOLUTION

To find orthologous non-milk proteins, 13-way mutual best BLASTP hits among human were identified, chimp, monkey macaque, mouse, rat, guinea pig, rabbit, cat, dog, horse, cow, opossum, and platypus. This method resulted in 1,412 sets of putative orthologs that were present among all 13 species. Each set of 13 proteins was aligned using ClustalW [19].

10.2.3 CALCULATING THE ISOELECTRIC POINT

First the signal peptide were cleaved off from each protein using a HMM search with SignalP-HMM [20]. The rest of the sequence was incorporated into an in-house perl script for the calculation of the *pI* that uses the Henderson-Hasselbach equation. The script searched for the number of R, K, Y, C, H, E, and D that are implicated in the *pI* of a protein. Each of the previous amino acids was assigned a pK_a value, 12.48, 10.54, 10.46, 8.18, 6.04, 4.07, and 3.9 respectively, 8.0 for the N-terminus, and 3.1 for the Cterminus. The charge due to arginine for example is the product of the corresponding pKa with the number of instances or R in the sequence. Then an estimated charge for the protein can be calculated at any particular *pH*. To determine the *pI* that is the *pH* value at which the estimated charge is zero, an initial *pH* was estimated at which the overall charge of the protein is positive and one where the charge is negative. Then a bisection method was used to estimate to a 10^{-2} precision the value that renders the overall charge null.

10.2.4 DEFINING SIGNIFICANT PI SHIFTING PROTEINS

A protein is considered as significantly shifting in, for example mouse, if the distance between its *pI* and that of its ortholog in human is higher than a threshold that is determined from the differences in *pI* of all orthologs between human and mouse. Setting a threshold of *pI* between two species is somewhat arbitrary because the data does not follow a known distribution, for this reason a non-parametric formula to define the threshold of significance. This threshold is calculated using the median, and third quartile of the absolute shift in *pI* between orthologous proteins this is: threshold = 2 × (3rd quartile - median).

10.2.5 ANCESTRAL RECONSTRUCTION AND AMINO ACID SUBSTITUTION RATE

To reconstruct the ancestral sequences of the current κ-casein protein, the κ-casein orthologs were aligned in the 12 species represented in Figure 2 using T-coffee [21]; this step was followed by a maximum likelihood reconstruction using codeml from the paml package [22].

To calculate the amino acid substitution the DNA coding sequences of κ-casein proteins were gathered from the ENSEMBL database. Good quality sequences could not be located for guinea pig, cat, and dog. The other nine κ-casein protein orthologs were aligned using T-coffee [21]. The DNA sequences were aligned based on the protein alignment. Codeml was implemented [22] on the DNA alignment to calculate the synonymous *dS* and non-synonymous *dN* substitutions.

10.2.6 DETECTING SELECTION IN THE CHARGED RESIDUES

To examine if the significant variation between human, chimp, mouse, and cow, in amino acid composition is due to selection, the DNA coding sequences of all milk-specific proteins were gathered from the ENSEMBL database. The proteins were aligned using T-coffee [21] and implemented a script that aligns to DNA based on the protein's alignment. Poorly aligned positions and divergent regions of a DNA alignment were removed using Gblocks [23]. The SLR method was used with the default parameters to detect positions that are likely to be under positive selection [7]. These positions are indicated on Figure 4.

10.3 RESULTS AND DISCUSSION

10.3.1 CALCULATION OF PI

To investigate if the milk proteins have experienced shifts in their *pI* between different mammalian species, nine milk proteins were selected that share three main conditions; firstly they are representative of one of the three components of milk (casein, whey, milk-fat-globules); secondly they are present in at least eight mammalian species allowing for comparative genomics; finally the proteins possess a well characterized protein and cDNA sequence. The *pI* of the milk proteins was calculated after removing defined signal peptides. Some proteins show quite strong evolutionary conservation of *pI* (Figure 1). α-S1-casein, β-casein, α-lactalbumin, and butyrophilin subfamily 1 member A1 have only changed slightly between species and remain acidic through the tree (Figure 1). Similarly, xanthine dehydrogenase/oxidase is maintained in the neutral range in all mammals (Figure 1).

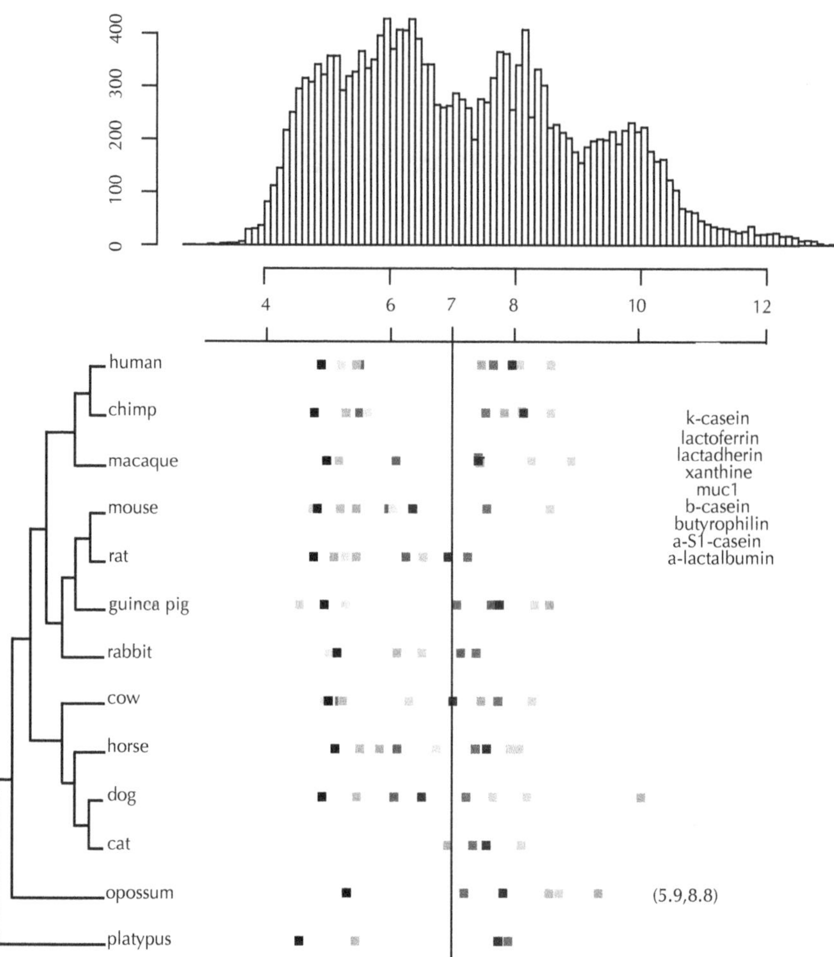

FIGURE 1 *pI* values for the nine major milk proteins in 13 mammalian species compared to the *pI* of the human proteome. The top histogram represents the *pI* distribution of the human proteome. The histogram's x-axis is shared with that of the major milk proteins' *pI* shown below. The Colors indicate the different milk proteins. The tree on the left is the mammalian species tree from Benton and Donoghue [23]. Two extra *pI* values are represented between brackets at the opossum level, these represent the *pI* of the reported extra copies of κ-casein this species possess [6]. The values are for the proteins with the accession numbers FJ548612 and FJ548626 respectively.

However, some proteins show more dramatic changes in one or multiple branches on the tree. Thus, κ-casein *pI* has apparently, under a parsimonious model, shifted from a basic ancestor to an acidic *pI* on the branch prior to the speciation of rodents, guinea pig, and rabbit (mouse *pI* = 4.75, rat *pI* = 6.53, guinea pig *pI* = 4.53, and rabbit *pI* = 6.51). Nevertheless, rat and rabbit are substantially less acidic than mouse and guinea pig, suggesting that more than one change in constraint on κ-casein *pI* during evolution in these lineages. κ-casein in cow has a much lower *pI* than horse, again suggesting an independent shift in constraint. Indeed, the most parsimonious scenario accounting for the current κ-casein *pI* values represented in Figure 1 and Figure 2 will require two changes in the ancestors of mouse and cow from an ancestral basic *pI* value to a more acidic observed value in both these species. In contradiction to this result, an ancestral reconstruction shows that the ancestor of κ-casein carried an acidic *pI*, and that further on in evolution this value shifted in a multitude of species to the current observed basic values (Figure 2 shows at least four independent shifts: in macaque, the ancestor of human and chimp, horse, and the ancestor of dog and cat). Besides according to this reconstruction all the current *pI* values are higher than the ancestral values (Figure 2), including the *pI* values of mouse and rat κ-casein (Figure 2). However it is known that ancestral reconstruction is somewhat unreliable especially at sites with alignment gaps. Thus it cannot argue for such a scenario, and from the current value a parsimonious scenario with fewer events is more likely to explain the current *pI* values in the κ-casein orthologs (Figure 1 and Figure 2).

It has been shown that platypus contains two extra copies of κ-casein [6]. These two copies have very different *pI* values ranging from acidic to basic, with *pI* = 5.9 for FJ548612, to *pI* = 8.8 for FJ548626 (Figure 1). Contrary to the other observed shifts in *pI* represented in Figure 1, the great shift in *pI* between the κ-casein copies cannot be explained by interspecies differences. It is noticeable that the pI of the current κ-casein orthologs is much higher than that of the ancestor values (Figure 2). However mouse and guinea pig seem to be an exception to this observation. It is unclear how much this is due to real *pI* shifts or to and artifact of the method of *pI* calculation.

Lactadherin has shifted at least twice on the tree. It is basic in the two outgroup species opossum (*pI* = 7.83) and platypus (*pI* = 7.75), in primates (*pI* = 7.96 human, *pI* = 8.17 chimp, *pI* = 7.42 macaque), and in guinea pig

(pI = 7.76), but seems to have shifted independently twice to acidic/neutral, once in rodents (pI = 6.36 in mouse, and pI = 6.9 in rat), and another time in dog (pI = 6.45 in dog).

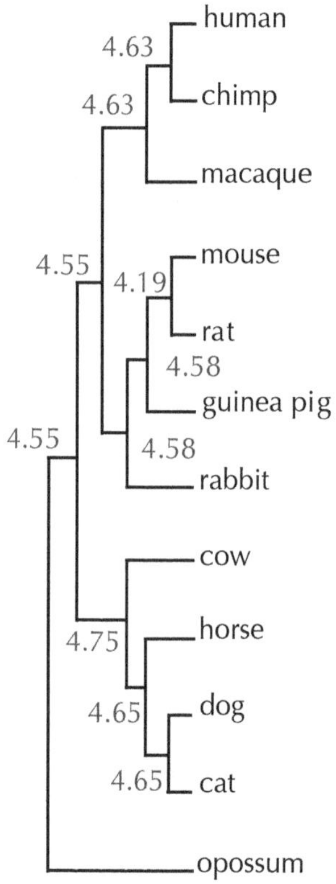

	pI	dN/dS	Stomach pH	
			Anterior	Posterior
human	9.32	0.49	1.5-5.0	5.0-7.0
chimp	8.82	---		
macaque	9.32	1.1	4.7-5.0	2.3-2.8
		0.41		
mouse	4.55	2.45	4.5	3.1
rat	6.51	0.7	4.3-5.1	2.3-4.0
guinea pig	4.22			
rabbit	6.71	1.07	1.9-2.2	1.9-2.2
cow	6.36	1.14		
horse	8.17	1.67		
dog	8.16		1.5-5.5	1.5-3.4
cat	8.16			
opossum	8.92	0.53		

FIGURE 2 Ancestral reconstruction of κ-casein and the representation of *pI, dN/dS* ratio and the stomach *pH* values. Ancestral values are represented in gray. The ratio *dN/dS* was calculated only for species with a well-defined cDNA (this is not the case for guinea pig, cat, and dog). When the ratio *dN/dS* is undefined due to extremely small *dS* values the symbol "--" was used. Despite the fact that the *pH* of other digestive compartments can show marked differences between mammals, only to represent the stomach *pH* values were chosen, as this compartment is the main first barrier for milk proteins. A review of the *pH* values is reported in Table 5 of the following reference [24] (p366), it could not found well-defined values for chimp, horse, cow, and guinea pig.

The pattern of *pI* change of muc1 protein shows a number of potential changes, shifting in two independent lineages to a lower *pI* in both rodents (mouse *pI* = 5.45, and rat *pI* = 5.09) and horse (*pI* = 5.83).

10.3.2 ARE THE SHIFTS IN THE PI OF SOME MILK PROTEINS IMPORTANT COMPARED TO WHOLE PROTEOME COMPARISON?

What appear as dramatic changes between the *pI*s of κ-casein, lactadherin, and muc1 orthologs, might not seem so dramatic compared to the changes across the entire proteome for non-milk protein orthologs.

To investigate this, all the orthologous proteins in the 13 mammals (human, chimp, monkey, mouse, rat, guinea pig, rabbit, cow, horse, dog, cat, opossum, platypus) were considered. A shift in *pI* between human and mouse were considered to be high if it was greater than 0.92, and an identical value between human and cow (See Methods for the rationale behind the choice of these cut-offs). Further it was tested for the significance of this threshold by randomly assigning *pI* values to proteins, and found that our set thresholds are in all cases significant ($p \leq 0.01$).

Figure 3 shows that κ-casein, lactadherin, and muc1 stand out on the figure as being part of a very small proportion of proteins that have shifted dramatically in *pI*, from being basic in man to being acidic in mouse (κ-casein, and lactadherin), from being basic/neutral in human to being acidic in mouse (muc1), or from being neutral/acidic in cow to being basic in man (κ-casein). Figure 3 also shows that most proteins conserve their *pI* despite the evolutionary distances separating human, mouse, and cow. These large shifts seen for certain milk proteins are therefore unexpected for typical proteins that have conserved their function in evolution.

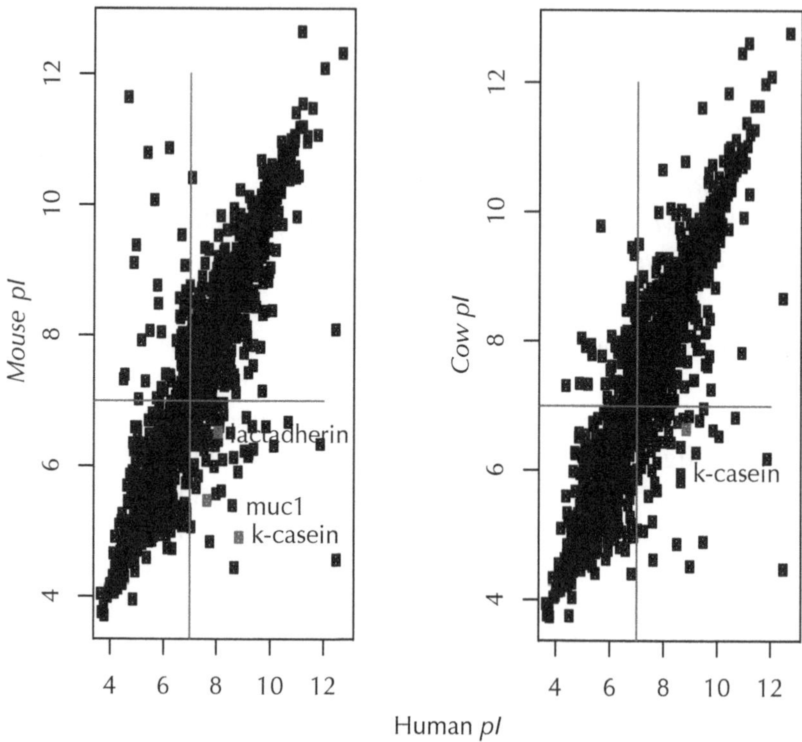

FIGURE 3 *pI* Values of all the Orthologous Proteins in Human, Mouse, and Cow. The vertical and horizontal lines represents the neutral $pH = 7$. Most proteins have a similar *pI* between species, with some exceptions lying out on both sides of the diagonal. The red dots represent the three milk proteins that have the highest shift (κ-casein, lactadherin, and muc1).

10.3.3 DIFFERENCES IN LENGTH BETWEEN ORTHOLOGS DUE TO INSERTIONS OR DELETIONS ARE ASSOCIATED WITH THE PISHIFT IN CERTAIN PROTEINS

The change in *pI* between the milk proteins may reflect amino acid replacement at a number of residues, or they might be due to large insertions or deletions that cause large changes in *pI*. This has been shown to be the major reason behind the shift in *pI* between mammal proteins carried out by Alendé and co-authors [5]. For κ-casein, the shifts do not appear to relate to size differences, since the sequence length between human and mouse is very similar, and the extra amino acid in human does not account for the difference (Table 2). However, noticeable changes were observed in length between lactadherin and muc1. For lactadherin, human is 76 residues shorter than mouse (Table 2). When the regions in mouse that are not aligned with those found in human are removed, the *pI* is 7.7, close to that of human *(pI* = 8.0). For muc1, the human protein is much longer than the mouse sequence. However, the *pI* of the human regions alignable with mouse muc1 was 7.12, broadly similar to the *pI* of the overall protein (7.47).

These results show that for Lactadherin the change in *pI* is mainly due to the mouse insertion. However this scenario does not account for the change in *pI* for muc1 and κ-casein where both shifts are accounted for by amino acid replacements between human and mouse.

10.3.4 SELECTION CAUSING PI CHANGE

Can selection have contributed to the change in *pI*? A recent study of the *pI* of mammalian proteins argues that selection has contributed to some of the *pI* shifts between orthologous proteins [5]. We searched for evidence of positive selection using the Sitewise Likelihood Ratio (SLR) method for the estimation of selection [7] in each site of the alignment of human, mouse, and cow for muc1 and κ-casein. SLR is a direct test of whether a particular site is evolving in a non-neutral fashion, inspecting the excess of non-synonymous over synonymous DNA changes; indicates which sites in the protein have strong evidence of positive selection, which correspond to sites that are unusually variable. For κ-casein evidence of 14 sites presenting positive selection were found ($p \leq 0.043$; Figure 4). Eleven of these sites change the *pI* of the protein, and 7 of those also change the overall charge of the protein at neutral *pH*. Only four positively selected sites have not affected the *pI* of the protein, and are not known to be implicated in any side modifications of the protein. It was found that there are significantly more sites that affect the *pI* that have undergone positive selection compared to all other sites that do not affect the *pI*. Thus, there are significantly more sites undergoing positive selection and that have an impact of the net charge of the protein compared to all other neutral sites

Table 2 Sequence lengths for eight milk proteins in human, mouse, and cow.

	Muc1	**lactadherin**	**κ-casein**	**β-casein**	**α-casein**	**butyrophilin subfamily 1 member**	**xanthine dehydrogenase/oxidase**	**lactoferrin**
Human	1255	387	182	226	185	526	1333	710
Mouse	630	463	181	231	313	524	1335	707
Cow	580	427	190	224	214	526	1332	708

(p = 0.03; 22% for charged residues versus 5% for neutral sites). Under a random distribution of the positively selected sites detected in the human κ-casein protein sequence, an average of 8.4% sites will be expected that undergo positive selection whether these are charged or neutral, which is less than the observed 22% charged sites that have undergone positive selection.

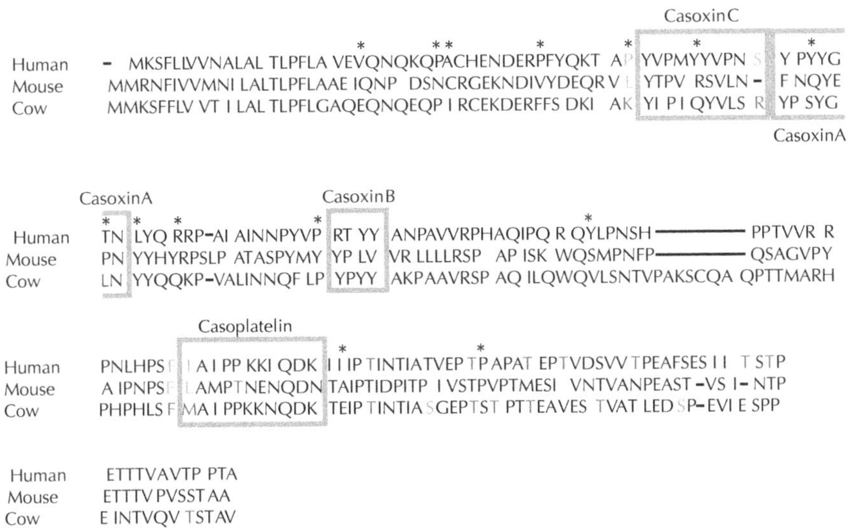

FIGURE 4 Alignment of κ-casein between human, mouse, and cow. The sequences of casoxin peptides A, B, and C are in the pink colored boxes. Cleavage sites are to the right of the red residues, while green residues are the corresponding residues that are not cleavable by the same enzyme in human and mouse. Casoxin-A and C are cleaved by a pepsin-trypsin digest for the former, and a trypsin digest for the later [12]. The peptide Casoplatelin [25] that inhibits ADP-induced platelet aggregation and fibrinogen binding is also represented on the figure together with the chymosin/rennin cleavage site between the F and M residues in red (while the same positions are in green in human and mouse). Horizontal lines represent gaps. Stars indicate sites that were predicted to be under positive selection (see results). Orange residues have been shown in the literature to undergo phosphorylation. Blue residues have been shown in the literature to undergo glycosylation. One potential phosphorylation site indicated in lavender in mouse.

Given that so many residues are experiencing adaptation in the human κ-casein and have a direct impact on the *pI* argues for adaptive changes in the *pI* of κ-casein. To further examine if positive selection has played a role in the evolution of κ-casein, the ratio of the rate of non-synonymous was calculated over synonymous substitutions (*dN/dS*). Figure 2 shows that the mouse κ-casein has undergone the greatest ratio indicating the action of positive selection on this protein in the mouse lineage. This also happens to correspond to the lineage undergoing the highest shift in *pI* (Figure 1, Figure 2). This positive selection seems to have consequently shifted the *pI* of mouse κ-casein. Two other orthologs seem to have also undergone some sort of fast evolutionary divergence (Figure 2 shows that horse and cow have $dN/dS > 1$) even though the *dN/dS* value might be too weak to speak about positive selection, the cow ortholog happens to have also diverged in its *pI* (Figure 1 and Figure 2, horse however seems to have diverged in sequence but retained a closer basic *pI* to the other mammals studied in this chapter.).

For muc1, 25 sites under positive selection ($p <= 4.7.10^{-2}$) were detected, 15 of these have changed the overall *pI* of the protein and also changed its net charge. Here again, it is found that there are significantly more sites undergoing positive selection and that these have an impact on the net charge of the protein compared to all other neutral sites ($p = 2.4 \times 10^{-9}$; 28% for charged residues versus 1.9% for neutral residues). Under a random distribution of the positively selected sites detected in the human muc1 protein sequence, an average of 4.4% sites will be expected that undergo positive selection whether these are charged or neutral, which is less than the observed 28% charged sites that have undergone positive selection.

Put together these results show that selection has played a part in the change of *pI* and consequently on the overall net charge of the protein.

10.3.5 SELECTION PRESSURES FOR CHANGES IN PI: THE ROLES OF DIETARY, MORPHOLOGICAL, AND INTRINSIC MILK PROTEIN FACTORS

What is driving this selection on the *pI*? Can it be the important differences in *pH* and compartmentalization between the digestive systems of different mammals? [8] Milk proteins travel down the digestive system. Some, such as the caseins, get broken down in the highly acidic conditions of the stomach, whereas others such as lactadherin and lactoferrin [9, 10] travel intact or partially intact to be broken down further down in the digestive tract. Given the very large shifts in *pI*, it would be anticipated that the processing and breakdown of milk proteins are likely to differ substantially. Thus, if it was to replace the human κ-casein with that of mouse, it seems unlikely that they will interact with their environment and function in an identical way, given that the mouse and human κ-casein *pI* is 4.75 in mouse, but 8.59 in human.

It might be imagined that the greatest shifts during evolution might occur when animals shift between largely carnivorous or omnivore diets and herbivore diets, since the more complex stomachs of some herbivores, and the more acid stomach *pH*s of some carnivores might alter functional constraints. However, inspection of Figure 1 indicates that many large shifts occur between species that have largely similar overall dietary strategies (dog and cat; mouse and rat). This suggests that the shifts in functional constraints may be associated with factors that are not linked with the gross morphology or diet of major clades. Similarly, the values of the posterior stomach *pH* in the different mammals represented in Figure 2 do not clearly argue for a stomach-*pH* change that is driving the shift in *pI* for κ-casein, including the significant *pI* shift observed in mouse (Figure 1, Figure 2, and Figure 3). Besides, the great difference observed between the *pI* values of the two extra copies of κ-casein in platypus (Figure 1; *pI* = 5.9 for FJ548612, to *pI* = 8.8 for FJ548626) does not argue for a stomach *pH* driven selection on milk proteins' *pI*.

It is interesting to speculate on how extrinsic factors, such as commensal and pathogenic bacteria, may exert selection pressures on milk protein function, but also of interest to consider how alterations in intrinsic milk protein functions may relate to adaptive changes. Milk proteins are known to yield many bioactive peptides that modulate and participate in various regulatory processes in the body [11]. These peptides are usually cleaved by digestive enzymes such as trypsin, pepsin, and chymotrypsin. Some proteases cleave near positively charged residues, such as trpysin, while others avoid positive charge in their substrate region (pepsin), and the adaptive requirements for the gain and loss of proteolytic cleavage sites in certain regions of the gut (e.g. the duodenum versus the stomach) may have some an impact on *pI*. In particular, when it is considered that the casoxins [12], known bioactive peptides released from bovine κ-casein that have opioid antagonist and anti-opioid activities- it is noted that although casoxin A, and C are released in cow, this is not the case in human and mouse, since the cleavage sites are not the same between the species (Figure 4). It is interesting to note that 3 residues of the 14 residues that are found to be positively selected on in κ-casein are found on the borders of the three peptides casoxin A, B, and C (Figure 4), indicating possible selection on the cleavage sites. Also, Figure 4 shows that three other positively selected sites are located within the peptides casoxin A, and B sequence, indicating adaptation of the individual peptides at least to cow. Thus, the shift in *pI* may be associated with divergence in functional requirements for either rates of digestion, or for functional components of the milk.

10.3.6 PHOSPHORYLATION AND GLYCOSYLATION

It is observed that all the proteins that have shifted dramatically are ones that also happen to be highly glycosylated and phosphorylated. Indeed the three proteins κ-casein, muc1, and lactadherin have more glycosylation sites than the other milk proteins with an average of seven glycosylations in human (nine glycosylations in muc1, seven in κ-casein, and five in lactadherin; these include referenced, probable, and potential sites), and in cow, as opposed to an average of 1.3 in the remaining six milk proteins in human, and in cow. Besides, also observed differences are done in phosphorylation sites, for example we have three referenced phosphorylations in cow κ-casein and none in human and mouse. Also, there are nine referenced phosphorylations in human muc1, while there are six and seven by similarity in cow and mouse respectively.

The analyzes of *pI* did not take into account these post-translational modifications. To examine if post-translational modifications can reduce the difference in the isoelectric point, experimentally validated phosphorylation and glycosylation sites were used, which are defined in cow, human and to a weaker extent in mouse. For κ-casein (Figure 4), the cow *pI* shifts from 5.93 to 5.34 when the two experimentally verified phosphorylations are added. Human remains the same $pI = 8.68$ (no experimentally validated phosphorylation so far), and mouse shifts from 4.67 to 4.52 (1 potential phosphorylation site; Figure 4.). The phosphorylation sites for muc1 in both cow and mouse are potential sites found with similarity rather than experimentally validated sites. These results show that despite shifting the *pI* of κ-casein and muc1 towards a more acidic *pH* as a result of phosphorylation in the three different species, the difference in *pI* remains very important between these two proteins.

The differences in glycosylation between human and cow for κ-casein might somewhat further reduce the *pI* shift between both these species. Indeed, in κ-casein (Figure 4) we have seven glycosylations in human as opposed to six in cow (none have been experimentally validated so far in mouse). For muc1, experimental validation is only available for human that has four O-linked, and five N-linked glycosylations. These might also narrow down the gap in the muc1 *pI* between the different species. Nonetheless, both cases where the *pI* difference is reduced or not are interesting. Indeed if the *pI* difference is reduced and becomes very close between both species, this reflects that the protein has adapted its *pI* so that the final product with the different number of glycosylations and phosphorylations becomes the same. Indeed, if the *pI* was initially not different, the addition of glycosylation will then further the gap between the *pI*s.

10.4 CONCLUSION

Although the production of milk is conserved between mammals for over 190 MA, our result argue that common proteins that have been shared by mammals are functionally diverging. Many humans consume cow's milk on a daily basis, and yet the *pI* of κ-casein in cow is very different from this κ-casein. It is shown that selection has acted on the residues that affect the protein's *pI*. The simplest explanation was the adaptation of the protein to the different digestive systems to accommodate reactions to changes in *pH* of the different compartments. However, it is found that the pattern of change did not correlate strongly with the greatest shifts in compartmentalization and *pH* during evolution, suggesting that other factors, potentially including milk proteins' functional features, may be associated with the adaptive changes.

Differences in the function of κ-casein between various species, raises the question of whether κ-casein of cow can functionally replace that of human. κ-casein is known to yield many bioactive peptides [12, 13] which, as it has been discussed, might have different affinities and functionalities between human and cow. Such functional changes may relate to regional positive selection seen within κ-casein in the family bovidae [14].

It is of interest to note that two of the proteins showing the most striking shifts in *pI* are also glycosylated extensively (κ-casein and muc1). It is not clear if this is merely coincidental, or whether glycosylated proteins play a particular role in the gut that is subjected to shifting selection pressures over evolutionary time. An obvious candidate function would be bacterial interactions, which are heavily influenced by glycosylated proteins, and κ-casein is known to play a role in altering *Helicobacter pylori* adhesion [15] (review [16]). Exactly how shifting the *pI* of these milk proteins might benefit the neonate is not entirely clear. However, given the ability of pathogens such as *H. pylori* to modify the host stomach *pH* [17], the ability of milk proteins to coat particular compartments or infected regions of altered *pH* is an obvious candidate factor to investigate. In this context, a specific question raised by this study is whether the muc1 and κ-casein in cow's milk provide optimal protection against bacterial infections of the stomach and intestine for human neonates.

KEYWORDS

- **Glycosylation**
- **Isoelectric Point**
- **κ-Casein**
- **Milk Proteins**
- **Orthologs**

ACKNOWLEDGMENT

This chapter was funded by the Irish Research Council for Science, Engineering & Technology (IRCSET), Marie Curie Actions under FP7, and Food for Health Ireland (FHI).

COMPETING INTERESTS

The authors declare that they have no competing interests.

AUTHORS' CONTRIBUTIONS

NK and DS conceived the study and wrote the paper. NK also carried out the analysis. Both authors read and approved the final manuscript.

REFERENCES

1. Andrade, M. A., O'Donoghue, S. I., and Rost, B. Adaptation of protein surfaces to subcellular location. *J Mol Biol* **276**(2), 517–525 (1998).
2. Nandi, S. Comparison of theoretical proteomes: identification of COGs with conserved and variable pI within the multimodal pI distribution. *BMC Genomics* **6,** 116 (2005).
3. Kiraga, J. The relationships between the isoelectric point and: length of proteins, taxonomy and ecology of organisms. *BMC Genomics* **8,** 163 (2007).
4. Purtell, J. N. Isoelectric point of albumin: effect on renal handling of albumin. *Kidney Int* **16**(3), 366–376 (1979).
5. Alendé, N. Evolution of the isoelectric point of mammalian proteins as a consequence of indels and adaptive evolution. *Proteins* **79**(5), 1635–1648 (2011).
6. Kraft, C., Reggiori, F., and Peter, M. Selective types of autophagy in yeast. *Biochim Biophys Acta* **1793**(9), 1404–1412 (2009).
7. Massingham, T. and Goldman, N. Detecting amino acid sites under positive selection and purifying selection. *Genetics* **169**(3), 1753–1762 (2005).
8. Martinez, M. Applying the biopharmaceutics classification system to veterinary pharmaceutical products. Part II. Physiological considerations. *Adv Drug Deliv Rev* **54**(6), 825–850 (2002).
9. Troost, F. J. Gastric digestion of bovine lactoferrin in vivo in adults. *J Nutr* (131), 2101–2104 (2001).

10. Peterson, J. A. Human milk fat globule (HMFG) glycoproteins: their association with lipid micelles in skim milk and survival in the stomach of milk-fed and pre-term infants. *Pediatric Research* **41**(4, Part 2), 87 (1997).
11. Meisel, H. Overview on Milk Protein-derived Peptides. *Int Dairy J* **8**(5–6), 363–373 (1998).
12. Chiba, H., Tani, F., Yoshikawa, M. Opioid antagonist peptides derived from kappa-casein. *J Dairy Res* **56**(3), 363–366 (1989).
13. Murray, B. A., FitzGerald, R. J. Angiotensin converting enzyme inhibitory peptides derived from food proteins: biochemistry, bioactivity and production. *Curr Pharm Des* **13**(8), 773–791 (2007).
14. Ward, T. J., Honeycutt, R. L., and Derr, J. N. Nucleotide sequence evolution at the kappa-casein locus: evidence for positive selection within the family Bovidae. *Genetics* **147**(4), 1863–1872 (1997).
15. Stromqvist, M. Human milk kappa-casein and inhibition of Helicobacter pylori adhesion to human gastric mucosa. *J Pediatr Gastroenterol Nutr* **21**(3), 288–296 (1995).
16. Lonnerdal, B. Nutritional and physiologic significance of human milk proteins. *Am Clin J Nutr* **77**(6), 1537S–1543S (2003).
17. Argent, R. H. Toxigenic Helicobacter pylori infection precedes gastric hypochlorhydria in cancer relatives, and H pylori virulence evolves in these families. *Clin Cancer Res* **14**(7), 2227–2235 (2008).
18. Rudolph, M. C., McManaman, J. L., Hunter, L., Phang, T., and Neville, M. C. Functional development of the mammary gland: use of expression profiling and trajectory clustering to reveal changes in gene expression during pregnancy, lactation, and involution. *J Mammary Gland Biol Neoplasia* **8**(3), 287–307 (2003).
19. Thompson, J. D., Higgins, D. G., and Gibson, T. J. CLUSTAL W.: improving the sensitivity of progressive multiple sequence alignment through sequence weighting, position-specific gap penalties and weight matrix choice. *Nucleic Acids Res* **22**(22), 4673–4680 (1994).
20. Nielsen, H. and Krogh, A. Prediction of signal peptides and signal anchors by a hidden Markov model. *Proc Int Conf Intell Syst Mol Biol* **6,** 122–130 (1998).
21. Notredame, C., Higgins, D. G., and Heringa, J. T-Coffee. A novel method for fast and accurate multiple sequence alignment. *J Mol Biol* **302**(1), 205–217 (2000).
22. Yang, Z. PAML 4: phylogenetic analysis by maximum likelihood. *Mol Biol Evol* **24**(8), 1586–1591 (2007).
23. Castresana, J. Selection of conserved blocks from multiple alignments for their use in phylogenetic analysis. *Mol Biol Evol* **17**(4), 540–552 (2000).
24. Benton, M. J. and Donoghue, P. C. Paleontological evidence to date the tree of life. *Mol Biol Evol* **24**(1), 26–53 (2007).
25. DeSesso, J. M. and Williams, A. L. Annual Reports in Medicinal Chemistry. In J. E. Macor (Ed.) *Contrasting the Gastrointestinal Tracts of Mammals*, Vol. 43. Wallingford CT, United States, pp. 353–371 (2008).
26. Jollès, P. Analogy between fibrinogen and casein. Effect of an undecapeptide isolated from kappa-casein on platelet function. *Eur Biochem J* **158**(2), 379–382 (1986).
27. Permyakov, E. A. and Berliner, L. J. alpha-Lactalbumin: structure and function. *FEBS Lett* **473**(2), 269–274 (2000).

28. Baker, E. N. and Baker, H. M. Molecular structure, binding properties and dynamics of lactoferrin. *Cell Mol Life Sci* **62**(22), 2531–2539 (2005).
29. Harmsen, M. C. Antiviral effects of plasma and milk proteins: lactoferrin shows potent activity against both human immunodeficiency virus and human cytomegalovirus replication in vitro. *J Infect Dis* **172**(2), 380–388 (1995).
30. Peterson, J. A. Structural and functional aspects of three major glycoproteins of the human milk fat globule membrane. *Adv Exp Med Biol* **501,** 179–187 (2001).
31. Vorbach, C., Scriven, A., and Capecchi, M. R. The housekeeping gene xanthine oxidoreductase is necessary for milk fat droplet enveloping and secretion: gene sharing in the lactating mammary gland. *Genes Dev* **16**(24), 3223–3235 (2002).
32. Silanikove, N. Role of xanthine oxidase, lactoperoxidase, and NO in the innate immune system of mammary secretion during active involution in dairy cows: manipulation with casein hydrolyzates. *Free Radic Biol Med* **38**(9), 1139–1151 (2005).

AUTHOR NOTES

CHAPTER 1

Acknowledgments

We thank Dr. Axel Schumacher for his help with drawing figures for this article. This research has been supported by the Special Initiative grant from the Ontario Mental Health Foundation and also by NARSAD, the Canadian Psychiatric Research Foundation, the Stanley Foundation, the Juvenile Diabetes Foundation International and the Crohn's and Colitis Foundation of Canada to A.P. The Epigenetic Perspective was originally published as "Phenotypic Differences in Genetically Identical Organisms: the Epigenetic Perspective" in Human Molecular Genetics 2004, 14 (supp. 1): R11-R18. Used with permission from Oxford University Press.

Author Affiliations

The Centre for Addiction and Mental Health, Departments of Psychiatry and Pharmacology, and the Institute of Medical Science at the University of Toronto, Ontario, Canada and Department of Psychiatry and Psychology, University of Minnesota, Minneapolis, MN, USA.

CHAPTER 2

Competing Interests

The author(s) declare that they have no competing interests.

Author Contributions

LS created the program and carried out the experiments. NC provided advice and resources to perform all of the required tests and design the algorithm and helped draft the manuscript. All authors read and approved the final manuscript.

Acknowledgments

Thank you to IBM for the use of the Blade Center computer system and to Steven Chen for the assistance in getting the experiments up and running. Correlating CpG Islands, Motifs, and Sequence Variants was originally published as "Correlating CpG Islands, Motifs, and Sequence Variants in Human Chromosome 21" in BMC Genomics 2011, 12(Suppl 2): S10. Used with permission.

CHAPTER 3

Competing Interests

The authors declare that they have no competing interests.

Author Contributions

DP: Conception and design, collection and assembly of data, data analysis and interpretation, manuscript writing. RJP: Collection and assembly of data, data analysis and interpretation. FMF: Data analysis and interpretation, manuscript writing, final approval of the manuscript. EEO, PAB, MCL, MJS and MSS: Provision of study material or patient samples, collection and assembly of data. ATC and NJM: Conception and design, manuscript writing, final approval of the manuscript. All authors read and approved the final manuscript.

Acknowledgments

We thank Dr. Simon Hayward for provision of BPH-1 cells, Dr. Johng Rhim for provision of RC-165N/hTERT and RC-92a/hTERT cells and Dr. David Hudson for provision of Bob and SerBob cells. We thank Dr. Pablo Navarro for his help with chromatin immunoprecipitation experiments. This work was supported by Yorkshire Cancer Research and The Freemasons' Grand Charity. Promoter Hypermethylation was originally published as "Regulation of the Stem Cell Marker CD133 Is Independent of Promoter Hypermethylation in Human Epithelial Differentiation and Cancer" in Molecular Cancer 2011, 10:94. Used with permission.

CHAPTER 4

Competing Interests

The authors declare that they have no competing interests.

Author Contributions

RKCY and WPR conceived the study. RKCY designed and performed the experiments. RJ prepared and karyotyped the samples. MSP performed the microarray experiment. RKCY analyzed the data. DEM contributed the tissue samples. RKCY and WPR wrote the paper. All authors read and approved the final manuscript.

Acknowledgments

We thank David Chai and Danny Leung for their technical assistance and advice, Alicia Murdoch and Jennifer Sloan for placenta donor recruitment, Dr. Angela Devlin for use of the Biotage PyroMark™ MD system, Dr. Michael Kobor for use of the Illumina array and Dr. Matthew Lorincz for the use of reagents for cloning. We also thank Dr. Louis Lefebvre for critical review of the manuscript. This work was funded by a grant from the Canadian Institutes for Health Research (to WPR) and by a graduate student scholarship from the Child & Family Research Institute (to RKCY). Genome-Wide Mapping was originally published as "Genome-Wide Mapping of Imprinted Differentially Methylated Regions by DNA Methylation Profiling of Human Placentas from Triploidies" in Epigenetics Chromatin 2011 July 13;4(1):10. Used with permission from the U.S. National Library of Medicine.

CHAPTER 5

Competing Interests

The authors declare that they have no competing interests.

Author Contributions

IM carried out most of the research. PG performed ChIP-qPCR. JB and JPB derived the KO mESCs. MB performed bioinformatics analysis. DM and PS contributed reagents. IM and ML designed the study, analyzed the data and wrote the manuscript. All authors read and approved the final manuscript.

Note Added in Proof

While this chapter was under review, an article published by Shang and colleagues (PNAS 2011, 108(18):7541-7546) revealed that the H3K9me3 demethylase JMJD2B greatly facilitates H3K4 methylation by purified MLL2 *in vitro* (demonstrating that H3K9 demethylation is required for efficient H3K4 methylation) and is required for transcription of MLL2 targets *in vivo*.

Acknowledgments

We thank Danny Leung, Sandra Lee, Lucia Lam and the University of British Columbia flow cytometry facility for technical support. We also thank Florian Lienert and Dirk Schübeler for providing the HA36 ES cell line, En Li for providing the *Dnmt1*-KO line, Yoichi Shinkai for providing the *Setdb1* KO line and Mark Bedford for helpful suggestions. This work was supported by CIHR grant 77805 (to ML) and CIHR grant 92090 (to ML and DM). This work was also supported by Biotechnology and Biological Sciences Research Council core strategic grants and Deutsche Forschungsgemeinschaft grant SI 1209/2-1 (to PS). ML is a Scholar of the MSFHR and a CIHR New Investigator. Gene Silencing was originally published as "H3K9me3 Binding Proteins Are Dispensable for SETDB1/H3k9me3-Dependent Retroviral Silencing" in Epigenetic Chromatin 2011 July 20;4(1):12. Used with permission from the U.S. National Library of Medicine.

CHAPTER 6

Competing Interests

The authors declare that they have no competing interests.

Author Contributions

AG designed the study, performed cell culturing and flow cytometry, and wrote most of the manuscript; BS performed RP-HPLC and HPCE, and wrote parts of the manuscript; HG performed flow sorting and wrote parts of the manuscript; AL isolated proteins; HL designed RP-HPLC and HPCE analysis, analyzed RP-HPLC and HPCE data, and helped supervise the project; and IR conceived and supervised the project, and wrote the final manuscript. All authors read and approved the final manuscript.

Acknowledgment

We are most grateful to Dr. Marie Larsson for expert help with T-cell purification and activation. This work, as part of the European Science Foundation EUROCORES Programme EuroDYNA, was supported by funds from the Austrian Science Foundation (I23-B03), the EC Sixth Framework Programme under contract number ERAS-CT-2003-980409, and the Swedish Cancer Society. Histone (H1) Phosphorylation was originally published as "Histone H1 Interphase Phosphorylation Becomes Largely Established in G_1 or Early S Phase and Differs in G_1 Between T-Lymphoblastoid Cells and Normal T Cells" in Epigentics Chromatin 2011 August 5;4:15. Used with permission from the U.S. National Library of Medicine.

CHAPTER 7

Competing Interests

The authors declare that they have no competing interests.

Author Contributions

SDB carried out the experiments, data analysis, and drafted the manuscript. SDT and AJT conceived of the study and participated in its design and coordination. AJT helped to draft the manuscript. All authors read and approved the final manuscript.

Acknowledgments

This work was funded by NIH R01DA025755, P20RR015569, P20RR016460 and F32GM093614. Post-Translational Modifications was originally published as "Quantitative Analysis of Histone Exchange for Transcriptionally Active Chromatin" in the Journal of Clinical Bioinformatics 2001, 1:17. Used with permission from the U.S. National Library of Medicine.

CHAPTER 8

Competing Interests

The authors declare that they have no competing interests.

Author Contributions

AK conceived of and designed the study. BJT, EDR and AK performed the experiments. PÓB, AAG, JMG and NK provided bioinformatics support and carried out the statistical analyzes. PW and KB contributed the samples. BJT, PW, AAG and AK drafted the paper. All authors read and approved the final manuscript.

Acknowledgments

We thank Carol Ware, Angel Nelson, Jennifer Hesson and Chris Cavanaugh at the Institute for Stem Cell and Regenerative Medicine for providing us with the stem cells used in this study. This work was supported by grants from the National Institutes of Health (National Cancer Institute grant CA109597), the US Department of Defense (grant W81XWH-08-1-0636) and the John H. Tietze Foundation (to AK) and by a Mary Gates Endowment scholarship (to BJT). Chromatin Signature was originally published as "Allele-Specific Transcriptional Elongation Regulates Monoallelic Expression of the *IGF2BP1* Gene" in Epigenetics Chromatin 2011 August 3;4:14. Used with permission from the U.S. National Library of Medicine.

CHAPTER 9

Competing Interests

The authors declare that they have no competing interests.

Author Contributions

MDG participated in the design of the study, carried out the molecular studies and drafted the manuscript; DG, ML, DV, NG, SL, KML, JAS-S, CP and RR carried out the molecular studies; DG, ML and RJG helped to draft the manuscript; JRH and ST participated in the bioinformatic and statistical analyzes; SS performed the statistical analysis; SEWJ, PV, RJG and TE participated in the design of the study and its coordination; DHR conceived the study and participated in its design and coordination and drafted the manuscript. All authors read and approved the final manuscript.

Acknowledgments

We thank the CBRG for bioinformatic support, I Dunham, C Langford and the Microarray Facility of the Wellcome Trust Sanger Institute for assistance with microarrays, K Clark for flow sorting, and T Milne for critical reading of the manuscript. This work was supported by the National Institute for Health Research Biomedical Research Centre Programme. Bivalent Chromatin Modification was originally published as "Generation of Bivalent Chromatin Domains During Cell Fate Decisions" in Epigenetics Chromatin 2011 June 6;4(1):9. Used with permission from the U.S. National Library of Medicine.

CHAPTER 10

Competing Interests

The authors declare that they have no competing interests.

Author Contributions

NK and DS conceived the study and wrote the paper. NK also carried out the analyses. Both authors read and approved the final manuscript.

Acknowledgments

This work was funded by the Irish Research Council for Science, Engineering & Technology (IRCSET), Marie Curie Actions under FP7, and Food for Health Ireland (FHI). Adaptive Divergence was originally published as "Shift in the Isoelectric-Point of Milk Proteins as a Consequence of Adaptive Divergence Between the Milks of Mammalian Species" in Biology Directions 2011 July 29:6:40. Used with permission from the U.S. National Library of Medicine.

INDEX

D

E

F

G

N

O

P